ASIC AND FPGA VERIFICATION:
A GUIDE TO COMPONENT MODELING

ABOUT THE AUTHOR

Richard Munden has been using and managing CAE systems since 1987. He has been concerned with simulation and modeling issues for as long.

Richard co-founded the Free Model Foundry *(http://eda.org/fmf/)* in 1995 and is its president and CEO. He has a day job as CAE/PCB manager at Siemens Ultrasound (previously Acuson Corp) in Mountain View, California. Prior to joining Acuson, he was a CAE manager at TRW in Redondo Beach, California. He is a well-known contributor to several EDA users groups and industry conferences.

His primary focus over the years has been verification of board-level designs.

ASIC AND FPGA VERIFICATION: A GUIDE TO COMPONENT MODELING

RICHARD MUNDEN

ELSEVIER

MORGAN KAUFMANN PUBLISHERS

AMSTERDAM • BOSTON • HEIDELBERG • LONDON
NEW YORK • OXFORD • PARIS • SAN DIEGO
SAN FRANCISCO • SINGAPORE • SYDNEY • TOKYO

Morgan Kaufmann Publishers is an imprint of Elsevier

The Morgan Kaufmann Series in Systems on Silicon

Series Editors: Peter Ashenden, Ashenden Designs Pty. Ltd. and Adelaide University, and Wayne Wolf, Princeton University

The rapid growth of silicon technology and the demands of applications are increasingly forcing electronics designers to take a systems-oriented approach to design. This has led to new challenges in design methodology, design automation, manufacture and test. The main challenges are to enhance designer productivity and to achieve correctness on the first pass. The Morgan Kaufmann Series in Systems on Silicon presents high quality, peer-reviewed books authored by leading experts in the field who are uniquely qualified to address these issues.

The Designer's Guide to VHDL, Second Edition
Peter J. Ashenden

The System Designer's Guide to VHDL-AMS
Peter J. Ashenden, Gregory D. Peterson, and Darrell A. Teegarden

Readings in Hardware/Software Co-Design
Edited by Giovanni De Micheli, Rolf Ernst, and Wayne Wolf

Modeling Embedded Systems and SoCs
Axel Jantsch

Multiprocessor Systems-on-Chips
Edited by Wayne Wolf and Ahmed Jerraya

Forthcoming Titles

Rosetta User's Guide: Model-Based Systems Design
Perry Alexander, Peter J. Ashenden, and David L. Barton

Rosetta Developer's Guide: Semantics for Systems Design
Perry Alexander, Peter J. Ashenden, and David L. Barton

Functional Verification
Bruce Wile, John Goss, and Wolfgang Roesner

Senior Editor   Denise E. M. Penrose
Publishing Services Manager   Andre Cuello
Project Manager   Brandy Palacios
Project Management   Graphic World
Developmental Editor   Nate McFadden
Editorial Assistant   Summer Block
Cover Design   Chen Design Associates

Composition   SNP Best-set Typesetter Ltd.
Technical Illustration   Graphic World
Copyeditor   Graphic World
Proofreader   Graphic World
Indexer   Graphic World
Printer   Maple Press
Cover printer   Phoenix Color

Morgan Kaufmann Publishers
An imprint of Elsevier.
500 Sansome Street, Suite 400
San Francisco, CA 94111
www.mkp.com

This book is printed on acid-free paper.

Library of Congress Cataloging-in-Publication Data: Application Submitted

ISBN: 0-12-510581-9

Printed in the United States of America
05   06   07   08   09   5   4   3   2   1

CONTENTS

Preface   xv

PART I   **INTRODUCTION**   **1**

CHAPTER 1   **INTRODUCTION TO BOARD-LEVEL VERIFICATION**   **3**

1.1   Why Models are Needed   3
       1.1.1   Prototyping   3
       1.1.2   Simulation   4
1.2   Definition of a Model   5
       1.2.1   Levels of Abstraction   6
       1.2.2   Model Types   7
       1.2.3   Technology-Independent Models   9
1.3   Design Methods and Models   10
1.4   How Models Fit in the FPGA/ASIC Design Flow   10
       1.4.1   The Design/Verification Flow   11
1.5   Where to Get Models   13
1.6   Summary   14

CHAPTER 2   **TOUR OF A SIMPLE MODEL**   **15**

2.1   Formatting   15
2.2   Standard Interfaces   17

	2.3	Model Delays	18
	2.4	VITAL Additions	19
		2.4.1 VITAL Delay Types	19
		2.4.2 VITAL Attributes	20
		2.4.3 VITAL Primitive Call	21
		2.4.4 VITAL Processes	22
		2.4.5 VitalPathDelays	24
	2.5	Interconnect Delays	25
	2.6	Finishing Touches	27
	2.7	Summary	31

PART II   RESOURCES AND STANDARDS   33

CHAPTER 3   VHDL PACKAGES FOR COMPONENT MODELS   35

	3.1	STD_LOGIC_1164	35
		3.1.1 Type Declarations	36
		3.1.2 Functions	37
	3.2	VITAL_Timing	37
		3.2.1 Declarations	37
		3.2.2 Procedures	38
	3.3	VITAL_Primitives	39
		3.3.1 Declarations	40
		3.3.2 Functions and Procedures	40
	3.4	VITAL_Memory	41
		3.4.1 Memory Functionality	41
		3.4.2 Memory Timing Specification	42
		3.4.2 Memory_Timing Checks	42
	3.5	FMF Packages	42
		3.5.1 FMF gen_utils and ecl_utils	43
		3.5.2 FMF ff_package	44
		3.5.3 FMF Conversions	45
	3.6	Summary	45

CHAPTER 4   AN INTRODUCTION TO SDF   47

	4.1	Overview of an SDF File	47
		4.1.1 Header	48

	4.1.2	Cell	50
	4.1.3	Timing Specifications	50
4.2	SDF Capabilities		52
	4.2.1	Circuit Delays	52
	4.2.2	Timing Checks	55
4.3	Summary		58

CHAPTER 5   **ANATOMY OF A VITAL MODEL**   **59**

5.1	Level 0 Guidelines		59
	5.1.1	Backannotation	60
	5.1.2	Timing Generics	60
	5.1.3	VitalDelayTypes	61
5.2	Level 1 Guidelines		63
	5.2.1	Wire Delay Block	63
	5.2.2	Negative Constraint Block	65
	5.2.3	Processes	65
	5.2.4	VITAL Primitives	70
	5.2.5	Concurrent Procedure Section	70
5.3	Summary		70

CHAPTER 6   **MODELING DELAYS**   **73**

6.1	Delay Types and Glitches		73
	6.1.1	Transport and Inertial Delays	73
	6.1.2	Glitches	74
6.2	Distributed Delays		75
6.3	Pin-to-Pin Delays		75
6.4	Path Delay Procedures		76
6.5	Using VPDs		82
6.6	Generates and VPDs		83
6.7	Device Delays		83
6.8	Backannotating Path Delays		88
6.9	Interconnect Delays		89
6.10	Summary		90

CHAPTER 7   **VITAL TABLES**   **91**

    7.1   Advantages of Truth and State Tables   91

    7.2   Truth Tables   92
        7.2.1   Truth Table Construction   92
        7.2.2   VITAL Table Symbols   92
        7.2.3   Truth Table Usage   93

    7.3   State Tables   97
        7.3.1   State Table Symbols   97
        7.3.2   State Table Construction   97
        7.3.3   State Table Usage   98
        7.3.4   State Table Algorithm   99

    7.4   Reducing Pessimism   100

    7.5   Memory Tables   101
        7.5.1   Memory Table Symbols   101
        7.5.2   Memory Table Construction   102
        7.5.3   Memory Table Usage   103

    7.6   Summary   105

CHAPTER 8   **TIMING CONSTRAINTS**   **107**

    8.1   The Purpose of Timing Constraint Checks   107

    8.2   Using Timing Constraint Checks in VITAL Models   108
        8.2.1   Setup/Hold Checks   108
        8.2.2   Period/Pulsewidth Checks   112
        8.2.3   Recovery/Removal Checks   114
        8.2.4   Skew Checks   117

    8.3   Violations   121

    8.4   Summary   122

PART III   **MODELING BASICS**   **123**

CHAPTER 9   **MODELING COMPONENTS WITH REGISTERS**   **125**

    9.1   Anatomy of a Flip-Flop   125
        9.1.1   The Entity   125
        9.1.2   The Architecture   129

9.1.3   A VITAL Process   131
9.1.4   Functionality Section   133
9.1.5   Path Delay   134
9.1.6   The "B" Side   135

9.2   Anatomy of a Latch   137
9.2.1   The Entity   138
9.2.2   The Architecture   140

9.3   Summary   146

CHAPTER 10   CONDITIONAL DELAYS AND TIMING CONSTRAINTS   147

10.1   Conditional Delays in VITAL   147

10.2   Conditional Delays in SDF   149

10.3   Conditional Delay Alternatives   150

10.4   Mapping SDF to VITAL   152

10.5   Conditional Timing Checks in VITAL   153

10.6   Summary   156

CHAPTER 11   NEGATIVE TIMING CONSTRAINTS   157

11.1   How Negative Constraints Work   157

11.2   Modeling Negative Constraints   158

11.3   How Simulators Handle Negative Constraints 176

11.4   Ramifications   177

11.5   Summary   178

CHAPTER 12   TIMING FILES AND BACKANNOTATION   179

12.1   Anatomy of a Timing File   179
12.1.1   Header   179
12.1.2   Body   181
12.1.3   FMFTIME   181

12.2   Separate Timing Specifications   182

12.3   Importing Timing Values   183

12.4   Custom Timing Sections   183

12.5   Generating Timing Files   184

12.6   Generating SDF Files   184

12.7   Backannotation and Hierarchy   185

12.8   Summary   187

PART IV   ADVANCED MODELING   189

CHAPTER 13   ADDING TIMING TO YOUR RTL CODE   191

13.1   Using VITAL to Simulate Your RTL   191

13.2   The Basic Wrapper   192

13.3   A Wrapper for Verilog RTL   206

13.4   Modeling Delays in Designs with Internal Clocks   206

13.5   Caveats   207

13.6   Summary   208

CHAPTER 14   MODELING MEMORIES   209

14.1   Memory Arrays   209
       14.1.1   The Shelor Method   210
       14.1.2   The VITAL_Memory Package   210

14.2   Modeling Memory Functionality   211
       14.2.1   Using the Behavioral (Shelor) Method   211
       14.2.2   Using the VITAL2000 Method   223

14.3   VITAL_Memory Path Delays   231

14.4   VITAL_Memory Timing Constraints   232

14.5   PreLoading Memories   235
       14.5.1   Behavioral Memory PreLoad   235
       14.5.2   VITAL_Memory PreLoad   237

14.6   Modeling Other Memory Types   238
       14.6.1   Synchronous Static RAM   238
       14.6.2   DRAM   241
       14.6.3   SDRAM   244

14.7   Summary   249

CHAPTER 15   CONSIDERATIONS FOR COMPONENT MODELING   251

15.1   Component Models and Netlisters   251
15.2   File Contents   253
15.3   Generics Passed from the Schematic   253
         15.3.1   Timing Generics   253
         15.3.2   Control Generics   253
15.4   Integrating Models into a Schematic Capture System   254
         15.4.1   Library Structure   254
         15.4.2   Technology Independence   255
         15.4.3   Directories   255
         15.4.4   Map Files   256
15.5   Using Models in the Design Process   256
         15.5.1   VHDL Libraries   257
         15.5.2   Schematic Entry   257
         15.5.3   Netlisting the Design   258
         15.5.4   VHDL Compilation   258
         15.5.5   SDF Generation   259
         15.5.6   Simulation   261
         15.5.7   Layout   261
         15.5.8   Signal Analysis   262
         15.5.9   Timing Backannotation   262
         15.5.10   Timing Analysis   262
15.6   Special Considerations   262
         15.6.1   Schematic Considerations   262
         15.6.2   Model Considerations   263
15.7   Summary   266

CHAPTER 16   MODELING COMPONENT-CENTRIC FEATURES   269

16.1   Differential Inputs   269
16.2   Bus Hold   279
16.3   PLLs and DLLs   282
16.4   Assertions   284
16.5   Modifying Behavior with the TimingModel Generic   285
16.6   State Machines   285
16.7   Mixed Signal Devices   288
16.8   Summary   294

CHAPTER 17   **TESTBENCHES FOR COMPONENT MODELS**   **295**

17.1   About Testbenches   295
       17.1.1   Tools   295
17.2   Testbench Styles   296
       17.2.1   The Empty Testbench   296
       17.2.2   The Linear Testbench   296
       17.2.3   The Transactor Testbench   296
17.3   Using Assertions   297
17.4   Using Transactors   298
17.5   Testing Memory Models   301
17.6   Summary   308

PREFACE

Digital electronic designs continue to evolve toward more complex, higher pincount components operating at higher clock frequencies. This makes debugging board designs in a lab with a logic analyzer and an oscilloscope considerably more difficult than in the past. This is because signals are becoming physically more difficult to probe and because probing them is more likely to change the operation of the circuit.

Much of the custom logic in today's products is designed into ASICs or FPGAs. Although this logic is usually verified through simulation as a standard part of the design process, the interfaces to standard components on the board, such as memories and digital signal processors, often go unsimulated and are not verified until a prototype is built.

Waiting to test for problems this late in the design process can be expensive, however. In terms of both time and resources, the costs are higher than performing up-front simulation. The decision not to do up-front board simulation usually centers around a lack of models and methodology. In *ASIC and FPGA Verification: A Guide to Component Modeling*, we address both of these issues.

Historical Background

The current lack of models and methodology for board-level simulation is, in large part, due to the fact that when digital simulation started to become popular in the 1980s, the simulators were all proprietary. Every Electronic Design Automation (EDA) vendor had their own and it was not possible to write models that were portable from one tool to another. They offered tools with names like HILO, SILO, and TEGAS. Most large corporations, like IBM, had their own internal simulators. At the ASIC and later FPGA levels each foundry had to decide which simulators they would support. There were too many simulators available for anyone to support them all. Each foundry had to validate that the models they provided worked correctly on each supported release of their chosen simulators.

At the board level, the component vendors saw it was impractical to support all the different simulators on the market. Rather than choose sides, they generally decided not to provide models at all. This led to the EDA vendors trying to provide models. After all, what good is a simulator if the customer has nothing to simulate?

So, each EDA vendor produced its own library of mostly the same models: 7400 series TTL, 4000 series CMOS, a few small memories, and not much else. In those days, that might be the majority of the parts needed to complete a design. But there were always other parts used and other models needed. Customers wanting to run a complete simulation had to model the rest of the parts themselves.

Eventually, someone saw an opportunity to sell (or rent) component models to all the companies that wanted to simulate their designs but did not want to create all the models required. A company (Logic Automation) was formed to lease models of off-the-shelf components to the groups that were designing them into new products. They developed the technology to model the components in their own internal proprietary format and translate them into binary code specific to each simulator.

Verilog, VHDL, and the Origin of VITAL

Verilog started out as another proprietary simulator in 1984 and enjoyed considerable success. In 1990, Cadence Design Systems placed the language in the public domain. It became an IEEE standard in 1995.

VHDL was developed under contract to the U.S. Department of Defense. It became an IEEE standard in 1987. Whereas Verilog is a C-like language, it is clear that VHDL has its roots in Ada. For many years there was intense competition between Verilog and VHDL for mind share and market share. Both languages have their strong points. In the end, most EDA companies came out with simulators that work with both.

Early in the language wars it was noted that Verilog had a number of built-in, gate-level primitives. Over the years these had been optimized for performance by Cadence and later by other Verilog vendors. Verilog also had a single defined method of reading timing into a simulation from an external file.

VHDL, on the other hand, was designed for a higher level of abstraction. Although it could model almost anything Verilog could, and without primitives, it allowed things to be modeled in a multitude of ways. This made performance optimization or acceleration impractical. VHDL was not successfully competing with Verilog-XL as a sign-off ASIC simulator. The EDA companies backing VHDL saw they had to do something. The something was named VITAL, the VHDL Initiative toward ASIC Libraries.

The VITAL Specification

The intent of VITAL was to provide a set of standard practices for modeling ASIC primitives, or macrocells, in VHDL and in the process make acceleration possible. Two VHDL packages were written: a primitives package and a timing package. The primitives package modeled all the gate-level primitives found in Verilog. Because these primitives were now in a standard package known to the simulator writers, they could be optimized by the VHDL compilers for faster simulation.

The timing package provided a standard, acceleratable set of procedures for checking timing constraints, such as setup and hold, as well as pin-to-pin propagation delays. The committee writing the VITAL packages had the wisdom to avoid reinventing the wheel. They chose the same SDF file format as Verilog for storing and annotating timing values.

SDF is the Standard Delay Format, IEEE Standard 1497. It is a textual file format for timing and delay information for digital electronic designs. It is used to convey timing and delay values into both VHDL and Verilog simulations. (SDF is discussed in greater detail in Chapter 4.)

Another stated goal of VITAL is model maintainability. It restricts the writer to a subset of the VHDL language and demands consistant use of provided libraries. This encourages uniformity among models, making them easily readable by anyone familiar with VITAL. Reabability and having the difficult code placed in a provided library greatly facilitate the maintainence of models by engineers who are not the original authors.

VITAL became IEEE Standard 1076.4 in 1995. It was reballoted in 2000. The 2000 revision offers several enhancements. These include support for multisource interconnect timing, fast path delay disable, and skew constraint timing checks. However, the most important new feature is the addition of a new package to support the modeling of static RAMs and ROMs.

The Free Model Foundry

In 1994 I was working at TRW in Redondo Beach California as a CAE manager. The benefits of board-level simulation were clear but models were not available for most of the parts we were using. I had written models for the Hilo simulator and then rewritten them for the ValidSim simulator and I knew I would have to write them again for yet another simulator. I did not want to waste time writing models for another proprietary simulator.

At this time VITAL was in its final development and a coworker, Russ Vreeland, convinced me to look at it. I had already tried Verilog and found it did not work well at the board level. Although the show-stopper problems were tool related, such as netlisting, and have since been fixed, other problems remain with the language itself. These include (but are not limited to) a lack of library support and the inability to read the strength of a signal. My personal opinion is that Verilog is fine for RTL simulation and synthesis but a bit weak at board- and system-level modeling. All that may be changed by SystemVerilog.

In 1994, VITAL seemed to have everything I needed to model off-the-shelf components in a language that was supported by multiple EDA vendors. Russ figured out how to use it for component models, developed the initial style and methodology, and wrote the first models. VHDL/VITAL seemed to be the answer to our modeling problem.

But TRW was in the business of developing products, not models. We felt that models should be supplied by the component vendors just as data sheets were. We suggested this to a few of our suppliers and quickly realized it was going to take a long time to convince them. In the mean time we thought we could show other engineers how our modeling techniques worked and share models with them.

In 1995, Russ Vreeland, Luis Garcia, and I cofounded the Free Model Foundation. Our hope was to do for simulation models what the Free Software Foundation had done for software: promote open source standards and sharing. We incorporated as a not-for-profit. Along the way the state of California insisted that we were not a "foundation" in their interpretation of the word. We decided we would rather switch than fight and renamed the organization the Free Model Foundry (FMF).

Today, FMF has models with timing covering over 7,000 vendor part numbers. All are free for download from our website at *www.eda.org/fmf/*. The models are generally copyrighted under the Free Software Foundation's General Public License (GPL). Most of the examples in this book are taken from the FMF Web site.

Structure of the Book

ASIC and FPGA Verification: A Guide to Component Modeling is organized so that it can be read linearly from front to back. Chapters are grouped into four parts: Introduction, Resources and Standards, Modeling Basics, and Advanced Modeling. Each part covers a number of related modeling concepts and techniques, with individual chapters building upon previous material.

Part I serves as an introduction to component models and how they fit into board-level verification. Chapter 1 introduces the idea of board-level verification. It defines component models and discusses why they are needed. The concept of technology-independent modeling is introduced, as well as how it fits in the FPGA and ASIC design flow. Chapter 2 provides a guided tour of a basic component model, including how it differs from an equivalent synthesizable model.

Part II covers the standards adhered to in component modeling and the many supporting packages that make it practical. Chapter 3 covers several IEEE and FMF packages that are used in writing component models. Chapter 4 provides an overview of SDF as it applies to component modeling. Chapter 5 describes the organization and requirements of VITAL models. Chapter 6 describes the details of modeling delays within and between components. Chapter 7 deals with VITAL truth tables and state tables and how to use them. In Chapter 8, the basics of modeling timing constraints are described.

Part III puts to use the material from the earlier chapters. Chapter 9 deals with modeling devices containing registers. Chapter 10 details the use of conditional delays and timing constraints. Chapter 11 covers negative timing constraints. Chapter 12 discusses the timing files and SDF backannotation that make the style of modeling put forth here so powerful.

Part IV introduces concepts for modeling more complex components. Chapter 13 demonstrates how to use the techniques discussed to build a timing wrapper around an FPGA RTL model so it can be used in a board-level simulation. Chapter 14 covers the two primary ways of modeling memories. Chapter 15 looks at some things to consider when writing models that will be integrated into a schematic capture system. Chapter 16 describes a number of different features encountered in commercial components and how they can be modeled. Chapter 17 is a discussion of techniques used in writing testbenches to verify component models.

Intended Audience

This book should be valuable to anyone who needs to simulate digital designs that are not contained within a single chip. It covers the creation and use of a particular type of model useful for verifying ASIC and FPGA designs and board-level designs that use off-the-shelf digital components. Models of this type are based on VHDL/VITAL and are distinguished by their inclusion of timing constraints and propagation delays. The numeric values used in the constraints and delays are external to the actual models and are applied to the simulation through SDF annotation.

The intent of this book is show how ASICs and FPGAs can be verified in the larger context of a board or system. To improve readability, the phrase "ASICs and FPGAs" will be abbreviated to just FPGAs. However, nearly everything said about FPGA verification applies equally to ASIC verification.

This book should also be useful to engineers responsible for the generation and maintenance of VITAL libraries used for gate-level simulation of ASICs and FPGAs. Component vendors that provide simulation models to their customers are able to take advantage of some important opportunities. The more quickly a customer is able to verify a design and get it into production, the sooner the vendors receive volume orders for their parts. The availability of models may even exert an influence over which parts, from which vendors, are designed into new products. Thus, the primary purpose of this book is to teach how to effectively model complex off-the-shelf components. It should help component vendors, or their contractors, provide models to their customers. It should also help those customers understand how the models work. If engineers are unable to obtain the models they need, this book will show them how to create their own models.

Readers of this book should already have a basic understanding of VHDL. This book will cover the details of modeling for verification of both logic and timing. Because many people must work in both Verilog and VHDL, it will show how to use VHDL component models in the verification of FPGAs written in Verilog.

The modeling style presented here is for verification and is not intended to be synthesizable.

Resources for Help and Information

Although this book attempts to provide adequate examples of models and tips on using published VHDL packages, most models and packages are too lengthy to be included in a printed text. All of the models discussed in this book are available in their entirety from the Free Model Foundry Web site (*www.eda.org/fmf/*). The full source code for the IEEE packages discussed should have been provided with your VHDL simulator. They may also be ordered from the IEEE at *standards.ieee.org*. Additional material may be found at *www.mkp.com/companions/0125105819*. Although I have been careful to avoid errors in the example code, there may be some that I have missed. I would be pleased to hear about them, so that I can correct them in the online code and in future printings of this book. Errata and general comments can be emailed to me at *rick.munden@eda.org*.

Acknowledgments

Very little in this book constitutes original thoughts on my part. I have merely applied other people's ideas. Russ Vreeland developed the concept of using VITAL for component modeling. That idea has formed the basis for not just this book but for the Free Model Foundry. Ray Steele took the idea, expanded it, and applied the notion of a rigorously enforced style. Yuri Tatarnikov showed us the basics of how to use VITAL to model complex components.

I would like to thank Peter Ashenden for publishing his *VHDL Cookbook* on the Internet. It was my introduction to VHDL back when there was nothing else available. Larry Saunders taught the first actual VHDL class I attended. I hope I do not ruin his reputation with this book.

Ray Ryan provided training on VITAL prior to it becoming a standard. His material was often referred to during the course of writing this book. His classes were instrumental in convincing Russ and I that VITAL would solve most of our technical problems regarding component modeling.

David Lieby patiently reviewed the first drafts of the book and weeded out all the really embarrassing errors. Additional valuable reviewers were Russ Vreeland, Ray Steele, Hardy Pottinger, Predrag Markovic, Bogdan Bizic, Yuri Tatarnikov, Randy Harr, and Larry Saunders.

Nate McFadden provided critical review of the logical structure of the text and smoothed the rough edges of my prose.

Finally, I thank my loving wife Mary, who fervently hopes I will never do anything like this again.

I Introduction

Part I provides a brief introduction to the board-level verification of FPGAs. The justification for the effort that goes into component modeling and the advantages of board-level simulation are discussed. Ideas for reducing the effort involved in component modeling are explored. In addition, we look at the different levels of abstraction at which models are written and their impact on simulation performance and accuracy.

Chapter 1 introduces board-level simulation. Component models are defined and the effort required to create them justified. Hints are also given regarding how to avoid having to create them all yourself. Technology-independent modeling is described and why it belongs in your FPGA design flow.

Chapter 2 observes a simple nand gate as it slowly evolves from a small synthesizable model to a full-fledged component model. It discusses the importance of consistent formatting and style in component modeling and how they affect maintenance. Basic concepts of modeling are introduced.

1

Introduction to Board-Level Verification

As large and complex as today's FPGAs are, they always end up on a board. Though it may be called a "system on a chip," it is usually part of a larger system with other chips. This chapter will introduce you to the concept of verifying the chip in the system.

In this chapter we discuss the uses and benefits of modeling and define component modeling. This is done in the context of verifying an ASIC or FPGA design. We also provide some historical background and differentiate the types of models used at different stages of digital design.

1.1 Why Models Are Needed

A critical step in the design of any electronic product is final verification. The designer must take some action to assure the product, once in production, will perform to its specification. There are two general ways to do this: *prototyping* and *simulation*.

1.1.1 Prototyping

The most obvious and traditional method of design verification is prototyping. A prototype is a physical approximation of the final product. The prototype is tested through operation and measurement. It may contain additional instrumentation to allow for diagnostics that will not be included in production. If the prototype performs satisfactorily, it provides proof that the design can work in production. If enough measurements are made, an analysis can be done that will provide insight into the manufacturing yield.

If the prototype fails to meet specifications, it will usually be examined to determine the source of its deficiency. Depending on the nature of the product, this may be easy or prohibitively difficult to do. An electronic doorbell built from off-the-shelf parts would lie on the easy end of the continuum; a high-end microprocessor would be prohibitively difficult. Almost all products get prototyped at least once during their design.

1.1.2   Simulation

The other method of design verification is simulation. Simulation attempts to create a virtual prototype by collecting as much information as is known or considered pertinent about the components used in the design and the way they are connected. This information is put into an appropriate set of formats and becomes a *model* of the board or system. Then, a program, the simulator, executes the model and shows how the product should behave. The designer usually applies a stimulus to the model and checks the results against the expected behavior. When discrepancies are found, and they usually are, the simulation can be examined to determine the source of the problem. The design is then changed and the simulation run again. This is an iterative process, but eventually no more errors are found and a prototype is built.

Simulation requires a large effort but in many situations it is worth the trouble for one or more of the following reasons:

- **Easier debugging** It is easier to find the source of a problem in a virtual prototype than in a physical prototype. In the model, all nodes are accessible. The simulator does not suffer from physical limitations such as bandwidth. Observing a node does not alter its behavior.

- **Faster, cheaper iterations** When a design error is identified, the model can be quickly fixed. A physical prototype could require months to rebuild and cost large sums of money.

- **Virtual components can be used** A virtual prototype can be constructed using models of components that are not yet available as physical objects.

- **Component and board process variations and physical tolerances can be explored** A physical prototype can embody only a single set of process variations. A virtual prototype can used to explore design behavior across a full range of variations.

- **Software development can begin sooner** Diagnostic and embedded software development can begin using the virtual prototype. The interplay between hardware and software development often shows areas where the design could be improved while there is still time to make changes.

For FPGA design, nearly everyone simulates the part they are designing. The FPGA vendor provides models of *simulation primitives* (cell models), the lowest-level structures in the design that the designer is able to manipulate. There are usually between 100 and 300 of these relatively simple primitives in an FPGA library. The silicon vendor supplies them because everyone agrees simulation is required and it is a reasonably sized task.

Every FPGA eventually ends up on a board, but for board-level design only the most dedicated design groups simulate. Most other groups would like to but it just seems too difficult and time consuming. The problem is few component vendors

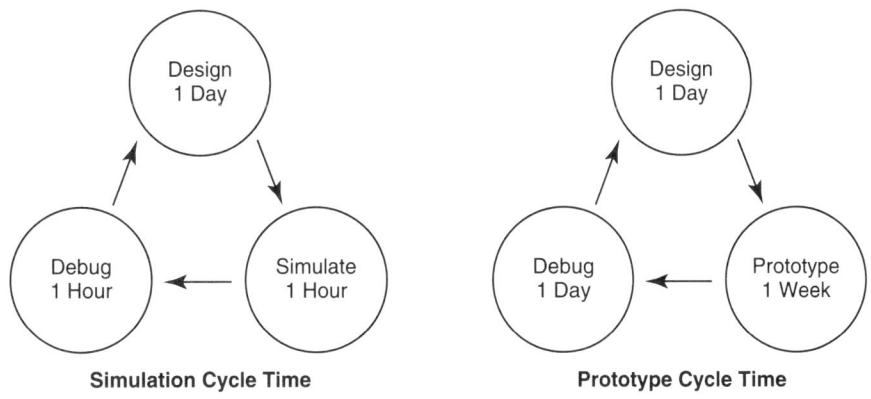

Figure 1.1 Relative iteration times

provide simulation models (although the number is slowly growing). Design groups must often write their own models. Unlike the FPGA primitives, each component needs to be modeled, and these models can be very large. In the end, many designers build prototypes. They then test them in the lab, as best they can, and build new prototypes to correct the errors that are found rather than performing the more rigorous simulations.

It has been said that the beauty of FPGAs is that you don't have to get them right the first time. This is true. However, you do have to get them right eventually. The iterative process of design and debug has a much faster cycle time when it is simulation based rather than prototype based. Not just the prototyping time is saved. The actual debug is much faster using a simulation than using a physical board, as illustrated in Figure 1.1. This is becoming even more true as boards incorporate larger, finer-pitched, ball grid array components.

1.2   Definition of a Model

For the purposes of this book, a model is a software representation of a circuit, a circuit being either a physical electronic device or a portion of a device. This book concentrates exclusively on models written in VHDL but it includes methods for incorporating Verilog code in a mixed-language verification strategy.

In modeling, there are different levels of abstraction and there are different types of models. The two are related but not quite the same thing. Think of a design being conceived of at a level of abstraction and modeled using a particular model type. It is always easier to move from higher to lower levels of abstraction than to move up. Likewise, it is difficult to model a component using a model type that is at a higher level than the data from which the model is being created.

1.2.1   Levels of Abstraction

All digital circuits are composed primarily of transistors. Modern integrated circuits contain millions and soon billions of these transistors. Transistors are analog devices. In digital circuits they are used in a simplified manner, as switches. Still, they must be designed and analyzed in the analog domain.

Analog simulation is very slow and computationaly intensive. To simulate millions of transistors in the analog domain would be exceedingly slow, complex, and not economically practical. Therefore, like most complex problems, this one is attacked hierarchically. Each level of hierarchy is a level of abstraction. Small groups of transistors are used to design simple digital circuits like gates and registers. These are small enough to be easily simulated in the analog domain and measured in the lab. The results of the analog simulations are used to assign overall properties, such as propagation delays and timing constraints to the gates. This is referred to as *characterization*.

The characterized gates can then be used to design more complex circuits, counters, decoders, memories, and so on. Because the gates of which they are composed are characterized, the underlying transistors can be largely ignored by the designer.

This process is extended so that still more complex products such as microprocessors and MPEG decoders can be designed from counters and instruction decoders. Computers and DVD players are then designed from microprocessors and MPEG decoders. This hierarchical approach continues into the design of global telecommunications networks and other planet-scale systems.

We will follow this process by discussing models starting at the gate level of complexity and going up through some of the more complex single integrated circuits.

Gate Level

Gate-level models provide the greatest level of detail for simulation of digital circuits and are the lowest level of abstraction within the digital domain. Gate-level simulation is usually a requirement in the ASIC design process. For FPGAs and ASICs, the library of gate-level models is provided by the component vendor or foundry. The gate-level netlist can be derived from schematics but more often is synthesized from RTL code.

Because it includes so much detail, gate-level simulation tends to be slower than register transfer level (RTL) or behavioral simulation. However, because gate-level models are relatively simple, several EDA vendors have created hardware or software tools for accelerating simulations beyond the speeds available from general-purpose HDL simulators.

RTL

The most discussed and practiced form of HDL modeling is RTL. RTL is used for designing chips such as ASICs and FPGAs. Its purpose is to describe design intent at a level less detailed than the gate level but detailed enough to be understood by a synthesis engine. The synthesis engine then decomposes the RTL description to a gate-level description that can be used to create an ASIC layout or to generate a program for an FPGA.

The person writing the RTL description is concerned primarily with the circuit's interior function. He can specify details such as the type of carry chain used in an adder or the encoding scheme used in a state machine. For the designer, the chip is the top level of the design. When the RTL is used in simulation, it will usually be the device under test (DUT).

RTL models can and should be simulated. This verifies the functionality of the code. However, the RTL code, as written for synthesis, does not include any delay or timing constraint information. Of course, delays could be included but until a physical implementation has been determined the numbers would not be accurate. Accurate timing comes from simulating the chip at the gate level.

Behavioral

In contrast to lower-level models, behavioral models provide the fewest details and represent the highest level of abstraction discussed in this book. The purpose of a behavioral model is to simulate what happens on the edge of a chip or cell. The user wants to see what goes in and what comes out and does not care about how the work is done inside. If delays are modeled, they are modeled as pin-to-pin delays.

The reduced level of detail allows behavioral models to generally run much faster than either gate-level or RTL models. They are also much easier to write, assuming a high-level description of the component is available.

Looking from this perspective, models of off-the-shelf components should be written at the behavioral level whenever possible. Besides being faster to write and faster to run, they give away little or no information about how a part is designed and built. They are inherently nonsynthesizable and do not disclose intellectual property (IP) beyond what is published in the component's data sheet.

Bus Functional

Bus functional models (BFMs) are usually created for very complex parts for which a full behavioral model would be too expensive to create or too slow to be of value. BFMs attempt to model the component interface without modeling the component function. They are not complete enough to simulate running software but they are adequate for verifying that the component is correctly designed into the larger system. Microprocessors and digital signal processors are candidates for bus functional models.

1.2.2   Model Types

There are three types of HDL models in common use: cell, RTL, and behavioral. Component models are usually a special case of behavioral models.

Cell

Cell models describe the functionality of the cells in the gate-level netlist. They are provided in libraries by FPGA vendors and are usually specific to a particular FPGA family. Cell models include propagation delays and timing constraints. However, the actual delay and constraint values are not coded directly into the models. Instead, these values are calculated by software usually provided by the FPGA vendor. The calculated values are then written to an SDF file and annotated into the simulation. Cell models are commonly of low to medium complexity.

RTL

The RTL model is used to describe a digital circuit a designer intends to synthesize. RTL models do not include timing. They are also used to simulate a design prior to synthesis. RTL models contain no information regarding propagation delays or timing constraints. A synthesis engine converts an RTL model into a gate-level netlist for a targeted FPGA. The gate-level netlist then instantiates a number of gates or cells and describes how they are connected to each other, as shown in Figure 1.2.

Behavioral

Behavioral models describe what a circuit or system does without attempting to explain how it does it. They are written with the fewest constraints: They may or may not be synthesizable, they may or may not include timing, and they may or may not be cycle accurate. They may be intended for use in architectural exploration, performance modeling, or hardware/software codesign.

Behavioral model development is not limited to VHDL and Verilog. They may be written in general computing languages such as C/C++ or any of the special system-level languages, such as Esterel or Rosetta.

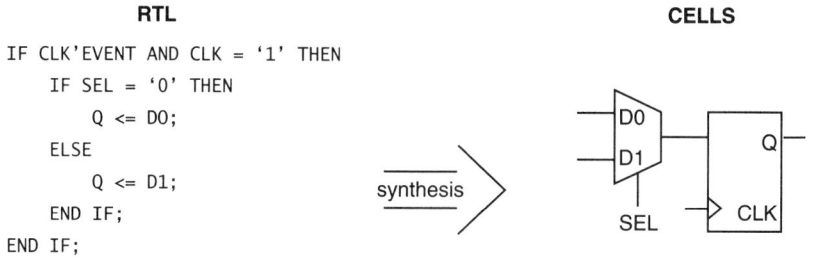

Figure 1.2 RTL produces cells

Component

Component models are models of off-the-shelf components used in board-level design. They use the same techniques for describing propagation delays and timing constraints as cell models. Models of simple components are constructed in the same manner as models of simple gates. Components, however, can be much more complex than gates. Complex components are modeled using a mixture of behavioral, RTL, and gate-level techniques with the intent of including as little detail as required to produce the correct behavior at the component interface. Sometimes, only their interfaces are modeled, in which case they may be referred to as BFMs.

It is possible to create a component model for an FPGA by embedding an RTL model in a wrapper that provides pin-to-pin propagation delays and timing constraint checks. Such a model can be used to accelerate the simulation of a board or system containing one or more large, user-designed components. A model constructed in this manner will provide much of the functionality of a gate-level model but execute at the speed of a RTL model. Figure 1.3 illustrates how the RTL code from Figure 1.2 might be incorporated into a component model.

1.2.3 Technology-Independent Models

Creating component models is work, and unless you work for a component vendor, it may not be your primary responsibility. Most often, you would like to write as few models as necessary to accomplish your verification goals. The best way to reduce the number of models needed is to make them technology (timing) independent.

The core concept is the separation of timing and behavior. In this method the VHDL (or Verilog) model describes component behavior but contains no timing

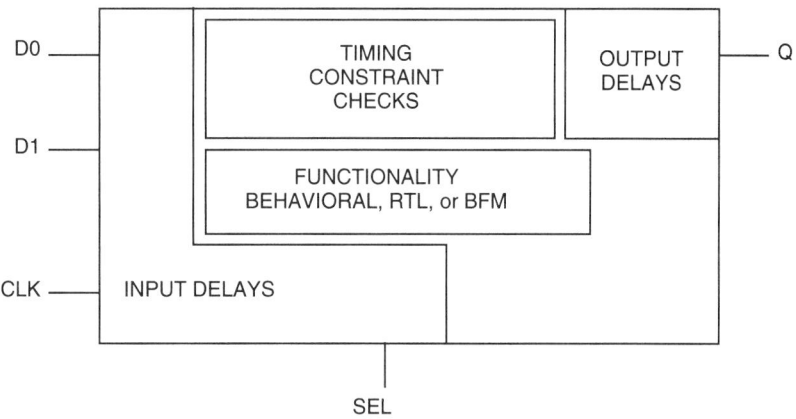

Figure 1.3 Component model of circuit from Figure 1.2

information. All timing values, for delays and constraints, reside in a separate ASCII file. A single model may represent many parts that differ only in timing. A single timing file contains all the different timings for that model. A tool is used to extract the desired timing from all the timing files for all the models and generate an SDF file for the entire design.

1.3   Design Methods and Models

There are many design methods that use the various type of models described here. One such method is the classic *top-down* style. In this method, a behavioral model of the system to be designed is written and simulated. It is modified until it adequately describes the desired product. It then becomes an executable specification for the design against which the design implementation can be compared. The design is then partitioned into sections that will be custom built with ASICs and FPGAs and sections that will be built with off-the-shelf (OTS) components (if any). There may be trade-off studies done to determine the optimum partitioning between custom and OTS hardware.

The custom section is further partitioned into as many different custom components as required and each of those is coded at the register-transfer level and synthesized to gates. The OTS section is designed using schematics. The custom parts are added to the schematic and, if models are available of the OTS parts, the system can be simulated to verify that all the components, including the custom ones, are correctly connected and will perform the desired functions. This method is shown in Figure 1.4.

The top-down method seems best suited for designs that have rigid performance requirements. These designs could be for defense or commercial markets that are performance driven and have very high volumes.

Another, more common approach may be called either *bottom-up* or *outside-in*. In this method performance goals are tempered by cost considerations. Instead of an executable specification, the design exploration may begin with the question, "How good can we make it and still meet our cost goals?" In such environments, custom components are designed only when they will be more cost effective than off-the-shelf components. Component availability can strongly influence the architecture of this type of product.

A representation of the bottom-up method is shown in Figure 1.5.

In both of the described methods, there is a point at which custom-designed components, ASICs and FPGAs, must be integrated with off-the-shelf components. Verifying correct integration is where component models come into the picture.

1.4   How Models Fit in the FPGA/ASIC Design Flow

Rarely does an FPGA become the only digital component in a product. Most have to interface with other FPGAs or off-the-shelf parts. These may be memories, microprocessors, digital signal processors, bus interface chips, or just glue logic. Although

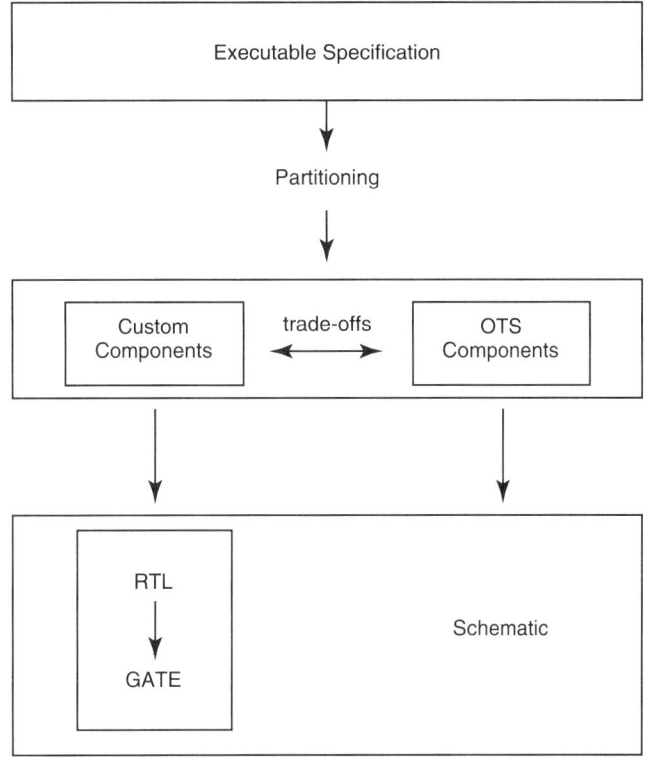

Figure 1.4   Classic top-down design method

most of the verification effort goes into proving that the internal logic of the custom part meets its specification, its interfaces with the rest of the system must be correct for it to contribute to a working product. Simulating these interfaces is most easily done by using models of the surrounding components.

Even when verifying the FPGA's internal logic, external component models may be used. A testbench may be more accurate and easier to construct if it incorporates models of peripheral components. These models can prove particularly helpful in uncovering errors in power up, reset, and boundary conditions.

1.4.1   The Design/Verification Flow

For FPGA-on-board verification, FPGAs are designed in VHDL or Verilog. They can be modeled at the behavioral level, RTL, or gate level. The boards they go into are designed using a schematic capture system. Schematics are still used for board design because they are a convenient and effective method of entering and conveying information about the logic and physical characteristics of a design. The

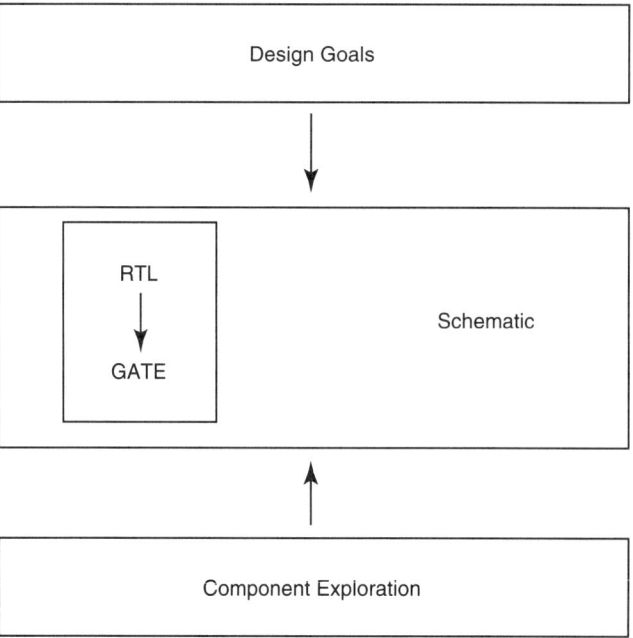

Figure 1.5 The bottom-up flow

schematic tool generates a VHDL netlist and other files needed to interface with a printed circuit board (PCB) layout tool. The components on the board are laid out and the connections between them routed in the PCB tool.

At this point, the FPGA has a model, the other components on the board have models, and the netlist describes how they are connected. All of these are fed to the HDL analyzer/compiler.

In addition to models describing their logical operation, these components have timing files. An SDF extraction tool reads the netlist and the timing files to create an SDF file with timing for all the components that are in the netlist. The FPGA may have its own SDF file, and still another SDF file can be generated by the PCB layout tool, or a signal integrity tool, to describe the interconnect delays.

The SDF file(s) and the compiled models are read by the simulator. A testbench provides the stimulus. The results are examined by the design engineer and the necessary changes are made to the board and/or FPGA design. This process is repeated until no more errors are found or it is otherwise determined to build the first prototype. This, of course, does not mean there are no more errors, just that you need to do something different to find them.

A diagram of this flow is shown in Figure 1.6. Keep it in mind as you learn to create and use component models as part of your ASIC/FPGA system verification strategy.

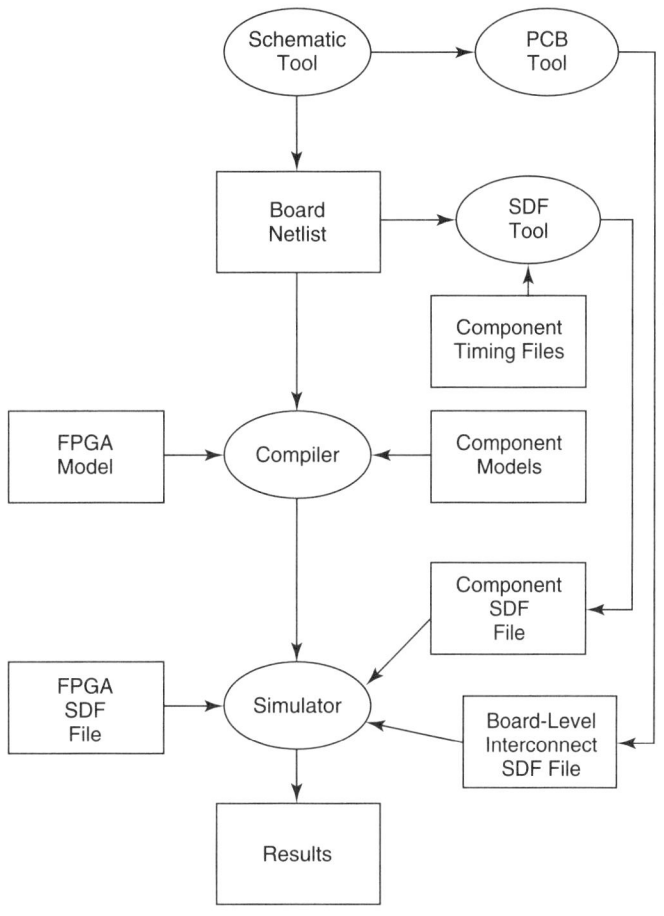

Figure 1.6 Simulation data flow

1.5   Where to Get Models

For a model to have maximum utility and portability, it must be available to the engineer as source code. There are three places to get such models. Some component vendors, mostly memory suppliers, provide source code models. Micron Technology was one of the pioneers in providing simulation models to its customers directly from its Web site. IDT and AMD memory divisions have both taken up the challenge and provide models of the style presented in this book. Intel flash memory division also offers some models. Although some of the models offered are quite good, others seem to have been written for the purpose of simulating a single, stand-alone part with no provision for verifying the component in a larger design.

There are also some EDA vendors who offer models. These vary widely and may be encrypted, requiring a license or special software, or they may be in source code form. Cost and usability also vary. Most models from EDA vendors and component vendors do not allow for backannotation of interconnect delays and may not allow for SDF backannotation at all.

Writing your own models is, of course, an option. This book provides the guidance you need to write complete and efficient models for use in ASIC/FPGA/board verification. This book also demonstrates how to incorporate your RTL or gate-level design into the board-level simulation. For the models you are able to find on the World Wide Web, it provides insight into how the models are constructed and how to use them.

In addition, as mentioned in the preface, another source of models is the Free Model Foundry. If you do write your own models of off-the-shelf components, you might consider sharing them with others. This can be done through the Free Model Foundry.

1.6   Summary

The verification of an FPGA is not complete until it has been simulated as a component at the board level. Doing this requires having models of the off-the-shelf components to which it is connected. These component models are quite different from the RTL models you write for synthesis. Some of them are available from vendor Web sites (if not, always ask for them) or from the Free Model Foundry, and others you will need to write yourself. Once you have the models, your FPGA design can be simulated at the board level to verify correct interfaces and system functionality.

The verification effort is worth the trouble because it can find errors that may be difficult to uncover in a prototype. It can also find them before the prototype is built, possibly saving board spins. Once a board-level model is available, it is often possible to begin some types of software development and reduce overall schedule.

2

Tour of a Simple Model

In this chapter we examine a very basic component model, a 2-input nand gate, in order to better understand the different goals of simulation and synthesis. This simple model allows us to review the basic requirements of a component model and see how such a model is different from the RTL models written for synthesis. The reason for beginning with such a trivial model is that it allows us to concentrate on the new concepts normally present in a VITAL component model that are not found in an RTL model. (Much more complex models will be discussed later in the book.)

The synthesizable 2-input nand gate we are to examine is shown in Figure 2.1. It is part of a larger synthesizable design. Written in VHDL code, the model has an active low output and is designated as such by appending a "neg" to the end of its name. (The reason for this particular convention is explained later.) This is a perfectly good model of a nand gate—if you are designing nand gates for synthesis only. On the other hand, if your job is to create a nand gate model that will be used as an FPGA simulation primitive or an off-the-shelf component to be used in a board-level simulation, you might find this model has some deficiencies. Let's look more closely at this model to see how it can be enhanced with simulation in mind, our goal being to create a VHDL model for the nand gate and the SDF to accompany it.

2.1   Formatting

Code formatting is often overlooked but it is one important way of improving any model that will be reused. When you are writing the RTL code for your latest chip, you may think no one else is going to spend much time trying to read your code. Perhaps not even you. You may feel that how you format your code is between you, your simulator, and your synthesis tool. Reuse is largely underexploited because too much code is written under constraints that do not allow the extra time to be taken to make it understandable and usable to other engineers. It is not written with other readers in mind. But when you are writing a component model, everybody wants to reuse your work. They just are not that excited about spending time modeling someone else's design. So everyone is going to read your code.

```
entity nandgate is
    port (a, b : in std_logic; yneg : out std_logic);
end nandgate;
architecture ex1 of nandgate is
begin
    yneg <= a nand b;
end;
```

Figure 2.1   A synthesizable 2-input nand gate

Also, because the parts you model are likely to reappear again in your next design, you are likely to read it too. Since you may also need to maintain it, you will want do what you can to make the model easy to edit and to understand.

Uniformity is important. If all your models are written in the same style and format, it becomes easy to navigate through them to find the section you want, and it will be easier to understand them. When you have a large number of models and a global change is required, having the models written in a consistent format may mean they can be updated using a script in batch mode instead of you having to plod through them one by one.

The first thing we can do is put a banner on the top so we can always see which file we are editing or which model we have printed. Let us make this banner 80 characters wide because 80 columns print reasonably well and we can set our window width to match the banner so we always know where to break a line.

```
-------------------------------------------------------------------------------
-- File Name: ex2_nand.vhd
-------------------------------------------------------------------------------
```

Next should come the library statements. Each model is standalone and in its own file, so each one needs its own library clauses.

```
LIBRARY IEEE; USE IEEE.std_logic_1164.ALL;
```

In the entity, we will use a separate line for each port. This takes up more space but is more readable and accessible by scripts. Besides, lines are cheap.

```
PORT (
    A    : IN    std_logic;
    B    : IN    std_logic;
    YNeg : OUT   std_logic
);
```

Let's add another banner to separate the entity and architecture sections.

```
-------------------------------------------------------------------------------
-- ARCHITECTURE DECLARATION
-------------------------------------------------------------------------------
```

Finally, we will capitalize key words and signal names so they stand out better. Some people prefer to make key words lowercase and capitalize everything else. Some are even passionate about whether key words are uppercase or lowercase. It is really just an arbitrary decision. I have chosen to use uppercase key words because that is

```
--------------------------------------------------------------------------------
-- File Name: ex2_nand.vhd
--------------------------------------------------------------------------------
LIBRARY IEEE; USE IEEE.std_logic_1164.ALL;

ENTITY nandgate IS
    PORT (
        A    : IN    std_logic;
        B    : IN    std_logic;
        YNeg : OUT   std_logic
    );
END nandgate;


--------------------------------------------------------------------------------
-- ARCHITECTURE DECLARATION
--------------------------------------------------------------------------------
ARCHITECTURE ex2 OF nandgate IS
BEGIN
    YNeg <= A nand B;
END;
```

Figure 2.2   Nand model with improved formatting

how it is done in the Institute of Electrical and Electronics Engineers (IEEE) packages. Figure 2.2 shows the nand model with the added formatting.

2.2   Standard Interfaces

Multichip or board-level simulation involves more than just ones and zeros. Signals can be strong or weak or high impedance. Drivers can have open collector outputs and require pull-up resistors. Realistic simulations require at least the 9-state logic found in the IEEE 1164 package, std_ulogic. For these reasons, except for mixed signal models, ports will always be of type std_ulogic. The difference between std_logic and std_ulogic types is that std_logic is a resolved subtype of std_ulogic. Using std_ulogic provides a slight performance improvement during simulation. Although the improvement is actually quite small, if there are thousands of instantiations of the model (and there could be many thousands) it could become significant. Vectored ports are not used because they would inhibit backannotation of interconnect delays. Interconnect delays are discussed later in this chapter.

It is good practice to explicitly specify default initial values for all ports. At the board level, sometimes an input pin may be left unconnected. When the design is netlisted, the unconnected pin is assigned to the key word OPEN. VHDL has a restriction that in order to be assigned to OPEN, an input port must have an explicit default value.

```
PORT (
    A    : IN    std_ulogic := 'U';
    B    : IN    std_ulogic := 'U';
    YNeg : OUT   std_ulogic := 'U'
);
```

In most cases the default value should be 'U' for uninitialized, as shown. However, some parts have inputs with internal pull up or pull down resistors so that unused inputs will be pulled to a known state and may be left unconnected if not needed. These pins are given initial values of '1' or '0' as appropriate.

There is at least one other case when an output is given an initial value other than 'U'. Some ECL logic parts have a VBB output. These pins are initialized to 'W' for reasons discussed in Chapter 16.

2.3   Model Delays

In Figure 2.2 we have a model that would function correctly as a nand gate but has zero delay. All physical parts have some delay. Sometimes we rely on that delay, other times we would like it to go away, but we always have to account for it. So how do we add delays to our models?

The simplest way of expressing a delay in VHDL is with an AFTER clause:

```
YNeg <= A nand B AFTER 6 ns;
```

This is fine if the part you are modeling happens to switch in 6 nanoseconds, in both directions, under all conditions. But then you would have to create another model when the new 4 nanosecond (ns) part came out.

VHDL has a stock solution for such problems: generics. Generics are used to pass information into a model. When a generic is used to pass information into a model, it describes a constant and can only be read. A generic is declared in the model entity and used in the architecture. Figure 2.3 shows our model with a generic named delay.

```
--------------------------------------------------------------------------------
-- File Name: ex3_nand.vhd
--------------------------------------------------------------------------------
LIBRARY IEEE; USE IEEE.std_logic_1164.ALL;

ENTITY nandgate IS
    GENERIC (
        delay : TIME := 10 ns
    );
    PORT (
        A    : IN    std_ulogic := 'U';
        B    : IN    std_ulogic := 'U';
        YNeg : OUT   std_ulogic := 'U'
    );
END nandgate;


--------------------------------------------------------------------------------
-- ARCHITECTURE DECLARATION
--------------------------------------------------------------------------------
ARCHITECTURE ex3 OF nandgate IS
BEGIN
    YNeg <= A nand B AFTER delay;
END;
```

Figure 2.3 Nand gate with delay generic

This is an improvement. Now only one model is needed to cover many possible nand gates. But the model still has symmetrical rise and fall delays that may not match the part used in your design. Most non-CMOS components use a totempole output structure that has different drive strengths depending on whether it is driving high or low. This causes the effective pin delay to be different for a high output than a low output.

2.4   VITAL Additions

The next several changes to the model are the addition of VITAL types, attributes, primitives, processes, or methods. By using these VITAL features, we make our models more uniform and improve their simulation performance. VITAL enables the technology independence we need to reuse models rather than constantly rewrite them.

2.4.1   VITAL Delay Types

Another deficiency in the model in Figure 2.3 is that the correct value for the delay must be written into the netlist, which can be inconvenient. Although there are many ways to overcome these limitations, in Figure 2.4 we use VITAL for pin-to-pin (path) delays, which is an IEEE standard. VITAL path delay generics are recognizable by their **tpd** prefix. There are a number of related changes made in Figure 2.4, and as you can see, the model has grown somewhat.

On lines 2 and 3 of Figure 2.4 are two new USE clauses calling out two VITAL packages. The first package contains the VITAL timing constraint and delay routines. The second package contains the VITAL accelerated primitives.

On lines 6 and 7 of Figure 2.4 are our delay generics. There is one for each input. Path delay generics (tpd) in VITAL are formed using a standardized formula. The names start with **tpd** and the names of the input port and output port are added in that order, all separated by underscores. Thus, the name of the generic to hold the value of the delay from pin A to pin YNeg is tpd_A_YNeg:

```
tpd_A_YNeg        : VitalDelayType01  :=  (1 ns, 1 ns);        -- 6
tpd_B_YNeg        : VitalDelayType01  :=  (1 ns, 1 ns)         -- 7
```

The type of these generics is VitalDelayType01. A generic of this type is used for paths that can cause the output to transition only between low and high. It takes two values, each is of type Time. The first is the delay for low to high transitions (LH), the second for high to low transitions (HL). The VITAL timing package also defines a type VitalDelayType01Z for paths that can cause an output to go high impedance ('Z'), and VitalDelayType01ZX for paths that can cause the output to go to 'X', the unknown. VitalDelayType01 Z takes 6 values: LH, HL, LZ, ZH, HZ, and ZL. VitalDelayType01ZX takes 12 values: LH, HL, LZ, ZH, HZ, ZL, LX, XH, HX, XL, XZ, and ZX. Definitions of the IEEE std_logic_1164 logic values are given in the next chapter.

```
--------------------------------------------------------------------------------
--  File Name: ex4_nand.vhd
--------------------------------------------------------------------------------
LIBRARY IEEE;     USE IEEE.std_logic_1164.ALL;                              -- 1
                  USE IEEE.VITAL_timing.ALL;                                -- 2
                  USE IEEE.VITAL_primitives.ALL;                            -- 3

ENTITY nandgate IS                                                         -- 4
    GENERIC (                                                              -- 5
        tpd_A_YNeg              : VitalDelayType01 := (1 ns, 1 ns);        -- 6
        tpd_B_YNeg              : VitalDelayType01 := (1 ns, 1 ns)         -- 7
    );                                                                     -- 8
    PORT (                                                                 -- 9
        A          : IN    std_ulogic := 'U';                             -- 10
        B          : IN    std_ulogic := 'U';                             -- 11
        YNeg       : OUT   std_ulogic := 'U'                              -- 12
    );                                                                     -- 13
    ATTRIBUTE VITAL_LEVEL0 of nandgate : ENTITY IS TRUE;                  -- 14
END nandgate;                                                             -- 15


--------------------------------------------------------------------------------
-- ARCHITECTURE DECLARATION
--------------------------------------------------------------------------------
ARCHITECTURE ex4 OF nandgate IS                                           -- 17
    ATTRIBUTE VITAL_LEVEL1 of ex4 : ARCHITECTURE IS TRUE;                -- 18

BEGIN                                                                     -- 20


    --------------------------------------------------------------------------
    -- Concurrent procedure calls
    --------------------------------------------------------------------------
    a_1: VitalNAND2 (                                                     -- 21
            q           => YNeg,                                          -- 22
            a           => A,                                             -- 23
            b           => B,                                             -- 24
            tpd_a_q     => tpd_A_YNeg,                                    -- 25
            tpd_b_q     => tpd_B_YNeg                                     -- 26
        );                                                               -- 27
END;                                                                     -- 28
```

Figure 2.4   Basic VITAL nand gate

As mentioned earlier, the nand gate has an active low output, and we indicate such by appending the **Neg** suffix to its name in accordance to Free Model Foundry convention. Many people like to use other conventions, such as an _L suffix. We cannot include underscores in port names in VITAL models because underscores are used as delimiters in the generic names.

2.4.2   VITAL Attributes

Because our enhancements to the model as shown in Figure 2.4 have made this a VITAL model, we need to notify the compiler of that fact. Line 14,

```
ATTRIBUTE VITAL_LEVEL0 of nandgate : ENTITY IS TRUE;
```

tells the compiler that the model is compliant with the level 0 VITAL specification. Level 0 pertains primarily to the entity part of a model. Its purpose is to promote

Table 2.1 Comparison of VITAL levels

	VITAL Level 0	VITAL Level 1
Provides:	SDF backannotation, negative timing constraints	acceleration of primitives and tables
Requires:	level 0 attribute, `std_ulogic` and `std_logic_vector` ports, no underscores in port names, special rules for timing generics	level 1 attribute, no shared variables, operators restricted to those in Standard and `std_logic_1164`, all outputs must be driven by a VitalPathDelay or a Vital primitive

the portability and interoperability of the model. It not only restricts the form and semantic content of the entity, but also standardizes the specification and processing of timing information. It enables the model to use the VITAL backannotation and timing check routines. The simulator will be able to read an SDF file and match it up with the model.

Line 18,

```
ATTRIBUTE VITAL_LEVEL1 of ex4 : ARCHITECTURE IS TRUE;
```

tells the compiler that we also claim VITAL level 1 compliance. Level 1 allows the compiler to optimize the compiled model for faster setup and simulation. To do this, we must restrict ourselves to certain VHDL constructs. We will discuss these restrictions as we come to them.

Level 1 compliance makes the most sense for smaller models that are best described in terms of gates. The execution speed of these models will be accelerated by the use of VITAL level 1. Larger and more complex models are better described in a behavioral style. They may not be practical or desirable to write at the gate level and will run faster as behavioral models anyway. Level 1 compliance is optional.

A brief comparison of the VITAL compliance levels is given in Table 2.1. More details are provided in Chapters 3 and 5.

2.4.3   VITAL Primitive Call

Lines 21 through 27 of Figure 2.4,

```
a_1: VitalNAND2 (                                          -- 21
        q          => YNeg,                                -- 22
        a          => A,                                   -- 23
        b          => B,                                   -- 24
        tpd_a_q    => tpd_A_YNeg,                          -- 25
        tpd_b_q    => tpd_B_YNeg                           -- 26
    );                                                     -- 27
```

are a concurrent procedure call to the VITAL primitive `VitalNAND2`. VITAL primitives are accelerated by the compiler and simulator for better simulation performance. The ports are mapped by name. The last two arguments are the two delays, one from each input to the output. Some components may specify identical delays from each input and others may not. The VITAL primitives always require that separate delays be specified. VITAL primitives do not offer as much flexibility in handling delays as another procedure, the `VitalPathDelay` (VPD), but VPDs must be called from within a process.

A further improved model incorporating a VITAL process is given in Figure 2.5.

2.4.4   VITAL Processes

A VITAL process consists of the following three sections: timing constraint checks, functionality, and path delays. The sections must be in the listed order. A more complete description is found in Chapter 5.

On line 20 of Figure 2.4,

```
VITALBehavior : PROCESS (A, B)
```

we begin a VITAL process. The use of a VITAL process gives us control over a number of behaviors that would be difficult to control otherwise. These will be pointed out as we walk through the code. Line 21,

```
VARIABLE YNeg_zd    : std_ulogic := 'U';
```

declares a functionality result variable. It will hold the zero delay result prior to it being scheduled for output. It must be a variable rather than a signal for the model to be level 1 compliant. The FMF convention, which follows examples given in the VITAL standard document, is to create the name for this variable by taking the name of the output port to which it refers and appending the characters _zd, for zero delay.

The next line, 22,

```
VARIABLE YNeg_GlitchData : VitalGlitchDataType;
```

declares the glitch variable for YNeg. Glitches and glitch handling will be covered in detail in the next chapter. For now, just be aware that a glitch variable is required for a path delay statement to be used.

The statement on line 24,

```
YNeg_zd := VitalNAND2 (a => A, b => B);
```

is the one (and only) statement that actually describes the functional behavior of the component being modeled. The description is in the form of a VITAL function. The variable YNeg_zd immediately gets the results of the VitalNAND2 function. The inputs are "A" and "B". The VitalNAND2 function comes from the VITAL_primitives package referenced on line 3.

```
--------------------------------------------------------------------------------
-- File Name: ex5_nand.vhd
--------------------------------------------------------------------------------
LIBRARY IEEE; USE IEEE.std_logic_1164.ALL;                         -- 1
            USE IEEE.VITAL_timing.ALL;                             -- 2
            USE IEEE.VITAL_primitives.ALL;                         -- 3

ENTITY nandgate IS                                                 -- 4
    GENERIC (                                                      -- 5
        tpd_A_YNeg   : VitalDelayType01 := (1 ns, 1 ns);          -- 6
        tpd_B_YNeg   : VitalDelayType01 := (1 ns, 1 ns)           -- 7
    );                                                             -- 8
    PORT (                                                         -- 9
        A    : IN    std_ulogic := 'U';                           -- 10
        B    : IN    std_ulogic := 'U';                           -- 11
        YNeg : OUT   std_ulogic := 'U'                            -- 12
    );                                                             -- 13
    ATTRIBUTE VITAL_LEVEL0 of nandgate : ENTITY IS TRUE;          -- 14
END nandgate;                                                      -- 15

--------------------------------------------------------------------------------
-- ARCHITECTURE DECLARATION
--------------------------------------------------------------------------------
ARCHITECTURE ex5 OF nandgate IS                                   -- 16
    ATTRIBUTE VITAL_LEVEL1 of ex5 : ARCHITECTURE IS TRUE;         -- 17
BEGIN                                                              -- 18
    VITALBehavior : PROCESS (A, B)                                -- 20

    -- Functionality Results Variables
    VARIABLE YNeg_zd    : std_ulogic := 'U';                      -- 21

    -- Output Glitch Detection Variables
    VARIABLE YNeg_GlitchData : VitalGlitchDataType;               -- 22

    BEGIN                                                          -- 23
    YNeg_zd := VitalNAND2 (a => A, b => B);                       -- 24

    --------------------------------------------------------------------------------
    -- Path Delay Section
    --------------------------------------------------------------------------------
    VitalPathDelay01 (                                           -- 25
        OutSignal      => YNeg,                                   -- 26
        OutSignalName  => "YNeg",                                 -- 27
        OutTemp        => YNeg_zd,                                -- 28
        Paths          => (                                       -- 29
          0 => (InPutChangeTime => A'LAST_EVENT,                  -- 30
                PathDelay       => tpd_A_YNeg,                    -- 31
                PathCondition   => TRUE ),                        -- 32
          1 => (InPutChangeTime => B'LAST_EVENT,                  -- 33
                PathDelay       => tpd_B_YNeg,                    -- 34
                PathCondition   => TRUE ) ),                      -- 35
        GlitchData   => YNeg_GlitchData );                        -- 36
    END PROCESS;                                                  -- 37
END;                                                              -- 38
--------------------------------------------------------------------------------
```

Figure 2.5   VITAL nand gate using VitalPathDelay

2.4.5   VitalPathDelays

For VITAL level 1 compliance, all output ports must be driven by a procedure call
to either a `VitalPathDelay` or a VITAL primitive.

`YNeg_zd` is not assigned directly to the output port. The assignment is
done through the `VitalPathDelay01` procedure beginning on Figure 2.4,
line 25,

```
VitalPathDelay01 (
```

If there were more than one output port, this procedure would be called for each
2-state output port. Let us examine the call, line by line:

```
OutSignal     => YNeg,                                              -- 26
OutSignalName => "YNeg",                                            -- 27
OutTemp       => YNeg_zd,                                           -- 28
```

`OutSignal` gets the name of the port to which the result is ultimately assigned.
`OutSignalName` gets a string that is the name of the port as you would like it stated
in any error messages that may be generated by the procedure. `OutTemp` gets the
variable that holds the temporary result of the functional simulation. Despite its
name, remember it is an input to the procedure.

Line 29,

```
Paths           => (                                               -- 29
    0 => (InPutChangeTime  => A'LAST_EVENT,                         -- 30
          PathDelay        => tpd_A_YNeg,                           -- 31
          PathCondition    => TRUE ),                               -- 32
    1 => (InPutChangeTime  => B'LAST_EVENT,                         -- 33
          PathDelay        => tpd_B_YNeg,                           -- 34
          PathCondition    => TRUE ) ),                             -- 35
```

is the beginning of the `Paths` section. For each possible path from each input to
the output there are three lines. The three lines constitute a VHDL record. The
records elements are:

InputChangeTime: The time of the last change on an input that may have
triggered the process we are in.

PathDelay: The set of delays to apply to the output, for this path.

PathCondition: The condition that must be met for this path to be considered
valid. It may be a boolean expression and must evaluate to true for this path to
be selected.

Together, all the paths become an array of records. The number at the beginning
of each path is its index number.

When the procedure is entered, all the paths are searched and the most valid is
selected. The most valid path will be one for which the `PathCondition` is true. If
there is more than one path for which the `PathCondition` is true, the one with
the most recent event will be chosen. If there were simultaneous events, the path

with the shortest delay will be chosen. (This may not always be what is desired. VPDs will be covered in depth in Chapter 6.) In the event none of the paths are valid, a default delay is applied. The default delay is zero but may be changed. This is discussed further in the next chapter.

Finally, on line 36,

```
GlitchData   => YNeg_GlitchData );
```

the `GlitchData` parameter is associated with its variable (discussed in Chapter 6). There are three versions of the `VitalPathDelay` procedure. Each has more parameters available than were shown in our example. They are discussed in Chapters 3 and 6.

2.5   Interconnect Delays

Although the model is now much improved, there is still something missing. At the board level, components are connected by copper printed circuit board traces. Depending on their lengths and the design's timing requirements, the delays introduced by these traces can be significant. Therefore, they need to be accounted for in the models.

Many PCB design tools and signal integrity analysis tools are capable of determining interconnect delays. The delay values can be exported to an SDF file that most simulators can read.

VITAL provides a method for handling interconnect delays. The method uses generics for holding the backannotated delay values and a `WireDelay` block for applying the delays to the input signals. Our nand gate model, shown in Figure 2.6, incorporates this and some other new features.

The first addition we find in Figure 2.6 is on line 4,

```
LIBRARY FMF;        USE FMF.gen_utils.ALL;
```

where we call out a new library and package. The library, named FMF, is from the Free Model Foundry and the package is called `gen_utils`. The FMF library has several packages that are discussed in Chapter 3. The `gen_utils` package is used in this model to supply default values to some generics, as shown later.

Lines 7 and 8,

```
tipd_A              : VitalDelayType01 := VitalZeroDelay01;
tipd_B              : VitalDelayType01 := VitalZeroDelay01;
```

declare the `tipd` generics, which are the interconnect delays between components on the PCB (or between boards). There should be one for each port of mode IN or INOUT in the port list. They are given default delay values of zero. Delay values for tipds must be nonnegative.

For each port with an associated `tipd` we declare a signal to hold the delayed value of that port. On lines 23 and 24 we have

```
SIGNAL A_ipd   : std_ulogic := 'U';
SIGNAL B_ipd   : std_ulogic := 'U';
```

```
----------------------------------------------------------------------
-- File Name: ex6_nand.vhd
----------------------------------------------------------------------
LIBRARY IEEE; USE IEEE.std_logic_1164.ALL;                      -- 1
             USE IEEE.VITAL_timing.ALL;                         -- 2
             USE IEEE.VITAL_primitives.ALL;                     -- 3
LIBRARY FMF;  USE FMF.gen_utils.ALL;                            -- 4

ENTITY nandgate IS                                              -- 5
    GENERIC (                                                   -- 6
        tipd_A         : VitalDelayType01 := VitalZeroDelay01;  -- 7
        tipd_B         : VitalDelayType01 := VitalZeroDelay01;  -- 8
        tpd_A_YNeg     : VitalDelayType01 := UnitDelay01;       -- 9
        tpd_B_YNeg     : VitalDelayType01 := UnitDelay01;       -- 10
        InstancePath   : STRING    := DefaultInstancePath;      -- 11
        TimingModel    : STRING    := DefaultTimingModel        -- 12
    );                                                          -- 13
    PORT (                                                      -- 14
        A    : IN   std_ulogic := 'U';                          -- 15
        B    : IN   std_ulogic := 'U';                          -- 16
        YNeg : OUT  std_ulogic := 'U'                           -- 17
    );                                                          -- 18
    ATTRIBUTE VITAL_LEVEL0 of nandgate : ENTITY IS TRUE;        -- 19
END nandgate;                                                   -- 20


----------------------------------------------------------------------
-- ARCHITECTURE DECLARATION
----------------------------------------------------------------------
ARCHITECTURE ex6 OF nandgate IS                                 -- 21
    ATTRIBUTE VITAL_LEVEL1 of ex6 : ARCHITECTURE IS TRUE;       -- 22

    SIGNAL A_ipd   : std_ulogic := 'U';                         -- 23
    SIGNAL B_ipd   : std_ulogic := 'U';                         -- 24

BEGIN                                                           -- 25
    WireDelay : BLOCK                                           -- 26
    BEGIN                                                       -- 27
        w_1: VitalWireDelay (A_ipd, A, tipd_A);                 -- 28
        w_2: VitalWireDelay (B_ipd, B, tipd_B);                 -- 29
    END BLOCK;                                                  -- 30

    VITALBehavior : PROCESS (A_ipd, B_ipd)                      -- 31
    -- Functionality Results Variables
    VARIABLE YNeg_zd    : std_ulogic := 'U';                    -- 32

    -- Output Glitch Detection Variables
    VARIABLE YNeg_GlitchData : VitalGlitchDataType;             -- 33

    BEGIN                                                       -- 34
    YNeg_zd := VitalNAND2 (a => A_ipd, b => B_ipd);             -- 35


    ----------------------------------------------------------------
    -- Path Delay Section
    ----------------------------------------------------------------
    VitalPathDelay01 (                                          -- 36
        OutSignal     => YNeg,                                  -- 37
        OutSignalName => "YNeg",                                -- 38
        OutTemp       => YNeg_zd,                               -- 39
        Paths         => (                                      -- 40
            0 => (InPutChangeTime  => A_ipd'LAST_EVENT,         -- 41
                  PathDelay        => tpd_A_YNeg,               -- 42
                  PathCondition    => TRUE ),                   -- 43
            1 => (InPutChangeTime  => B_ipd'LAST_EVENT,         -- 44
                  PathDelay        => tpd_B_YNeg,               -- 45
                  PathCondition    => TRUE ) ),                 -- 46
        GlitchData  => YNeg_GlitchData );                       -- 47
    END PROCESS;                                                -- 48
END;                                                            -- 49
```

Figure 2.6 VITAL nand gate model with interconnect delays

The names are the same as the two input port names with the _ipd (interconnect path delay) suffix added. The _ipd suffix is an FMF convention. The signals must be of type std_ulogic. We initialize them to 'U'.

A WireDelay block begins on line 26:

```
WireDelay : BLOCK                                          -- 26
BEGIN                                                      -- 27
    w_1: VitalWireDelay (A_ipd, A, tipd_A);               -- 28
    w_2: VitalWireDelay (B_ipd, B, tipd_B);               -- 29
END BLOCK;                                                 -- 30
```

The label WireDelay is mandatory for this block. The only thing allowed in a WireDelay block is calls to the VitalWireDelay procedure. The VitalWire Delay procedure delays an input by the delay value specified by its tipd_ generic using a transport delay. A port with an associated wire delay should be read only in the WireDelay block. Elsewhere in the model, only the delayed (_ipd) signal should be read. It is recommended (another FMF convention) that each call to the VitalWireDelay procedure be given a label beginning with w.

The VITALBehavior process beginning on line 31,

```
VITALBehavior : PROCESS (A_ipd, B_ipd)                    -- 31
```

has changed only in that whereas ports A and B were referenced in the prior example, we now use the delayed signals A_ipd and B_ipd. This is a VITAL level 1 requirement. Otherwise, this process is the same as in Figure 2.4.

The default delays assigned on lines 9 and 10 of Figure 2.5,

```
tpd_A_YNeg    : VitalDelayType01 := UnitDelay01;          --  9
tpd_B_YNeg    : VitalDelayType01 := UnitDelay01;          -- 10
```

are defined in the gen_utils package as 1 nanosecond for both rising and falling outputs. If there is no timing backannotation to the netlist, all models will exhibit a propagation delay of 1 nanosecond. It was originally assumed that a unit delay simulation would run much faster than one with realistic timing. Experience has shown this not to be the case. There is little to be gained from not using actual delays.

The overall structure of a VITAL model can be seen in Chapter 5.

2.6   Finishing Touches

The examples so far have been for a generic nand gate. Now lets look at a model for a real part family. We will use the 54/74xx01 family. Our models are technology independent, so one model may represent any number of parts that have identical functionality but differ only in timing.

We call the model in Figure 2.7 STD01. STD01 is also a nand gate, but in this case it has an open collector output. The example comes from the Free Model Foundry library. Line numbers have been added for reference.

```
--------------------------------------------------------------------------------
-- File Name: std01.vhd
--------------------------------------------------------------------------------
-- Copyright (C) 1998 Free Model Foundry
--
-- This program is free software; you can redistribute it and/or modify
-- it under the terms of the GNU General Public License version 2 as
-- published by the Free Software Foundation.
--
-- MODIFICATION HISTORY:
--
-- version | author  | mod date | changes made
--   V1.0  R. Munden  98 APR 02  Initial release
--------------------------------------------------------------------------------
-- PART DESCRIPTION:
--
-- Library:    STND
-- Technology: 54/74XXXX
-- Part:       STD01
--
-- Description:  2-input positve-NAND gate with open-collector output
--------------------------------------------------------------------------------
```

```
LIBRARY IEEE;    USE IEEE.std_logic_1164.ALL;                     -- 1
                 USE IEEE.VITAL_timing.ALL;                       -- 2
                 USE IEEE.VITAL_primitives.ALL;                   -- 3
LIBRARY FMF;     USE FMF.gen_utils.ALL;                           -- 4
```

```
--------------------------------------------------------------------------------
-- ENTITY DECLARATION
--------------------------------------------------------------------------------
ENTITY std01 IS                                                   -- 5
    GENERIC (                                                     -- 6
        -- tipd delays: interconnect path delays
        tipd_A            : VitalDelayType01 := VitalZeroDelay01;  -- 7
        tipd_B            : VitalDelayType01 := VitalZeroDelay01;  -- 8
        -- tpd delays
        tpd_A_YNeg        : VitalDelayType01 := UnitDelay01;       -- 9
        -- generic control parameters
        MsgOn             : BOOLEAN  := DefaultMsgOn;              -- 10
        XOn               : Boolean  := DefaultXOn;               -- 11
        InstancePath      : STRING   := DefaultInstancePath;      -- 12
        -- For FMF SDF technology file usage
        TimingModel       : STRING   := DefaultTimingModel        -- 13
    );                                                            -- 14
    PORT (                                                        -- 15
        B         : IN    std_ulogic := 'U';                      -- 16
        A         : IN    std_ulogic := 'U';                      -- 17
        YNeg      : OUT   std_ulogic := 'U'                       -- 18
    );                                                            -- 19
    ATTRIBUTE VITAL_LEVEL0 of std01 : ENTITY IS TRUE;             -- 20
END std01;                                                        -- 21
```

```
--------------------------------------------------------------------------------
-- ARCHITECTURE DECLARATION
--------------------------------------------------------------------------------
ARCHITECTURE vhdl_behavioral of std01 IS                          -- 22
    ATTRIBUTE VITAL_LEVEL1 of vhdl_behavioral : ARCHITECTURE IS TRUE;  -- 23

    SIGNAL B_ipd      : std_ulogic := 'U';                        -- 24
    SIGNAL A_ipd      : std_ulogic := 'U';                        -- 25
```

Figure 2.7 Complete VITAL model of a nand gate component

```
BEGIN                                                                    -- 26
----------------------------------------------------------------------
-- Wire Delays
----------------------------------------------------------------------
WireDelay : BLOCK                                                        -- 27
BEGIN                                                                    -- 28

    w_1: VitalWireDelay (B_ipd, B, tipd_B);                             -- 29
    w_2: VitalWireDelay (A_ipd, A, tipd_A);                             -- 30

END BLOCK;                                                               -- 31


----------------------------------------------------------------------
-- VITALBehavior Process
----------------------------------------------------------------------
    VITALBehavior1 : PROCESS(A_ipd, B_ipd)                              -- 32

    -- Functionality Results Variables
    VARIABLE YNeg_zd        : std_ulogic := 'U';                        -- 33

    -- Output Glitch Detection Variables
    VARIABLE Y_GlitchData : VitalGlitchDataType;                        -- 34
BEGIN                                                                    -- 35

    ------------------------------------------------------------------
    -- Functionality Section
    ------------------------------------------------------------------
        YNeg_zd := VitalNAND2(a=> A_ipd, b => B_ipd,                   -- 36
                              Resultmap => STD_wired_and_rmap);        -- 37


    ------------------------------------------------------------------
    -- Path Delay Section
    ------------------------------------------------------------------
VitalPathDelay01 (                                                      -- 38
    OutSignal       =>  YNeg,                                          -- 39
    OutSignalName   =>  "YNeg",                                        -- 40
    OutTemp         =>  YNeg_zd,                                       -- 41
    XOn             => XOn,                                            -- 42
    MsgOn           => MsgOn,                                          -- 43

    Paths           => (                                              -- 44
        0 => (InputChangeTime   => A_ipd'LAST_EVENT,                  -- 45
              PathDelay         => tpd_A_Yneg,                        -- 46
              PathCondition     => TRUE ),                            -- 47
        1 => (InputChangeTime   => B_ipd'LAST_EVENT,                  -- 48
              PathDelay         => tpd_A_Yneg,                        -- 49
              PathCondition     => TRUE ) ),                          -- 50
    GlitchData      => Y_GlitchData );                                -- 51

    END PROCESS;                                                       -- 52

END vhdl_behavioral;                                                   -- 53
```

Figure 2.7 Complete VITAL model of a nand gate component *(continued)*

The first thing to note about the model in Figure 2.7 is that we have taken 114 lines to describe a simple nand gate for which a synthesizable model could have been written in just one line. Of these 114 lines, only 53 are actual VHDL/VITAL code. The rest are comments and white space. The different requirements of component models and synthesis models justify this huge disparity. Lets look at all the details.

The first 22 lines are comments. We start by documenting the name of the file as discussed earlier. Then comes a copyright statement. You are going to want to share your models and your coworkers are going to want to use them. The copyright states who owns the model and what rights are reserved. Even if you intend to retain no rights at all to the model, you should explicitly say so here.

Models change over time. As with any other software, there needs to be a way to determine the revision of a particular copy. It is also helpful to have some idea what changed from one revision to the next. Therefore, in the header, we have a MODIFICATION HISTORY block. This describes the version number, the author or editor of that version, the release date, and a brief summary of the changes made. If somebody contacts you about a problem with a model, the first thing you will want to know is the version number.

FMF models track history through comments in the header. This is for documentation purposes only. During development, you could use a software revision control system such as RCS or CVS to make it easier to record the actual code changes. This could be particularly beneficial during the development of very large models where there are multiple developers involved.

The part description section of the header states in which library the part has been placed. It tries to indicate the technology, if relevant, and it gives the name of the model. Finally, there is a one-line description of the part's function.

Major sections of the model are separated by comment banners. These may be indented but are extended to the full 80 character width of the format. As mentioned, models should always be limited to 80 characters in width to facilitate printing and online viewing. The banners indicate the nature of the following section of code. Indented comments describe subsections of code or groups of generics or variables.

This model in Figure 2.6 has three lines of generic control parameters on lines 10, 11, and 12:

```
-- generic control parameters
MsgOn                 : BOOLEAN  := DefaultMsgOn;              -- 10
XOn                   : BOOLEAN  := DefaultXOn;               -- 11
InstancePath          : STRING   := DefaultInstancePath;      -- 12
```

The generic MsgOn is a boolean that controls message generation should a glitch be detected in the VitalPathDelay procedure. If there were timing constraint procedures, they would also read the value of MsgOn. Because a model may contain several VitalPathDelay procedures, it is convenient to control all of them through a generic.

The next generic, XOn, controls the generation of 'X's on the output if a glitch is detected.

The last of the three generics, InstancePath, is a string that becomes part of the warning or error message produced by a timing constraint procedure when a violation is detected. Although this model does not have any timing constraints, we still include the generic.

The last line in the generic list, 13,

```
-- For FMF SDF technology file usage
TimingModel         : STRING   := DefaultTimingModel                   -- 13
```

declares a very important generic. The `TimingModel` generic is used for selecting
the section of the timing file that corresponds to the part number used in the design
for each instance of each component. A `TimingModel` property or attribute is
attached to the component symbol in the schematic. The schematic's VHDL netlister will copy the value of the attribute to the generic map of the instance in the
netlist. When the `mk_sdf` program (a free utility used to generate the SDF file,
described in more detail in Chapter 12) reads the netlist, it picks out the value
and uses it to match a section in the timing file for the model. It then uses the
information from the timing file to create an entry for the instance in the SDF
file for the design. We will look at SDF in Chapter 4 and the formats for timing
files Chapter 12.

Port, signal, and variable declarations are made one per line. Indentations are
set to four spaces. We always use spaces instead of tabs. Tabs may be set to anything by the reader or printer. By using spaces, we can control the formatting and
be sure the model will print legibly.

Except for `WireDelay` blocks, always try to use named associations. For example,
on lines 36 and 37 we write

```
YNeg_zd:= VitalNAND2(a=> A_ipd, b => B_ipd,                          -- 36
                     Resultant => STD_wired_and_rmap);               -- 37
```

in which we specify that `A_ipd` is associated with a, and so on. Although this makes
the model easier to understand, there is another important reason. Many of the
VITAL functions and procedures have default parameters set in them. Because not
all parameters always need to be passed during the call, named association is
required to ensure the values given are passed to the right parameters. In Figure
2.6, the `VitalNAND2` function was called without the `Resultmap` parameter. The
function defaulted to outputting one of ('U', 'X', '0', '1'), which is normal for
an output that can drive both high and low. In Figure 2.7, `Resultmap` gets
`STD_wired_and_rmap` from the `FMF.gen_utils` package and the output is
mapped to one of ('U', 'X', '0', 'Z'). This is correct for an open collector output
that can drive low but not high.

2.7   Summary

A component model is quite different than a synthesizable model. Component
behavior is modeled at as high a level of abstraction as practical. Formatting and
readability are more important in a component model because it is likely to see
wider circulation and have a longer useful life. The use of standard interfaces,
specifically `std_ulogic`, is required to ensure that all component models can easily
be integrated into board-level simulations.

Unlike synthesizable models, component models are used to verify timing. They include the simulation of propagation delays and interconnect delays and the checking of timing constraints. Taking advantage of the VITAL standard allows the models to use generics to bring in the actual timing values through SDF files. By maintaining all timing values external to a model, it can be technology independent. As processes evolve and new speed grades become available, the timing file can be updated without the need to modify a tested model.

II Resources and Standards

In Part II we examine the standards adhered to in component modeling and the many supporting packages that make life easier for the component modeler. These standards are from the IEEE. They include VHDL, VITAL, and SDF. The packages covered are the IEEE VITAL packages and some packages written expressly for component modeling from the Free Model Foundry.

Chapter 3 covers several IEEE and FMF packages that are used in writing component models. The Standard Logic 1164 package is discussed. Particular attention is given to the VITAL packages. These include VITAL Timing, VITAL Primitives, and VITAL Memory. Four packages from FMF are also reviewed.

Chapter 4 provides a basic tutorial on the Standard Delay Format as it applies to component modeling. The overall file format is described. The capabilities of SDF and its syntax are explored.

Chapter 5 describes the organization and requirements of VITAL models. The different requirements for level0 and level1 models are provided. The mapping of SDF to VITAL generics is explained. The sections of a VITAL model are described along with the order in which they must appear.

Chapter 6 is a detailed examination of modeling delays within and between components. The use of VITAL path delay procedures is unraveled. Different delay modes and how they relate to glitch detection are explained. The trade-offs between distributed delays and pin-to-pin delays are discussed.

Chapter 7 discloses the truth behind VITAL truth tables and state tables and their employment in component modeling. VITAL memory tables are not forgotten. This chapter reveals the differences between truth tables and state tables, how to create them, and when each is appropriate. It also touches on memory tables.

In Chapter 8, timing constraints are defined and the essentials of constraint modeling are described. Each type of timing constraint is explained, along with its usage.

3 VHDL Packages for Component Models

VHDL packages are used to simplify models and facilitate code reuse. In this chapter, we are going to take a brief look at eight packages from two libraries that are frequently used in modeling board-level digital components.

A package is a design unit much like a model. The difference is a package contains code intended for use in other models. Packages are organized and compiled into libraries. One of the great strengths of VHDL is the way it works with libraries and packages. The purpose of these packages is to make modeling faster and easier by taking commonly used functions, procedures, and declarations and packaging them so they may be referenced by many models rather than copied or reinvented. Every model will require at least one of these packages. Few component models will use less than four of them.

The first library we will look at is the IEEE library, and the packages are `std_logic_1164` and the three VITAL packages. All the packages in this library are balloted standards of the IEEE. There are other IEEE packages that are not described here because they have not been used in component models.

The second library is the FMF library from the Free Model Foundry. The four packages we will explore are all written explicitly for component modeling. They are not an official standard but are available for public use as open source under the Free Software Foundation GPL.

Details of specific features of these packages will be given in later chapters as we get to models that use them. The source code for the IEEE packages should be available in your simulator installation tree. The FMF packages are available from the FMF Web site. You are encouraged to read the code in these packages for a better understanding of how they work and how to use them.

3.1 STD_LOGIC_1164

The `std_logic_1164` package, also called `std_logic`, is at the foundation of component modeling and board-level simulation. Every model and every package we write uses this package.

When the VHDL language was being written, the designers decided not to define all the logic levels that would be needed for years to come. They took the approach of allowing users to define the logic levels that met their needs. Although this allows the maximum flexibility it also encourages the minimum interoperability, so an IEEE committee created the `std_logic_1164` standard to define a set of logic values and functions that would work under most circumstances.

3.1.1   Type Declarations

The most used contribution of `std_logic` to the world of component modeling and board-level simulation is the definition of the 9-value logic system called `std_logic`. While inside your FPGA you maybe able to get by with type `bit` and `bit_vector`, or maybe even `signed` and `unsigned`, on the outside you have to deal with tri-state signals and resistive drivers.

`std_logic`'s 9 values are, in order,

'U', Uninitialized;

'X', Forcing Unknown;

'0', Forcing 0;

'1', Forcing 1;

'Z', High Impedance;

'W', Weak Unknown;

'L', Weak 0;

'H', Weak 1;

'-', don't care.

There are two flavors of this system, `std_logic` and `std_ulogic`. `std_logic` is a resolved type and `std_ulogic` is not. This means a signal of type `std_logic` can have multiple drivers and a resulting signal value will be determined by a resolution function. For example, for a signal that has an open collector driver and a pull-up resistor, the OC output can drive either a 'Z' or a '0' and the pull-up always drives an 'H'. The resolution function in `std_logic_1164` causes 'Z' and 'H' to resolve to 'H', '0' and 'H' resolve to '0'.

It is worth noting here that although we can speak of a model driving a 'Z', it represents no drive at all in a physical component. It is an output driver that is turned off.

Every assignment of a value to a signal or variable of type `std_logic` results in a call to the resolution function. To save this overhead, signals with a single driver can be of type `std_ulogic`. `std_ulogic` has the same list of values as `std_logic` but without resolution.

We usually represent bused signals as type `std_logic_vector` or `std_ulogic_vector`. Many times we will have a signal that should never have a weak value inside a model. We can write more compact and efficient code by typing such a signal using one of the std_logic subtypes X01 or UX01. However, caution must be exercised, as many VITAL functions and procedures expect arguments of std_logic and some compilers will complain if a subtype is used instead.

3.1.2   Functions

The std_logic_1164 package also contains several functions. These are the ones we will use most often in component modeling:

TO_X01 and TO_UX01 perform a type conversion from std_ulogic to either type X01 or UX01. These functions will cause 'H' and 'L' values to be translated to '1' and '0'.

RISING_EDGE detects a rising transition on a signal of type std_ulogic.

FALLING_EDGE likewise detects a falling transition.

IS_X detects a value of 'U', 'X', 'Z', 'W', or '-' on a signal of type std_ulogic or std_ulogic_vector.

The use of each of these functions is illustrated in later examples.

3.2   VITAL_Timing

The VITAL_Timing package is the root of the VITAL library. It is called by all the other VITAL packages. It provides the facilities for specifying propagation delays and timing constraints as well as SDF backannotation. Every VITAL model must reference this package.

3.2.1   Declarations

The VITAL_Timing package declares many types that we will use extensively. Here are the ones with which you will become most familiar:

VitalDelayType is a subtype of TIME. It is used to hold simple delays.

VitalDelayType01 is an array of TIME. It holds two values for rising (tr01) and falling (tr10) transition delays. It is used to hold delays for 2-state outputs.

VitalDelayType01Z is an array of TIME. It holds six values for tr01, tr10, tr0Z, trZ1, tr1Z, and trZ0 transitions, in that order. It is used to describe delays through 3-state devices.

VitalDelayType01ZX is an array of TIME. It holds twelve values for tr01, tr10, tr0Z, trZ1, tr1Z, trZ0, tr0X, trX1, tr1X, trX0, trXZ, and trZX transitions, in that order. It is used to describe output delays on a device that can traverse an unknown state, such as many memory devices.

VitalResultMapType is an array (UX01) of std_ulogic. It is used to map the outputs of VITAL primitives to other output strengths, such as open collector.

VitalTableSymbolType is an enumerated list of symbols used to represent signal transitions or steady state conditions. It is used in timing constraint procedures and in truth and state tables. The procedures for truth tables and state tables are defined in the VITAL_Primitives package. The symbols are (in order of enumeration) defined as follows:

'/'—$0 \rightarrow 1$
'\'—$1 \rightarrow 0$
'P'—Union of '/' and '∧' (any edge to 1)
'N'—Union of '\' and 'v' (any edge to 0)
'r'—$0 \rightarrow X$
'f'—$1 \rightarrow X$
'p'—Union of '/' and 'r' (any edge from 0)
'n'—Union of '\' and 'f' (and edge from 1)
'R'—Union of '∧' and 'p' (any possible rising edge)
'F'—Union of 'v' and 'n' (any possible falling edge)
'∧'—$X \rightarrow 1$
'v'—$X \rightarrow 0$
'E'—Union of 'v' and '∧' (any edge from X)
'A'—Union of 'r' and '∧' (rising edge to or from X)
'D'—Union of 'f' and 'v' (falling edge to or from X)
'*'—Union of 'R' and 'F' (any edge)
'X'—Unknown level
'0'—low level
'1'—high level
'-'—don't care
'B'—0 or 1
'Z'—high impedance
'S'—steady value

The reader should note that this list contains both uppercase and lowercase characters. As with any enumerated type, the values are case sensitive. Also note that although some of the symbols are also used in the type std_logic, they do not have the same meaning. Other types are also defined in this package and are discussed as they are encountered in models.

3.2.2   Procedures

The VITAL_Timing package has a number of important procedures used for controlling output delays and defining timing constraints. Here are some of the procedures you will call directly:

VitalPathDelay01 is used for assigning a delayed value to an output signal that can only have a high or low value. It can take up to 13 parameters for precisely controlling its behavior.

`VitalPathDelay01Z` is like `VitalPathDelay01` but is for outputs that can be put in a high impedance state. Both these procedures will be explained in detail in Chapter 6.

`VitalWireDelay` is used to delay an input signal to simulate interconnect delays. Its use is detailed in Chapter 6.

`VitalSignalDelay` is used in models that have negative timing constraints. The topic of negative timing constraints is taken up in Chapter 11.

`VitalSetupHoldCheck` detects a setup or hold violation. The test signal may be either a scalar or a vector. It can take up to 22 arguments.

`VitalRecoveryRemovalCheck` detects the presence of a recovery or removal violation. It has 21 parameters. Recovery and removal usually refer to asynchronous signals such as the preset and clear functions on a 7474 flip-flop.

`VitalPeriodPulseCheck` is used to test for a minimum pulse width, either high or low, and a maximum periodicity (1/freq.) of a signal. The types of signals most commonly tested are clocks, resets, and write enables. This procedure can accept up to 13 parameters.

`VitalInPhaseSkewCheck` detects an in-phase skew violation between two input signals.

`VitalOutPhaseSkewCheck` detects an out-of-phase skew violation between two signals. Each of the skew check procedures can take up to 18 parameters.

There are several more procedures in the `VITAL_Timing` package that are used internal to the package and will not be discussed. All of the timing constraint procedures are examined in depth in Chapter 8.

3.3 VITAL_Primitives

Primitives are used in simulation languages to describe the most basic logical functions. They are used in structural descriptions or netlists of logic designs. The physical implementation of these functions are called *gates*. These include the *and*, *or*, *invert*, and *buffer* functions and their variations (*nand*, *nor*, etc.). Verilog defines 14 logical primitives along with 12 switches. The VHDL language does not define any primitives. Switches are bidirectional primitives and have no corresponding elements in VHDL.

The primary purpose of the `VITAL_Primitives` package is to provide the primitives that were not defined in the VHDL language and to accelerate the simulation of ASIC gate-level netlists. It contains a set of functions and procedures that roughly correspond to Verilog's gate-level simulation primitives. It also contains the declarations, functions, and procedures required to enable writing truth tables and state tables similar to Verilog's user-defined primitives (UDPs). Placing this code in an IEEE standard package has enabled simulator developers to write compilers that can better optimize their compiled code for efficient simulation. This generally means smaller memory requirements and faster simulations.

3.3.1   Declarations

Few of the declarations made in this package are used outside the package in the normal course of modeling. They will not be listed here.

3.3.2   Functions and Procedures

The first group of functions and procedures you will encounter in the VITAL_Primitives package are the simulation primitives. Each primitive is provided as both a function and a procedure. The functions are called from inside VHDL processes where they will be executed serially. Their inputs will often be variables. The procedures are called from outside VHDL processes and are executed concurrently. The procedures take only signals for their logical inputs. In addition to signals, they can accept constants for specifying input to output delays. Both types of primitives accept result maps to control the strengths of their output signals. VITAL outdid Verilog in defining primitives. There are 39 primitives in the VITAL_Primitives package. Their names are self-explanatory. The functionality of some of these primitives goes well beyond single gates:

```
VitalIDENT

VitalBUF: VitalBUF, VitalBufIf0, VitalBufIf1

VitalINV: VitalINV, VitalInvIf0, VitalInvIf1

VitalAND: VitalAND, VitalAND2, VitalAND3, VitalAND4

VitalNAND: VitalNAND, VitalNAND2, VitalNAND3, VitalNAND4

VitalOR: VitalOR, VitalOR2, VitalOR3, VitalOR4

VitalNOR: VitalNOR, VitalNOR2, VitalNOR3, VitalNOR4

VitalXOR: VitalXOR, VitalXOR2, VitalXOR3, VitalXOR4

VitalXNOR: VitalXNOR, VitalXNOR2, VitalXNOR3, VitalXNOR4

VitalMux: VitalMux, VitalMux2, VitalMux3, VitalMux4

VitalDecoder:   VitalDecoder,   VitalDecoder2,   VitalDecoder4,
VitalDecoder8
```

The VITAL_Primitives package defines the procedures for VitalTruthTables and VitalStateTables. The primary input to one of these procedures is a constant in the form of a table. Figure 3.1 is an example of a VitalStateTable describing the functionality of D register.

Truth and state tables are worthy of their own chapter. They are explored in more detail in Chapter 7.

```
----------------------------------------------------------------
-- Simple register without previous states
----------------------------------------------------------------
CONSTANT DREG_tab : VitalStateTableType  := (

    ----INPUTS-------|-OUTPUT--
    -- Viol CLK    D | Qi     --
    ----------------|---------
    ( 'X', '-', '-', 'X'), -- timing violation
    ( '-', 'X', '-', 'X'), -- clk unknown
    ( '-', '/', '0', '0'), -- active clock edge
    ( '-', '/', '1', '1'), -- active clock edge
    ( '-', '/', '-', 'X'), -- active clock edge
    ( '-', '-', '-', 'S')  -- default

); -- end of VitalStateTableType definition
```

Figure 3.1 StateTable for a D register

3.4   VITAL_Memory

The VITAL_Memory package was added to the VITAL standard in the 2000 release. It was created to support the modeling of static memory (SRAM) used in ASIC designs. It is the largest of the VITAL packages, about the size of the VITAL_Timing and VITAL_Primitives packages combined.

This package can be thought of as having three sections: memory functionality procedures, memory timing specifications, and memory timing check procedures. Parts of it can be used only in VITAL level 1 models.

3.4.1   Memory Functionality

The VITAL_Memory package provides one function and three procedures to aid in the modeling of static memory.

VitalDeclareMemory is a function to declare and initialize memory. It establishes a storage mechanism that is very efficient but only for word sizes that are multiples of eight bits. It can automatically initialize the contents of memory from a file but only at simulation elaboration time.

VitalMemoryTable is a procedure to perform memory read, write, and corruption. It uses a compact, table-based approach to modeling similar to a state table. Such tables encourage the creation of packages for reuse.

VitalMemoryCrossPorts is a procedure for implementing multiport contention and crossport reads in multiport memories. Fortunately, it is rarely needed for component modeling.

VitalMemoryViolation is a procedure supporting the modeling of memory violation actions. It takes a table-based approach to memory corruption policies.

3.4.2   Memory Timing Specification

The VITAL_Memory package timing section contains three procedures for specifying propagation delays. The path delay procedures together support complex output scheduling and include support of output retain specifications.

VitalMemoryInitPathDelay is used to initialize the data structure for the output path delay schedule. It takes a maximum of three arguments.

VitalMemoryAddPathDelay is used to specify a delay path from an input to an output port. While there is a maximum of only nine arguments to this procedure, it is overloaded 24 times, which can make it difficult to debug.

VitalMemorySchedulePathDelay schedules the functional output value on the output port. It includes a particularly useful feature known as output retention that is explained in Chapter 14.

3.4.3   Memory Timing Checks

The VITAL_Memory package timing section contains two procedures for specifying timing constraints.

VitalMemorySetupHoldCheck is similar to the VitalSetupHoldCheck procedure. It has two additional arguments for timing arcs and memory subword size.

VitalMemoryPeriodPulseCheck detects periodicity and pulse width violations. It differs from the VitalPeriodPulseCheck procedure in that it can check both minimum and maximum periods.

The VITAL_Memory package is given more attention in Chapter 7 and examples are provided as we investigate memory modeling techniques in Chapter 14.

3.5   FMF Packages

The next few packages we will survey are from the Free Model Foundry. Although not standards in the sense of IEEE standards, they have been written expressly for use in modeling components for ASIC/FPGA and board-level design verification. They are freely available from the FMF Web site.

Another difference is the FMF packages do not come precompiled with your simulator. To use these packages, you must create a library with the logical name FMF someplace in your file system where they will be visible to any user who needs them. Then compile all the packages into that library. Finally, your simulation tool must be configured to know where to find the FMF library. How this is done is different for every simulator. Using Modelsim as an example, the system "modelsim.ini" file should be edited to associate the FMF logical library with the location in the file system where the FMF packages have been compiled.

3.5.1   FMF gen_utils and ecl_utils

The gen_utils package is the most used of the FMF packages. It contains a number of declarations for constants and functions:

STD_wired_and_rmap is a constant of type VitalResultMapType. It is used to map the outputs of VITAL primitives and path delay procedures to open collector drive strengths ('U', 'X', '0', 'Z').

diff_rec_tab is a constant of type VitalStateTableType. It is used to convert a differential data input to a single-ended signal for use internal to the model.

The constants UnitDelay, UnitDelay01, UnitDelay01Z, and UnitDelay01ZX are all used to provide default timing values of 1 nanosecond to VITAL timing generics.

GenParity is a function that will generate an odd or even parity bit for each 8 bits of a std_logic_vector and return it as the ninth bit.

CheckParity is a function that will check for correct parity in 9-bit words.

To_UXLHZ is a function that converts strong signal values to weak values. Its primary application has been in writing VITAL wrappers for RTL models with bidirectional ports.

The ecl_utils package is much like the gen_utils package except it is tailored to modeling ECL components. It also has VitalStateTables and VitalTruth Tables for reading differential inputs and BB outputs as described in Chapter 16. Most of its contents are described here:

ECL_wired_or_rmap is a constant of type VitalResultMapType. It is used to map the outputs of VITAL primitives and path delay procedures to open emitter drive strengths ('U', 'X', 'Z', '1').

The constants ECLUnitDelay, ECLUnitDelay01, and ECLUnitDelay01Z are all used to provide default timing values of 1 nanosecond to VITAL timing generics.

ECLVbbValue is a constant of type std_logic. It used to set the value for ECL VBB output pins. It is set to 'W'.

The constant ECL_diff_mode_tab is a look-up table used to determine whether an input pair is differential or single ended. It senses which, if either, of the inputs is connected to the VBB output. This table's output is used as an input to another table that decodes the differential inputs.

The constant ECL_s_or_d_inputs_tab is another table for determining the value of a nonclock differential input pair.

The constant ECL_clk_tab is the final table for computing a single signal from a differential ECL clock pair. One of the inputs to this table is the output from the ECL_diff_mode_tab table.

3.5.2   FMF ff_package

The **ff_package** is the largest of the FMF packages. It contains VitalStateTables for 37 different types and configurations of latches and flip-flops. It is widely used, particularly in glue logic models. Should any new types of latches or flip-flops be discovered, they will be added to this package.

Examples of how to use this package are given as we discuss VITAL state tables in Chapter 7 and modeling register devices in Chapter 9.

Table 3.1 Supported conversion functions

From	*To*	*Conversion*
std_logic_vector	natural	to_nat
std_logic_vector	integer	to_int
std_logic_vector	string	to_time_str
std_logic_vector	hex string	to_hex_str
std_logic_vector	decimal string	to_int_str
std_logic_vector	octal string	to_oct_str
std_logic_vector	binary string	to_bin_str
std_logic	natural	to_nat
std_logic	string	to_time_str
std_logic	binary string	to_bin_str
natural	std_logic_vector	to_slv
natural	std_logic	to_sl
natural	time	to_time
natural	string	to_time_str
natural	hex string	to_hex_str
natural	decimal string	to_int_str
natural	octal string	to_oct_str
natural	binary string	to_bin_str
time	natural	to_nat
time	string	to_time_str
hex string	std_logic_vector	h
hex string	natural	h
decimal string	std_logic_vector	d
decimal string	natural	d
octal string	std_logic_vector	o
octal string	natural	o
binary string	std_logic_vector	b
binary string	natural	b

3.5.3   FMF Conversions

The conversions package was contributed to FMF by SEVA Technologies. It is used in most complex models. It provides functions to perform type conversions between various signal types (`std_logic`, `std_logic_vector`), the numeric type `NATURAL`, `TIME`, and several string types (binary, octal, decimal, and hex). It is well documented internally.

Table 3.1 shows the various supported conversions along with the name of the function to use for each conversion.

The primary reason for creating this package was a proliferation of incompatible arithmetic packages provide by simulator vendors. The use of any one of those packages was likely to prevent a model from being used by any other brand of simulator. FMF has furnished the conversions package so that people may write portable component models.

The problem of incompatible arithmetic packages has been somewhat alleviated by the IEEE packages `NUMERIC_STD` and `NUMERIC_BIT`. However, both of these packages are designed for use with synthesis tools. This may reduce simulation performance. Perhaps more important, companies that provide models of their standard components usually prefer those models to be nonsynthesizable.

3.6   Summary

Packages contain proven code that can be used to make your models simpler and more robust. They also make model development less work because they allow you to incorporate functionality that others put much thought and effort into devising. The IEEE standard packages are usually precompiled and optimized by the tool vendors to maximize simulation performance. However, the FMF packages must be downloaded and compiled on your system and entered into the configuration of your simulator.

4

An Introduction to SDF

SDF is a file format used to convey timing information to the simulator. Understanding it will help you write correct model entities. In this chapter we present the basics of SDF as they apply to the verification of ASICs and FPGAs at the board level.

The Standard Delay Format is based on IEEE Standard 1497. [1] The specification describes an ASCII file format that contains propagation and interconnect delays and timing constraints. SDF can be read by a simulator to supply values for propagation and interconnect delays and timing constraints. SDF is the standard format for backannotating timing into a VHDL/VITAL or Verilog simulation. All major simulators can read this format. When you simulate your ASIC or FPGA at the gate level, you will usually read in an SDF file generated by your tool chain. Likewise, SDF is generated and used to backannotate values for component delays, interconnect delays, and timing constraints at the board level.

The component modeling method described in this book results in technology-independent models. For example, a single model of a specific memory type can be used for all speed grades of that memory without editing the model. This is possible because timing values are stored external to the model. The model expresses only the behavior of the part. As speed grades are made available in the future, they are added to the external timing file. Examples of timing files will be given in later chapters. For now, all you need to know is that they are composed of sections of SDF encapsulated in eXtensible Markup Language (XML).

4.1   Overview of an SDF File

Let's look at the structure of an SDF file. At the top level, an SDF file can be divided into two sections: a header and a cell list. A simple SDF file is presented in Figure 4.1. This SDF file is for a testbench referencing a single instance of an std646. The timing is for an SN74BCT646NT. A simplified schematic of an std646 is shown in Figure 4.2.

It is important to understand that an SDF file can only be created for a testbench or netlist, never for a bare component model. This is because SDF backannotation always applies timing values to instances:

```
(DELAYFILE
 (SDFVERSION "3.0")
 (DESIGN "tbstd646")
 (DATE "Sun Feb  9 13:12:08 2003")
 (VENDOR "Free Model Foundry")
 (PROGRAM "SDF timing utility(tm)")
 (VERSION "2.0.3")
 (DIVIDER /)
 (VOLTAGE)
 (PROCESS)
 (TEMPERATURE)
 (TIMESCALE 1ns)
 (CELL
  (CELLTYPE "std646")
  (INSTANCE std646_1)
  (DELAY (ABSOLUTE
    (IOPATH A B (3.1:6:9.5) (3.7:6.8:10.5))
    (IOPATH SAB A (3.9:8.8:13.8) (3.3:8.3:912.9))
    (IOPATH CLKAB B (3.6:7:11.2) (3.9:7:10.6))
    (IOPATH DIR A () () (3.2:7.3:11.8) (2.8:7.8:13.1) (3.8:8.4:12.6) (3.8:8.9:14.6))
    (IOPATH OENEG A () () (3.4:7:10.5) (4:7.9:13.2) (4:7.2:10.9) (4.6:8.8:14.4))
  ))
  (TIMINGCHECK
   (SETUP A CLKAB (6:6:6))
   (SETUP B CLKBA (6:6:6))
   (HOLD A CLKAB (.5:.5:.5))
   (HOLD B CLKBA (.5:.5:.5))
   (WIDTH (posedge CLKAB) (6:6:6))
   (WIDTH (posedge CLKBA) (6:6:6))
   (WIDTH (negedge CLKAB) (6:6:6))
   (WIDTH (negedge CLKBA) (6:6:6))
   (PERIOD (posedge CLKAB) (12.1:12.1:12.1))
   (PERIOD (posedge CLKBA) (12.1:12.1:12.1))
  )
 )
)
```

Figure 4.1 Sample SDF file

```
(CELL
 (CELLTYPE "std646")
 (INSTANCE std646_1)
```

A model does not contain an instance of itself, so backannotation is not possible.

4.1.1   Header

The first 12 lines in Figure 4.1 constitute the header:

```
(DELAYFILE
  (SDFVERSION "3.0")
  (DESIGN "tbstd646")
  (DATE "Sun Feb 9 13:12:08 2003")
  (VENDOR "Free Model Foundry")
  (PROGRAM "SDF timing utility(tm)")
  (VERSION "2.0.3")
```

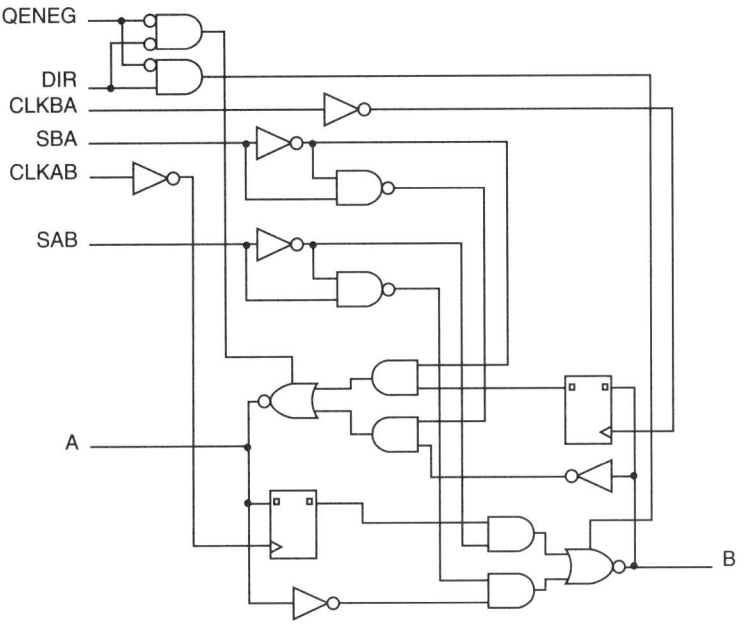

Figure 4.2   Simplified schematic of an STD646

```
(DIVIDER /)
(VOLTAGE)
(PROCESS)
(TEMPERATURE)
(TIMESCALE 1ns)
```

It begins with the key word DELAYFILE. After that there are fields specifying the following:

SDFVERSION, which version of the SDF standard we are using. The current version is 4.0. You should never see anything older than 2.1.

DESIGN, the name of the design for which the file is written. This field is optional and for documentation purposes only.

DATE, when the file was created. An optional field for documentation purposes only.

VENDOR, another optional field listing the name of the company manufacturing the device (if the file is for an ASIC or FPGA) or the originator of the program that created the file. In this case, the file was created by a perl script provided by the Free Model Foundry.

PROGRAM, the name of the program that created the file. Small SDF files may be created by hand but usually you will want to use a program. This field is optional.

VERSION, the version of the program. Again, this field is optional.

DIVIDER, the hierarchy divider. It separates elements of the hierarchical path to each cell. It can have one of two possible values: "." or "/" (without the quotes). It is an optional field. If omitted the separator defaults to "."

VOLTAGE, PROCESS, and TEMPERATURE, optional fields that apply to ASICs but not board-level netlists.

TIMESCALE, specifies the units for all time values in the SDF file. It is optional and defaults to 1 nanosecond. For SDF files discussed in this book and files generated by FMF programs, we will always use 1 nanosecond as the unit for our SDF files.

Because your SDF files will usually be generated by a program, you will rarely need to create a header.

4.1.2   Cell

The rest of the SDF files is a list of cells. The cells need not be in any particular order. However, a path will be given to each cell in the netlist.

```
(CELL
(CELLTYPE "std646")
(INSTANCE std646_1)
```

For Figure 4.1, the cell elements are as follows:

CELL, the key word signifying the beginning of a cell entity. There will be a cell for each component in the design for which you will backannotate timing values.

CELLTYPE, the name of the component model as it appears in the HDL netlist. It could also be the name of the hierarchical region if the netlist is hierarchical.

INSTANCE, identifies the particular instance of the cell, including the hierarchical path. In our example, the cell is at the top level so there is no path. In a more complex design it could look like "top/memory/idt7016_3." If an instance is not in the SDF file it will not be backannotated, but there will be no error.

4.1.3   Timing Specifications

Next come the timing specifications. They are divided into delays and constraints.

```
(DELAY (ABSOLUTE
  (IOPATH A B (3.1:6:9.5) (3.7:6.8:10.5))
  (IOPATH SAB A (3.9:8.8:13.8) (3.3:8.3:912.9))
  (IOPATH CLKAB B (3.6:7:11.2) (3.9:7:10.6))
  (IOPATH DIR A () () (3.2:7.3:11.8) (2.8:7.8:13.1) (3.8:8.4:12.6)
    (3.8:8.9:14.6))
  (IOPATH OENEG A () () (3.4:7:10.5) (4:7.9:13.2) (4:7.2:10.9)
    (4.6:8.8:14.4))
))
(TIMINGCHECK
  (SETUP A CLKAB (6:6:6))
  (SETUP B CLKBA (6:6:6))
  (HOLD A CLKAB (.5:.5:.5))
  (HOLD B CLKBA (.5:.5:.5))
  (WIDTH (posedge CLKAB) (6:6:6))
  (WIDTH (posedge CLKBA) (6:6:6))
  (WIDTH (negedge CLKAB) (6:6:6))
  (WIDTH (negedge CLKBA) (6:6:6))
  (PERIOD (posedge CLKAB) (12.1:12.1:12.1))
  (PERIOD (posedge CLKBA) (12.1:12.1:12.1))
```

DELAY, the key word signaling the beginning of the delay section. There are four delay types possible. They are PATHPULSE, PATHPULSEPERCENT, ABSOLUTE, and INCREMENT. We are only concerned with ABSOLUTE in this book. See the IEEE-1497 standards document for details on the others.

IOPATH, is an input–output path delay. It is followed by the names of the input and output ports and an ordered list of delays. The name of the input can have an optional edge identifier. The ordered list of delays consists of a number of triplets. Each triplet specifies minimum, typical, and maximum values for its particular transition, as shown later. Ports A and B are tristate outputs. A can drive B high or low. Therefore, there are two triplets: tr01 and tr10. However, OENEG is an output enable. It can only cause A to switch between high impedance and low impedance. It requires a set of six triplets: tr01, tr10, tr0Z, trZ1, tr1Z, and trZ0. Because OENEG cannot cause a transition between high and low, the first two triplets can be left empty.

```
(IOPATH A B (3.1:6:9.5) (3.7:6.8:10.5))
(IOPATH OENEG A () () (3.4:7:10.5) (4:7.9:13.2) (4:7.2:10.9) (4.6:8.8:14.4))
```

In the std646 model, the delays from A to B are the same as from B to A. Therefore, a single generic is used for both paths and a single line in the SDF file supplies the needed values. This is also true of other path delays.

TIMINGCHECK, the key word to signal the beginning of the timing constraint section.

SETUP, specifies a setup constraint value. It is followed by the names of two input ports. The first is the test port, usually address or data, and the second is the reference port, usually a clock. Finally there is a triplet. Because we are only interested in worst-case constraints, not min, typ, or max, the three values are identical. Alternatively, only a single value could be supplied.

```
(SETUP A CLKAB (6:6:6))
```

HOLD, identical to SETUP except it specifies a hold constraint value.

```
(HOLD A CLKAB (.5:.5:.5))
```

WIDTH, defines a minimum pulse width value. A single port name is preceded by an edge specifier. Finally the constraint value is given. The posedge specification indicates the high phase of the pulse, negedge the low phase.

```
(WIDTH (posedge CLKAB) (6:6:6))
(WIDTH (negedge CLKAB) (6:6:6))
```

PERIOD, specifies the minimum value for a period constraint. A single port name is preceded by an optional edge specifier. Finally the constraint value is given.

```
(PERIOD (posedge CLKAB) (12.1:12.1:12.1))
```

This was a relatively simple example. Next we will look at a wider range of what is possible with SDF.

4.2   SDF Capabilities

SDF was originally intended for use in a chip design process. Things can be done with it that are beyond those needed for board-level simulation. In this discussion we will limit ourselves to those features that we know apply to component modeling.

As we saw in our example, SDF includes constructs for modeling both circuit delays and timing checks. All of the constructs we are about to examine have mappings to VITAL. Those mappings are shown in later chapters.

4.2.1   Circuit Delays

Circuit delays can be either interconnect delays or path delays. SDF files describing interconnect delays are generated by (nearly all) vendor-supplied tools that generate interconnect information, such as printed circuit board layout tools or signal integrity tools. We will not cover the interconnect portion of the SDF standard.

Path delays are delays within or through a component. In most cases we specify pin-to-pin path delays, but in modeling some complex parts we must also model internal delays deep inside the component.

Each delay definition we are about to study requires a set of delay values. The set is formally referred to as a *delval_list*. That is to say it is a list of *delval* tokens.

Each of these tokens consists of either one, two, or three real numbers. If the token has one number it is simply interpreted as a delay. If it has two numbers, they are taken as minimum and maximum delays. If there are three numbers, they will represent minimum, typical, and maximum delays.

The formal syntax for *delval_list* is as follows:

```
delval_list ::=
    delval
| delval delval
| delval delval delval
| delval delval delval delval delval delval
| delval delval delval delval delval delval delval delval delval delval
    delval delval
```

The number of tokens in the *delval_list* can be one, two, three, six, or twelve. For component modeling,

A single token is specified if a single value is sufficient for all transitions.

Two tokens are specified for 2-state drivers with $0 \to 1$, $1 \to 0$ transitions.

Three tokens are specified for $0 \to 1$, $1 \to 0$, $? \to Z$ (1 or 0 to Z) transitions.

Six tokens are specified for normal 3-state drivers that have $0 \to 1$, $1 \to 0$, $0 \to Z$, $Z \to 1$, $1 \to Z$, and $Z \to 0$ transitions.

Twelve tokens are specified for drivers that can drive low impedance with unknown values. Memories often have this property. The transitions are $0 \to 1$, $1 \to 0$, $0 \to Z$, $Z \to 1$, $1 \to Z$, $Z \to 0$, $0 \to X$, $X \to 1$, $1 \to X$, $X \to 0$, $X \to Z$, and $Z \to X$.

The delay section of an SDF cell description begins with the key word DELAY. Delays can be ABSOLUTE or INCREMENTAL. We will only need ABSOLUTE. A delay section might look like this:

```
(DELAY (ABSOLUTE
  (IOPATH A Y (.440:.620:.810) (.440:.620:.810))
))
```

After opening the absolute delay section we can list one or more delay definitions. The most commonly used definition specifies a delay from an input pin to an output pin. Its key word is IOPATH. The IOPATH key word must be followed by the names of the two ports for which the delay is being given. The order in which the ports are listed is input first then output. These are usually followed by a list of delays, as shown. The formal syntax from the standard document is

```
iopath_def ::=
        (IOPATH port_spec port_instance { retain_def } deval_list )
retain_def ::=
        (RETAIN retval_list )
```

In this syntax,

IOPATH is the key word.

port_spec is the input port.

port_instance is the output port.

retain_def will be discussed shortly.

delval_list is the delay data.

In our definition, RETAIN specifies the time an output port retains its previous logic value after a change at a related input, as shown in Figure 4.3.

```
(IOPATH ADDR0 DOUT0
          (RETAIN (2:3:4) (3:4:5))              // RETAIN delays
          (5:10:15) (6:12:18))                  // IOPATH delays
```

The pin-to-pin delay through a component may vary depending on the state of another pin. SDF accommodates this situation by providing the conditional path delay. The formal syntax for conditional path delay is

```
cond_def ::=
   ( COND [ qstring ] conditional_port_expr iopath_def )
```

Here, the fields are interpreted as follows:

COND is the key word.

qstring is an optional symbolic name. Its mapping in VITAL is not well documented, so we shall avoid using it.

conditional_port_expr is the description of the state dependency of the path delay. A particular conditional path delay will be used only if the condition is TRUE. Only expressions using ports are legal. VITAL offers additional,

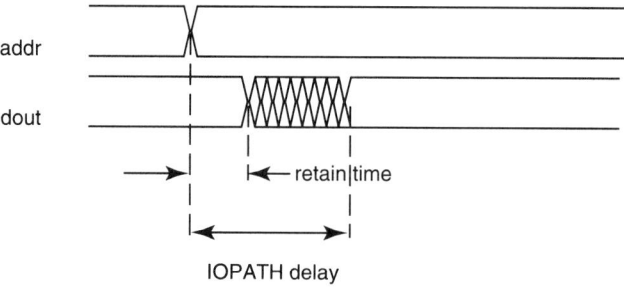

Figure 4.3 Retain time

more flexible methods of selecting among competing path delays. The mapping of condition expressions will be covered in a later chapter.

iopath_def has the same meaning as described earlier.

Here is an example of conditional path delay excerpted from the timing file for a 74GTL1655:

```
(DELAY (ABSOLUTE
  (COND VERC == 1 (IOPATH A0 B0 (3.0:4.5:5.1) (2.9:5.5:6.5)))
  (COND VERC == 0 (IOPATH A0 B0 (2.3:3.3:4.3) (2.0:3.4:4.4)))
```

Sometimes we need to define a delay that is internal to a component and not related to any port. This cannot be done with an IOPATH delay. For this situation SDF provides a device delay.

Although the device delay was devised for gate-level models using distributed timing, it is also applicable in component modeling for things like expressing the refresh interval in a dynamic memory. There is an example of using device delays in a model in Chapter 6.

The formal syntax for device delays is

```
device_def ::=
  (DEVICE [ port_instance ] delval_list )
```

Where the fields mean

DEVICE is the key word.

port_instance is an optional field specifying the output port to which the delay will be applied.

delval_list is the delay data.

Here is an example of a device delay used in a DRAM model:

```
(CELL (CELLTYPE "VITALbuf" )
  (INSTANCE U1/REF) (DELAY (ABSOLUTE ( DEVICE (15625) ) ) )
```

A complete explanation is given in Chapter 6.

4.2.2   Timing Checks

An important part of the board-level verification of your FPGA is ensuring that the timing requirements of all the components with which it interfaces are met. To do this we must build timing constraint checks into our component models. VITAL provides us with the procedures for performing the constraint checks, but SDF gives us the means to pass in the actual values.

The formal syntax for an SDF timing specification is

```
tc_spec ::=
   (TIMINGCHECK tchk_def { tchk_def } )
```

where

TIMINGCHECK is the key word marking the beginning of the timing specification.

Any number of *tchk_def* constructs can be listed in a *tc_spec*.

The *tchk_def* syntax here is not a complete list of all the constructs available in SDF, but it lists all the constructs used in component modeling:

```
tchk_def ::=
          setup_timing_check
        | hold_timing_check
        | recovery_timing_check
        | removal_timing_check
        | skew_timing_check
        | bidirectional_skew_timing_check
        | width_timing_check
        | period_timing_check
        | nochange_timing_check

setup_timing_check :: =
      ( SETUP port_tchk port_tchk value )

hold_timing_check
      ( HOLD port_tchk port_tchk value )

recovery_timing_check
        ( RECOVERY port_tchk port_tchk value )

removal_timing_check
        ( REMOVAL port_tchk port_tchk value )

skew_timing_check
        ( SKEW port_tchk port_tchk value )

bidirectional_skew_timing_check
        ( BIDIRECTSKEW port_tchk port_tchk value value )

width_timing_check
        ( WIDTH port_tchk value )

period_timing_check
        ( PERIOD port_tchk value )

nochange_timing_check
        ( NOCHANGE port_tchk value )
```

Waveforms illustrating these constraint checks are provided in Chapter 8.

```
(SETUP DQO CLK (1.0))
(HOLD  DQO CLK (1.0))
(RECOVERY CLRNeg CLK (2.0))
(WIDTH  (posedge DQS) (2.6))
(PERIOD (posedge CLK) (7.5))
```

Figure 4.4   Example usage of SDF constructs

Figure 4.4 shows how some of these constructs appear in an SDF file.

As you may have guessed, just as conditions can be applied to an IOPATH specification, so may they be applied to a timing constraint specification. The syntax
is similar:

```
port_tchk ::=
      port_spec
    | ( COND [ qstring ] timing_check_condition port_spec )
```

where

COND is the key word

qstring is an optional symbolic name that we will not use.

timing_check_condition is the description of the state dependency of the
timing check.

conditionport_spec is the input port.

Here is an example of a conditional timing check taken from the STDH1655
timing file:

```
(SETUP (COND CLK == 0 AO) LEAB (2.6))
(SETUP (COND CLK == 1 AO) LEAB (2.8))
```

Any *port_spec* can be further qualified with an edge identifier:

```
port_spec ::=
      port_instance
    | port_edge
port_edge ::=
      ( edge_indentifier port instance )
```

A list of legal edge identifiers is given in Chapter 10. Because VITAL also provides the capability to specify edge conditions for timing checks, we really only
need to use edge identifiers in SDF to specify separate values for a constraint, as
illustrated here:

```
(WIDTH (posedge CLK) (3.0))
(WIDTH (negedge CLK) (3.5))
```

The subject of timing constraints in component models is discussed in greater
detail in Chapter 8.

4.3   Summary

SDF is a convenient way to annotate timing values into a simulation. It is supported by both VHDL/VITAL and Verilog simulators. SDF has the capability to annotate circuit delays as pin-to-pin delays or as device delays. It is also able to annotate values for a variety of timing constraint checks. In both cases conditions can be included to allow multiple values to be specified and have the correct value selected by the simulator.

5

Anatomy of a VITAL Model

VHDL component models used in FPGA and board-level verification are based on the VITAL standard and methodology. This methodology can also be used to add delays and timing constraints to your RTL models. In this chapter we examine the architecture of a VITAL model.

VITAL provides two levels of support to better achieve its goals because ASIC libraries, and more important to us, component libraries, must often accommodate a wide range of models. VITAL level 0 facilitates portability and interoperability. It also specifies the method for bringing timing information into the simulation. VITAL level 1 defines a usage model for constructing complete cell (component) models. It also facilitates compiler optimization and accelerated execution of the models.

5.1   Level 0 Guidelines

VITAL level 0 compliance is a prerequisite for level 1 compliance. Both levels restrict the form and semantic content of the model. The level 0 guidelines ensure that all the functionality and timing semantics are defined only within the VHDL-1993 language. Foreign language interfaces and the use of vendor-supplied attributes are prohibited, as are directives or metacomments that affect the behavior or timing characteristics of a model. This is to ensure that models are interoperable across all VITAL compliant simulators.

Level 0 provides a standard interface specification that addresses ports, generic timing parameters, and types. In this way it guarantees that models will work together and that backannotation using SDF can be done.

To accomplish its objectives, level 0 makes several restrictions. Most of the restrictions pertain to the model's entity. Port names may not contain underscore (_) characters. Ports may not be of mode linkage. Scalar ports must be of type `std_ulogic` or a subtype. Vector ports, if needed, must be of type `std_logic_vector`.

VITAL timing generics must conform to a specific naming convention. A detailed explanation will be given shortly. A set of control generics are provided. Other generics may be defined without restriction.

A level 0 model must include a VITAL_Level0 attribute to facilitate compliance checking.

```
ATTRIBUTE VITAL_LEVEL0 of ⟨VitalCompliantEntity⟩ : ENTITY IS TRUE;
```

Level 0 compliance also places a few restrictions on the model's architecture. All functionality and timing must be expressed in VHDL. The VITAL_Level0 attribute must be present.

```
ATTRIBUTE VITAL_LEVEL0 of ⟨VitalCompliantArchitecture⟩ : ARCHITECTURE IS TRUE;
```

Routines defined in the VITAL packages should be used where possible to improve compiler optimization.

5.1.1   Backannotation

Backannotation is used to add timing information to a simulation. The information may pertain to propagation delays and timing constraints. The simulator reads SDF files to obtain timing information and modify the values of a model's timing generics. Separate SDF files may be used to convey model delays and interconnect delays. Additional SDF files may be used to convey FPGA timing information. It is also possible that all the timing information could be combined into a single SDF file.

In all cases, timing data are backannotated to instances in a testbench or a netlist. They cannot be backannotated to a bare cell or model. Figure 5.1 illustrates how the statements map to a netlist.

Delay calculations are performed external to the model and prior to simulation. Backannotation occurs immediately after elaboration and just before negative constraint calculation (see Chapter 11). Once the timing generics have been updated by backannotation, they remain constant during simulation.

5.1.2   Timing Generics

Timing backannotation to a VITAL model is performed through one or more timing generics. For the SDF annotator to determine which generics are to receive which values, VITAL timing generic names must be constrained by a predefined naming convention. The declaration of a VITAL timing generic,

```
tpd_A_YNeg              : VitalDelayType01 := UnitDelay01;
```

includes the predefined generic name, a VITAL-defined type, and an optional default value. The generic name is composed of three parts:

Prefix, denotes the kind of parameter (propagation delay, setup time, etc.).

Signal(s), or path to which the timing value applies. VITAL requires the use of actual port names.

Condition(s), and/or edge designation(s) associated with the indicated signal(s).

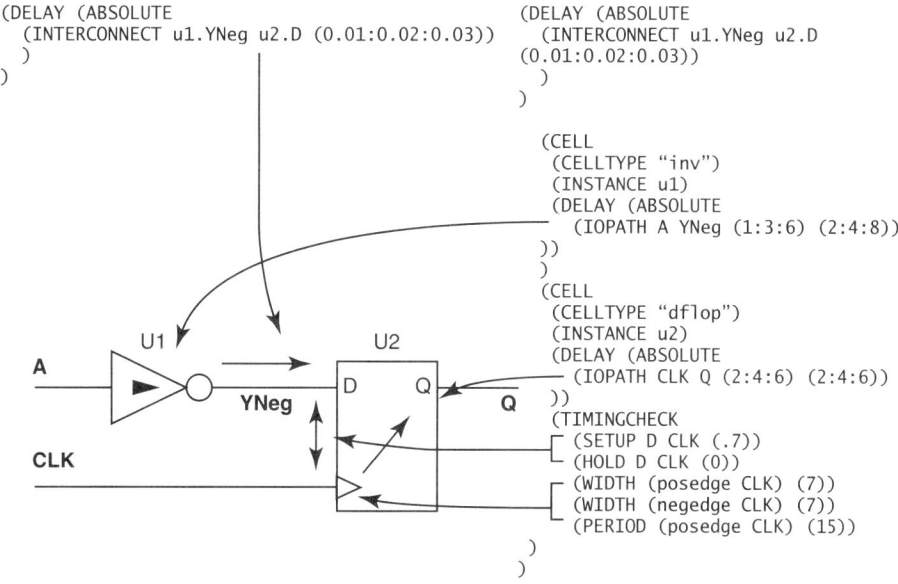

Figure 5.1 SDF to netlist mapping

For example,

```
tipd_A
tpd_OENeg_Y
tpw_CLK_posedge
```

Each generic to which timing information will be annotated is mapped to an SDF key word, as shown in Table 5.1.

5.1.3   VitalDelayTypes

Each VITAL timing generic is declared to be of a particular VitalDelayType as appropriate. VITAL defines an enumerated type to represent twelve possible types of signal transitions:

```
TYPE VitalTransitionType IS (tr01, tr10, tr0z, trz1, tr1z, trz0, tr0X,
    trx1, tr1x, trx0, trxz, trzx);
```

This type is used in all of the VITAL timing procedures. It is usually referenced through one of the VITALDelayTypes.

There are four VITAL-defined scalar delay types:

VitalDelayType, a subtype of TIME used to specify a single delay value.

```
SUBTYPE VitalDelayType IS TIME;
```

Table 5.1 SDF key words and their corresponding VITAL prefixes

SDF Key word	VITAL prefix	Description
INTERCONNECT	tipd	interconnect path delay
		delay between components
IOPATH	tpd	propagation delay
		pin-to-pin delay within a component
DEVICE	tdevice	device delay
		delay not associated with a pin pair
SETUP	tsetup	input setup time
HOLD	thold	input hold time
RECOVERY	trecovery	input recovery time
REMOVAL	tremoval	input removal time
NOCHANGE	tncsetup	no change setup time
	tnchold	no change hold time
SKEW	tskew	input skew time
BIDIRECTSKWEW	tskew	input skew time (1st generic)
	tskew	input skew time (2nd generic)
WIDTH	tpw	pulse width
PERIOD	tperiod	cycle period
	tbpd	biased propagation delay
	tisd	internal signal delay
	ticd	internal clock delay

VitalDelayType01, an array type used to specify two delay values (indexed by tr01 and tr10). It is used with 2-state drivers.

```
TYPE VitalDelayType01 IS ARRAY (VitalTransitionType RANGE tr01 to tr10) OF
    TIME;
```

VitalDelayType01Z, an array type used to specify six delays (indexed by tr01 through trz0). It is used with 3-state drivers.

```
TYPE VitalDelayType01Z IS ARRAY (VitalTransitionType RANGE tr01 to trz0) OF
    TIME;
```

VitalDelayType01ZX, an array type used to specify 12 delays (indexed by tr01 through trzx). It is used with drivers that can have unknown states.

```
TYPE VitalDelayType01ZX IS ARRAY (VitalTransitionType RANGE tr01 to trzx)
    OF TIME;
```

Timing constraints are specified with the simple VitalDelayType. Delays can be specified with any of the delay types as required.

5.2   Level 1 Guidelines

The intent of the level 0 specification is to enable portability and interoperability. The intent of the level 1 specification is to facilitate optimization of compilation and execution of the models, and allow higher levels of simulation performance. While the level 0 guidelines focused on the model's entity, level 1 focuses on its architecture.

Level 1 is more restrictive than level 0. As we examine the structural elements of a level 1 architecture, keep in mind they are all available in a level 0 architecture too. Indeed, most of the models you will create will be level 0.

Level 1 allows a model to have multiple processes. However, no two processes may drive the same signal. The use of subprogram calls and operators in a level 1 model is limited to those declared in packages Standard, `Std_Logic_1164`, and the VITAL packages.

In the declarative part of the architecture, the `Vital_Level1` attribute must appear:

```
ATTRIBUTE VITAL_LEVEL1 of ⟨VitalCompliantArchitecture⟩ : ARCHITECTURE IS TRUE;
```

All signals declared must be of type `std_ulogic`, `std_logic_vector`, or a subtype. Alias declarations may appear but no other declarations are allowed.

A diagram of the structure of a level 1 model (from the IEEE standard) is shown in Figure 5.2. Although we are discussing level 1 models here, this general structure will also apply to level 0 models.

5.2.1   Wire Delay Block

Following the declarative portion of the model we come to the wire delay block. Every model of a digital component should utilize this block. It is where the interconnect timing values are applied to the input signals. There may be at most one wire delay block in a VITAL architecture, and it must have the label `WireDelay`. A wire delay block is shown in Figure 5.3.

In the statement labeled `w_1`, the internal signal `D1_ipd` is driven to the value of input port `D1` after a delay of `tipd_D1`.

A wire delay block contains one or more calls to `VitalWireDelay`. This routine may be called only once for each port of mode `IN` or `INOUT`. It is FMF's convention to label each call to `VitalWireDelay` with a w_N label. The `VitalWireDelay` routine may not be called from anywhere in the model outside of this block.

A GENERATE statement may be used for vector ports. A wire delay block incorporating a GENERATE statement is show in Figure 5.4.

Before using vectored ports in a model, verify that the tool that will be generating your interconnect SDF files is compatible with such models.

A port read by the `VitalWireDelay` routine may not be read anywhere else in the architecture. The output of the call must be an internal signal, not a port. The author prefers to name such signals by appending `_ipd` to the port name, as shown in Figure 5.4.

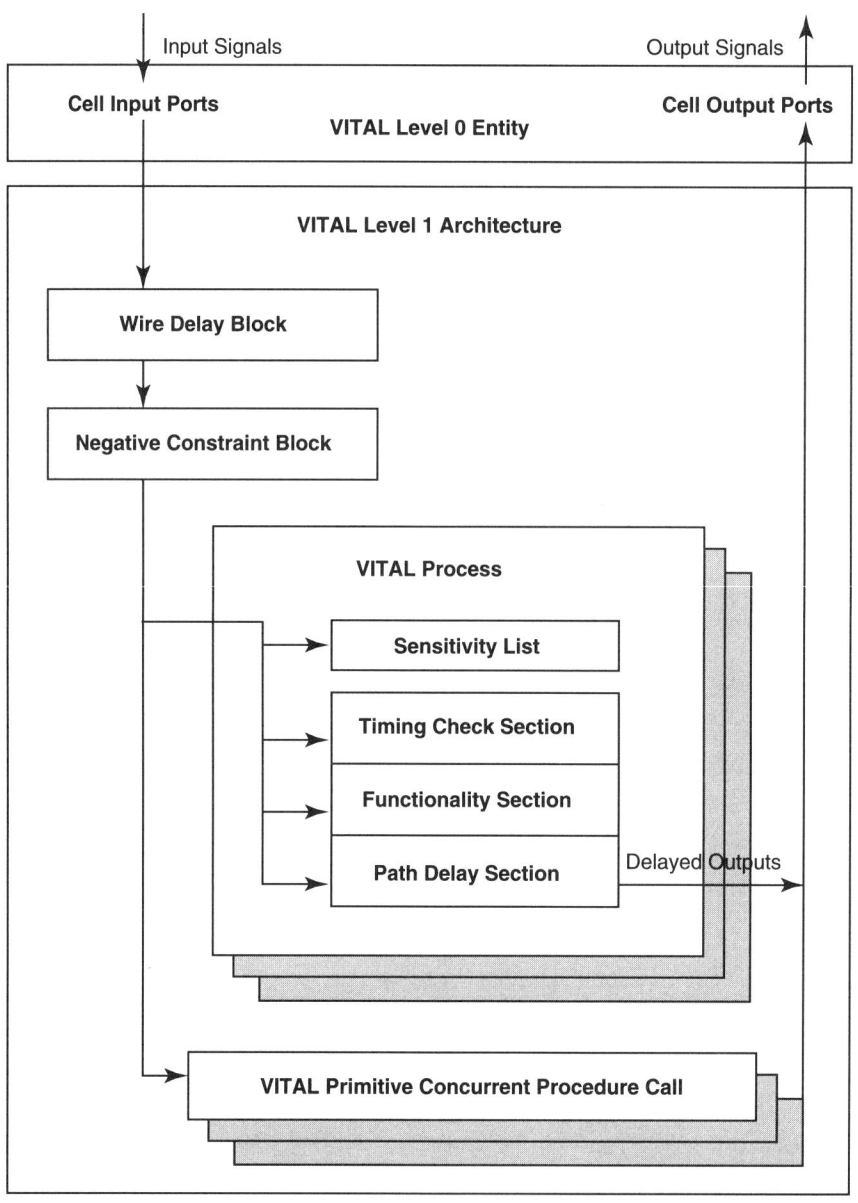

Figure 5.2 Structure of a VITAL model

```
BEGIN

    -------------------------------------------------------------------------
    -- Wire Delays
    -------------------------------------------------------------------------
    WireDelay : BLOCK
    BEGIN

        w_1: VitalWireDelay (D1_ipd, D1, tipd_D1);
        w_2: VitalWireDelay (D0_ipd, D0, tipd_D0);

    END BLOCK;
```

Figure 5.3   Example wire delay block

```
    WireDelay : BLOCK
    BEGIN
        w_1 : VitalWireDelay (CLRNeg_ipd, CLRNeg, tipd_CLRNeg);
        w_2 : VitalWireDelay (OE1Neg_ipd, OE1Neg, tipd_OE1Neg);
        wdgen : FOR i IN 7 downto 0 GENERATE
            VitalWireDelay (IO_ipd(i), IO(i), tipd_IO(i));
        END GENERATE;

    END BLOCK;
```

Figure 5.4   Wire delay block with GENERATE statement

5.2.2   Negative Constraint Block

The negative constraint block is also called the signal delay block. It is used in models with negative timing constraints. Because a component can have a negative timing constraint only if a signal has an internal delay, we can model such a component by implementing a similar internal delay. This is done using calls to VitalSignalDelay. A timing diagram illustrating negative setup and hold constraints is given in Chapter 11.

A schematic of a component that would have a VitalSignalDelay is shown in Figure 5.5. There may be at most one signal delay block in a VITAL architecture, and it must have the label SignalDelay. An example of a signal delay block is shown in Figure 5.6. A signal delay block contains exactly one call to

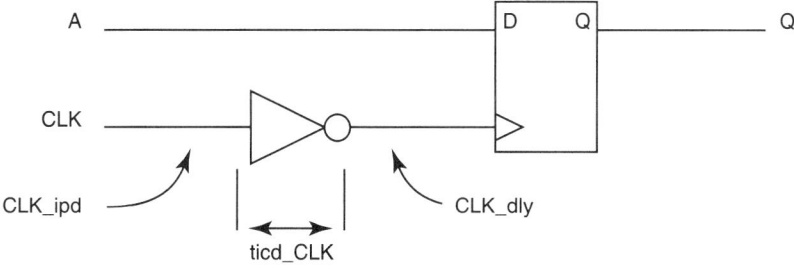

Figure 5.5   Component with VitalSignalDelay

```
--------------------------------------------------------------------------------
-- Negative Timing Constraint Delays
--------------------------------------------------------------------------------
SignalDelay : BLOCK
BEGIN

    s_1: VitalSignalDelay (CLK_dly, CLK_ipd, ticd_CLK);

END BLOCK;
```

Figure 5.6 Signal delay block

VitalSignalDelay for each timing generic representing an internal clock delay or internal signal delay. Negative timing constraints are the topic of Chapter 11.

5.2.3   Processes

A VITAL model may contain as many processes as required. There are four possible sections in a VITAL process. Most are optional:

- Declarative section

- Timing check section

- Functionality section

- Path delay section

The sequence of the various sections in the process is significant. They must appear in the order shown here.

All VITAL processes must have a sensitivity list. Each and every signal that is read in the process must appear in the sensitivity list of that process. This is due to the model's tracking of timing for all its signals. If a signal were read in the process without it being in the sensitivity list, the model could register the signal transition at the wrong time. Timing checks and delays based on the signal would become inaccurate.

Declarative Section

The declarative section of a VITAL process may include constants, variables, aliases, and attributes. In a level 1 model, variables must be of type std_ulogic, std_logic_vector, boolean, time, or one of the VITAL internal (restricted) types. Keep in mind most component models will not be level 1 models. Instead, they will be behavioral models that are easier to write and faster to run.

The restricted types you will declare in a VITAL process and their uses are:

VitalGlitchDataType is used with VitalPathDelay procedures to store timing data used for glitch detection.

VitalTimingDataType is used with VitalSetupHoldCheck and Vital-RecoveryRemovalCheck procedures to store timing data used to detect timing violations.

VitalPeriodPulseDataType is used with VitalPeriodPulseCheck procedures to determines pulse widths and periods.

PreviousDataIn is of type std_logic_vector and is used with VitalStateTable calls to store previous state information.

VitalSkewDataType is used with VitalInPhaseSkewCheck and VitalOutPhaseSkewCheck procedures to store timing data used to detect skew violations.

Variables of these restricted types are used to store persistent data for private use by certain VITAL procedures. They may not be modified by the model itself. Figure 5.7 shows a VITAL process declarative section.

Timing Check Section

The timing check section performs timing constraint checks through predefined timing check procedures. These procedures can generate timing violation messages to the user and set violation flags that can be read by the functionality section of the process. The predefined procedures available are as follows:

- VitalSetupHoldCheck

- VitalRecoveryRemovalCheck

- VitalInPhaseSkewCheck

```
-----------------------------------------------------------------------------
-- Main Behavior Process
-----------------------------------------------------------------------------
VitalBehavior : PROCESS (CLKint, D_ipd, MR_ipd)

    -- Timing Check Variables
    VARIABLE Tviol_D_CLK     : X01 := '0';
    VARIABLE TD_D_CLK        : VitalTimingDataType;

    VARIABLE Rviol_MR_CLK    : X01 := '0';
    VARIABLE TD_MR_CLK       : VitalTimingDataType;

    VARIABLE Pviol_CLK       : X01 := '0';
    VARIABLE PD_CLK          : VitalPeriodDataType := VitalPeriodDataInit;

    VARIABLE Pviol_MR        : X01 := '0';
    VARIABLE PD_MR           : VitalPeriodDataType := VitalPeriodDataInit;

    VARIABLE Violation       : X01 := '0';

    -- Functionality Results Variables
    VARIABLE Q_zd            : std_ulogic;
    VARIABLE PrevData        : std_logic_vector(0 to 3);

    -- Output Glitch Detection Variables
    VARIABLE Q_GlitchData    : VitalGlitchDataType;
```

Figure 5.7   Example of VITAL process declarative section

- `VitalOutphaseSkewCheck`

- `VitalPeriodPulseCheck`

The timing check section is constructed of a single IF statement. The `Timing-ChecksOn` generic must be used as the only condition. The timing check procedures are called from within the IF statement. No `ELSE` or `ELSIF` clauses are allowed:

```
BEGIN
    -----------------------------------------------------------------
    -- Timing Check Section
    -----------------------------------------------------------------
    IF (TimingChecksOn) THEN
-- as many calls to timing check procedures as required
...
        END IF;
```

The timing check section only detects timing violations. The model must function correctly when `TimingChecksOn` is FALSE. Timing constraint violations may be translated to 'X' outputs or corrupted memory locations in the functionality section. Signal assignments may not be made in the timing check section. All timing checks must be independent of one another. Details of the various timing check procedures are given in Chapter 8.

Functionality Section

The function of a VITAL model may be coded by utilizing the VITAL functionality section, concurrent procedure calls, or both. The functionality section defines the logical function of the model. It computes new output values based on the input values, but without timing.

The operation of a component is modeled in the VITAL level 1 process through a sequence of variable assignments and/or calls to the `VitalTruthTable` or `VitalStateTable` procedures. Right-hand-side expressions may include

- Function calls to VITAL primitives

- Operators and functions defined in the `std_logic_1164` package

- Function calls to `VitalTruthTables`

- Concatenation and aggregate forms

- Variables, constants, signals (`_ipd`), and ports

The following are not allowed in a level 1 process (but are allowable in a level 0 process):

- `IF`, `CASE`, `LOOP`, `NEXT`, `EXIT`, and `RETURN` statements

- `WAIT` statements

- Signal assignments

- Assertion statements

- Procedure calls to other than VITAL procedures

These restrictions limit the usefulness of level 1 processes to modeling relatively simple components. The majority of models will be compliant with VITAL level 0. A level 1 functionality section might look like this:

```
-------------------------------------------------------------------------
-- Functionality Section
-------------------------------------------------------------------------
Violation := Tviol_D_CLK OR Pviol_CLK OR Rviol_MR_CLK OR Pviol_MR;

VitalStateTable (
     StateTable          => DFFR_tab,
     DataIn              => (Violation, CLKint, D_ipd, MR_ipd),
     Result              => Q_zd,
     PreviousDataIn   => PrevData
);
```

We will see level 0 functionality sections in later chapters.

Path Delay Section

The path delay section receives the undelayed computed output values from the functionality section and uses them to drive ports or internal signals after applying the appropriate delays. It accomplishes this task through calls to the `VitalPath-Delay` procedures. (A more in-depth discussion of path delays is given in Chapter 6.) Figure 5.8 shows the path delay section from the eclps151 model:

```
-------------------------------------------------------------------------
-- Path Delay Section
-------------------------------------------------------------------------
VitalPathDelay01 (
    OutSignal        => Qint,
    OutSignalName    => "Qint",
    OutTemp          => Q_zd,
    GlitchData       => Q_GlitchData,
    XOn              => XOn,
    MsgOn            => MsgOn,
    Paths            => (
        0 => (InputChangeTime    => CLKint'LAST_EVENT,
               PathDelay          => tpd_CLK1_Q,
               PathCondition      => TRUE),
        1 => (InputChangeTime    => MR_ipd'LAST_EVENT,
               PathDelay          => tpd_MR_Q,
               PathCondition      => TRUE)
    )
);
```

Figure 5.8 Example path delay

5.2.4   VITAL Primitives

The VITAL primitives afford a convenient and efficient method for coding a number of simple logical functions. They provide the basic functional support for level 1 models but are also useful in level 0 models. In a process, they are called as functions. For example,

```
Y_zd := VitalOR2 (a => D_ipd, b => ENeg_ipd, ResultMap => ECL_wired_or_rmap);
```

Y_zd, the zero delay output variable, receives the result of an OR operation between D_ipd and ENeg_ipd, both delayed input signals. The output is further modified by the result map. In this case, the output simulates an open emitter which can drive a '1' but not a '0'. It is always wise to call a primitive using named association. Positional notation is difficult to maintain and is a common source of errors. The 39 available primitives are listed in Chapter 3.

5.2.5   Concurrent Procedure Section

A concurrent procedure section may be placed after the delay block(s) and either before or after the processes. This placement is a VITAL convention, not a VHDL requirement. In a VITAL level 1 model, only procedure calls to VITAL primitives are allowed. Other types of assignments are permitted in a level 0 model. Procedure calls to VITAL primitives may include delay specifications, as show in Figure 5.9. It is preferred to use path delay procedure calls for reasons explained in Chapter 6. Providing a label for each VITAL primitive procedure call is recommended for debug purposes.

5.3   Summary

The VITAL standard and its packages provide us with a uniform modeling methodology and a tool box of types, functions, and procedures. There are two levels of VITAL compliance. Level 0 enables the use of path delays, timing constraints, and the backannotation of timing values through SDF. It can be used for modeling complex components at the behavioral level of abstraction. Level 0 primarily affects the model entity.

Level 1 provides a set of simulation primitives that can be used for modeling simple components at the gate level. It also allows compiler optimization for faster

```
a_7: VitalINV (
        q => DOUTNeg,
        a => DINint,
        tpd_a_q => tpd_DIN_DOUT,
        ResultMap => ECL_wired_or_rmap
    );
```

Figure 5.9   Procedure call to VITAL primitive

simulation, but at the price of modeling restrictions. Level 1 primarily affects the model architecture.

VITAL timing generics are named using a formula. There is a one-to-one mapping between the VITAL generics and SDF statements that is required for backannotation.

VITAL compliant simulators can use VITAL procedures to enable negative timing constraint checks. Values for the timing generics used for negative timing constraints are not backannotated but are calculated by the simulator prior to the start of simulation.

A VITAL process must have a sensitivity list that includes every signal read within the process. VITAL path delay procedures are used to add propagation delays to output signals in a VITAL process.

6 Modeling Delays

Timing is an important aspect of system verification. Timing includes delays through components and between components as well as timing constraints. In this chapter we examine the ways delays within a component can be modeled. We also discuss delays between components.

There are a number of ways in which component delays can be modeled. Delays can be distributed among the various parts of the model. They can also be modeled as pin-to-pin delays, or a combination of the two approaches can be taken.

6.1 Delay Types and Glitches

In a physical component, it takes a finite time for a transition on an input pin to have an effect on an output pin. In modeling, this is called a path delay. It also takes a finite time for the change on the output pin to be felt on the input pin of the next component. This is called interconnect delay.

In VHDL, a signal is an object that consists of a list of values and the times at which those values are scheduled to take effect. When a signal assignment statement is executed in VHDL, it is scheduled to become effective at some time in the future. The difference between the current time and the time the signal is scheduled to become effective is the delay.

6.1.1 Transport and Inertial Delays

There are two types of delays in VHDL and they correspond with two types of digital circuits. Which one you use should depend on which logic family you are modeling.

The simplest is *transport* delay. It is applicable to wire delays and to nonsaturating logic families such as emitter coupled logic (ECL). A feature of transport delay, as illustrated in Figure 6.1, is that any signal transition on the input appears on the output after a delay.

Most logic families do not behave this way. Most digital components use a saturated transistor technology in which a gate capacitance must be charged or

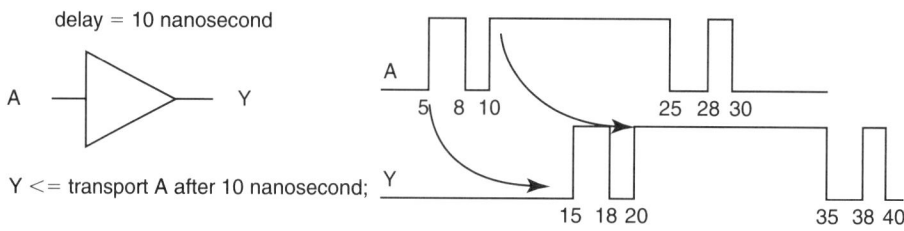

Figure 6.1   Transport delay

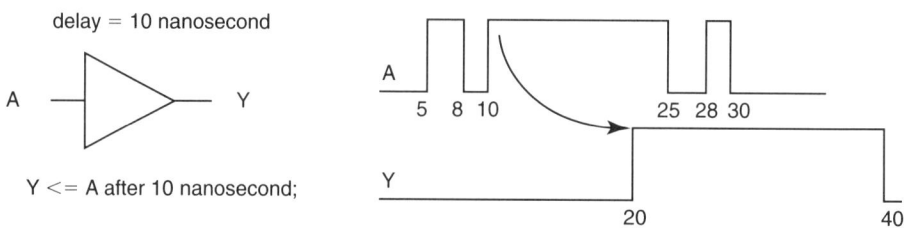

Figure 6.2   Inertial delay

discharged before the output begins to change. The output transistors are either completely shut off or completely turned on (saturated). CMOS (Complimentary Metal Oxide on Silicon) is the most common example of this today. In CMOS parts, a resistance capacitance (RC) time constant of the drive circuit and the gate capacitance controls how fast an input can be charged or discharged. If an input pulse is much longer than that time constant, the output switches as expected. If the pulse is much shorter than the time constant, the output is unaffected because before the input gate can be charged up enough to begin switching the output, it is discharged again. If an input pulse is of an intermediate duration, the result is uncertain. This type of propagation delay is called *inertial* delay and is illustrated in Figure 6.2.

6.1.2   Glitches

The short pulses that do not result in an output transition are referred to as *glitches*. Very short glitches are "swallowed" by the input circuit without affecting the output. However, because it is not known exactly how long a glitch must be to produce an event on the output, the effect of any input pulse shorter than the propagation delay is conservatively regarded as indeterminate. Under such conditions it is often prudent to have the model output an 'X'. The VITAL path delay procedures provide a means of specifying and controlling such behavior.

6.2   Distributed Delays

Just as your FPGA can be viewed as being composed of a number of connected components, each with its own delays, so too can other board-level digital parts. The data sheets for many 7400 devices include schematics depicting their logical construction. It is easy to model such a part as a netlist of VITAL primitives (but beware, these schematics are often inaccurate).

The schematic of one section of a SN74LS245 as shown in a 1988 Texas Instruments data book [2] is presented in Figure 6.3. One way to describe the timing of this device would be to assign a delay to each gate. This is similar to the way delays are modeled in a gate-level netlist. Each gate would be coded as a VITAL concurrent primitive procedure call with a delay parameter. Delays would be read from SDF DEVICE statements or IOPATH statements into VITAL tdevice or tpd generics. Separate generics would be required for U1, U2, U3, and U4. In Figure 6.3, the propagation delay from G# to B1 would be the sum of the delays through U1 and U3.

Such an approach has several drawbacks. The most significant is that the model must effectively be a gate-level netlist. Modeling at that level is about as much work as designing the device in the first place, and simulation speed is poor. The schematic provided in a component's data sheet is frequently inaccurate. It may not be possible to achieve correct timing with a gate-level model based on such a schematic. It may also be difficult or impossible to determine the gate-level delays because the component's timing is usually specified on a pin-to-pin basis.

6.3   Pin-to-Pin Delays

Modeling delays as pin-to-pin delays is much less work. It is also more consistent with the way IC vendors characterize their products. Because the model can be

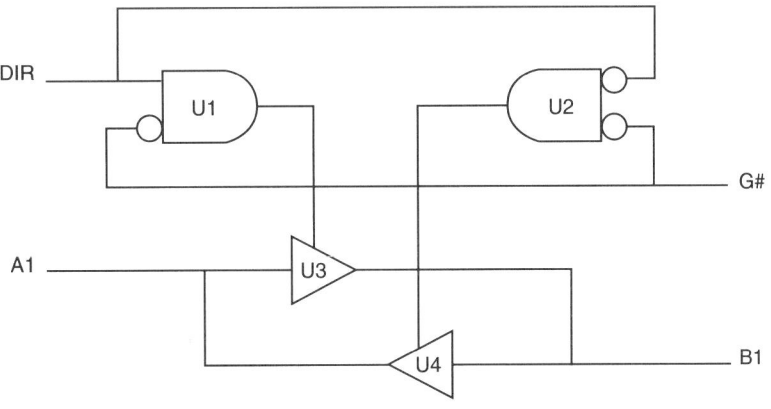

Figure 6.3   Schematic of 74LS245 section

written at a higher level of abstraction, it is less tedious to code and has better
simulation performance. The component functionality is modeled without delays.
Delays are added using calls to VITAL path delay procedures. IOPATH statements
in SDF would be mapped to VITAL tpd generics in the model.

Besides being easier to understand, the behavioral model with pin-to-pin delays
is more flexible and better lends itself to technology independence or, more precisely, to timing independence. A greater degree of control over delay issues is available when a model is written using the VITAL path delay procedures.

Yet another advantage of modeling with pin-to-pin delays is independence
between timing and functionality implementation. In a gate-level model with distributed timing, a change to the netlist, perhaps to make a correction, is likely to
change the timing and necessitate and modification to the timing file. In the model
incorporating pin-to-pin delays, the delay timing and the functionality are largely
separate. One can be changed with minimal impact on the other.

Even in cases in which a gate-level representation of functionality makes sense,
pin-to-pin delays are effective. Figure 6.4 presents the behavior process from the
Free Model Foundry model of the 74LS245.

6.4   Path Delay Procedures

The VitalPathDelay procedure comes in three flavors: VitalPathDelay,
VitalPathDelay01, and VitalPathDelay01 Z. Each supports

- Transition-dependent delays

- Path delay selection

- User-controlled glitch detection

- User-controlled 'X' generation

- User-controlled error reporting

- Output signal scheduling

- and more

Of the three varieties of VPDs, VitalPathDelay is for simple path delays of type
VitalDelayType that are not transition dependent. The VitalPathDelay01 procedure is for delays of type VitalDelayType01 that require different rise/fall transition delay values. Finally, VitalPathDelay01Z is for scheduling path delays on
signals for which there is a transition to 'Z'. All three procedures take the same
set of parameters,

OutSignal, output port or internal signal to be driven. Type std_logic.

OutSignalName, name of OutSignal to use when generating messages. Type string

OutTemp, the value to be applied to the output; computed in the functionality
section (_zd). This is an input of type std_ulogic.

```
------------------------------------------------------------------------
-- VITALBehavior Process
------------------------------------------------------------------------
VITALBehavior1 : PROCESS(A_ipd, B_ipd, DIR_ipd, ENeg_ipd)

    -- Functionality Results Variables
    VARIABLE A_zd        : std_ulogic := 'X';
    VARIABLE B_zd        : std_ulogic := 'X';
    VARIABLE Aen_int     : std_ulogic := 'X';
    VARIABLE Ben_int     : std_ulogic := 'X';

    -- Output Glitch Detection Variables
    VARIABLE A_GlitchData : VitalGlitchDataType;
    VARIABLE B_GlitchData : VitalGlitchDataType;

BEGIN

    --------------------------------------------------------------------
    -- Functionality Section
    --------------------------------------------------------------------
    Aen_int := VitalAND2(a=> NOT(DIR_ipd), b => NOT(ENeg_ipd));
    Ben_int := VitalAND2(a=> DIR_ipd, b => NOT(ENeg_ipd));

    A_zd := VitalBUFIF1 (data => B_ipd, enable => Aen_int );
    B_zd := VitalBUFIF1 (data => A_ipd, enable => Ben_int );

    --------------------------------------------------------------------
    -- Path Delay Section
    --------------------------------------------------------------------
    VitalPathDelay01Z (
        OutSignal       => A,
        OutSignalName   => "A",
        OutTemp         => A_zd,
        Paths           => (
            0 => (InputChangeTime   => B_ipd'LAST_EVENT,
                    PathDelay       => VitalExtendToFillDelay(tpd_B_A),
                    PathCondition   => (Aen_int = '1')),
            1 => (InputChangeTime   => DIR_ipd'LAST_EVENT,
                    PathDelay       => tpd_DIR_A,
                    PathCondition   => TRUE ),
            2 => (InputChangeTime   => ENeg_ipd'LAST_EVENT,
                    PathDelay       => tpd_ENeg_A,
                    PathCondition   => TRUE ) ),
        GlitchData      => A_GlitchData );

    VitalPathDelay01Z (
        OutSignal       => B,
        OutSignalName   => "B",
        OutTemp         => B_zd,
        Paths           => (
            0 => (InputChangeTime   => A_ipd'LAST_EVENT,
                    PathDelay       => VitalExtendToFillDelay(tpd_A_B),
                    PathCondition   => (Ben_int = '1')),
            1 => (InputChangeTime   => DIR_ipd'LAST_EVENT,
                    PathDelay       => tpd_DIR_B,
                    PathCondition   => TRUE ),
            2 => (InputChangeTime   => ENeg_ipd'LAST_EVENT,
                    PathDelay       => tpd_ENeg_B,
                    PathCondition   => TRUE ) ),
        GlitchData      => B_GlitchData );

END PROCESS;
```

Figure 6.4   Behavioral process for 74LS245

Paths, a list of paths of VitalPathArrayType. Each path gives the delays and conditions for driving the output.

DefaultDelay, delay used when no delay path applies. If absent, defaults to zero.

IgnoreDefualtDelay, a boolean. If FALSE, the default delay will be used. If TRUE, no event will be visible if no paths are selected. If absent, defaults to FALSE.

GlitchData, a variable (local to the behavior process) used by the VPD to store data required for glitch detection. Must not be referenced elsewhere.

MsgOn, control for message generation on glitch detection. When TRUE, glitches are reported. Defaults to TRUE.

MsgSeverity, the severity level at which the message is reported.

XOn, glitch 'X' generation control. When TRUE, 'X's are scheduled for glitches. If absent, defaults to TRUE.

Mode, selects the type of glitch detection. Value should be one of OnEvent (default), OnDetect, VitalInertial, on VitalTransport (described later in this chapter).

OutputMap, an array that allows mapping of output strength. For Vital-PathDelay01Z only.

NegPreemptOn, if TRUE, enables negative preemptive glitch handling.

RejectFastPath, if TRUE, enables rejection of fast signal path.

Some of the parameters listed require further explanation and are covered in later sections.

Paths

The Paths parameter is an array of records. There may be one or more elements in the array. The order of the elements is not significant. Each record is made up of three elements:

InputChangeTime, the amount of time since the input changed. This should be of the form <signal>'LAST_EVENT.

PathDelay, a VitalDelayType, VitalDelayType01, or VitalDelayType01Z value specifying the delay time (from a specific input). Most often, this will be a tpd generic.

Condition, a boolean expression controlling whether this delay path is to be considered.

Three examples of paths taken from three FMF models are displayed in Figure 6.5. The VitalDelayType in the PathDelay element of the record must be consistent with the VITAL path delay procedure call. However, when using Vital-

```
    1 => (InputChangeTime => CLKIn'LAST_EVENT,
          PathDelay => tpd_CLK_DQ2,
          PathCondition   => CAS_Lat = 2),
    2 => (InputChangeTime => CLKIn'LAST_EVENT,
          PathDelay => tpd_CLK_DQ3,
          PathCondition   => CAS_Lat = 3)

  0 => (InputChangeTime    => CLK_ipd'LAST_EVENT,
        PathDelay          => tpd_CLK_Q,
        PathCondition      => ((CLRint = '0') AND (PREint = '0'))),


  0 => (InputChangeTime    => A_ipd'LAST_EVENT,
        PathDelay          => VitalExtendToFillDelay(tpd_A_Y),
        PathCondition      => TRUE),
```

Figure 6.5   Example path parameters

PathDelay01Z, the function VitalExtendtoFillDelay may be employed to
extend a constant of VitalDelayType or VitalDelayType01 to VitalDelay-
Type01Z, as demonstrated in the third example.

DefaultDelay

If none of the paths are selected and IgnoreDefaultDelay is FALSE, the
DefaultDelay is used for the output delay. If no default value is supplied by the
model, a zero delay will be used. In a properly written model, every path and condition is anticipated, so use of the DefaultDelay should never happen. However,
when testing a model, a zero delay can be very handy because it signals that no
path was selected and an unanticipated condition occurred, an indication that the
model is not yet perfected.

IgnoreDefualtDelay

If IgnoreDefualtDelay is set to TRUE and no path is selected, the output event
is scheduled for TIME'HIGH. From a simulation perspective, this is the end of time.
If there is a requirement that transitions not covered by any of the paths never
reach the output port, this is the way to do it.

MsgOn

The MsgOn parameter controls the emission of messages in the event of a glitch
detection. In most cases you will want to see these messages. However, should you
prefer to ignore them, the value of this parameter can be set through generics and
controlled on an instance-by-instance basis or for the entire schematic.

MsgSeverity

Messages emitted by the VPD have a severity level. Most simulators can be configured to pause or abort a simulation based on the severity level of a message.

Alternatively, most simulators can be configured to hide messages of low severity. The MsgSeverity parameter allows the user to set the severity level of messages emitted upon glitch detection. The value must be note, warning, error, or failure. If no value is supplied, the value will default to warning.

XOn

If a glitch occurs on an input, the VPD will, by default, schedule an 'X' on the output. Although this may be a good thing for regression tests, the 'X' propagation can be inconvenient during initial debug. Fortunately, it is easy to control through the XOn generic routinely included in your model entity.

Mode

Given that a glitch is detected and XOn is TRUE, there are four modes for propagating the 'X': VitalInertial, VitalTransport, OnEvent, and OnDetect. They are shown in Figure 6.6.

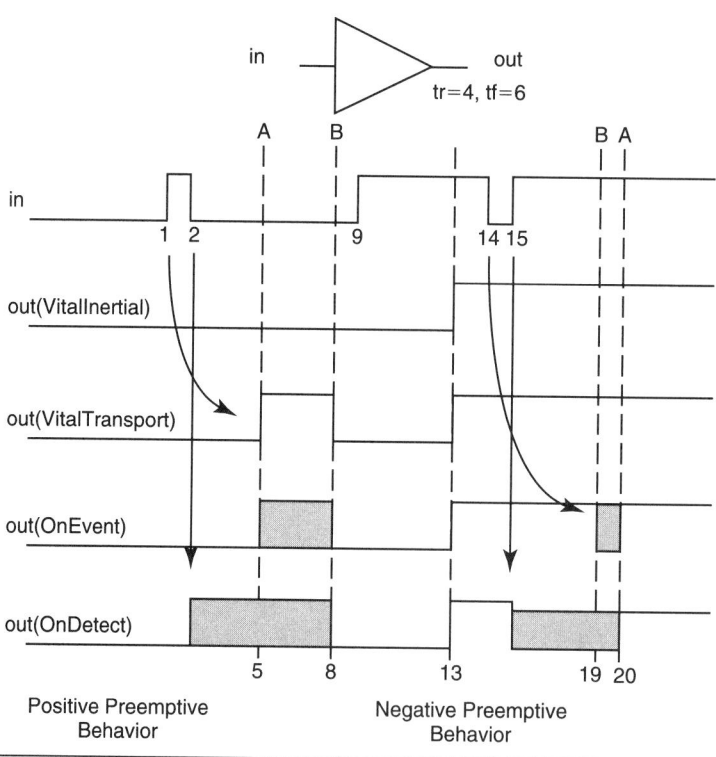

Figure 6.6 The four glitch propagation types

VitalInertial and VitalTransport are identical to VHDL inertial and transport modes. If either of these modes is selected, glitches will not be detected. Otherwise if the duration of an input pulse is less than the propagation delay for that pulse, a glitch is detected at the end of the pulse. OnEvent and OnDetect are special glitch handling modes. OnEvent outputs 'X' from the time an output event is scheduled. OnDetect mode outputs 'X' from the time the glitch is detected (when the glitch ends). "A" is the first scheduled event, "B" is the second scheduled event.

NegPreemptOn

In Figure 6.6, the buffer has a longer delay for a falling output than for a rising one. Therefore, the first positive going glitch has its pulse width extended. The second glitch shown, with its negative going pulse, would normally produce no output. This is because the shorter delay of the rising edge would preempt and thus cancel the falling edge. That is positive preemption. If we are concerned with such pulses and want them to cause 'X' outputs, we can set the NegPreemptOn parameter to TRUE in the VitalPathDelay procedure call.

RejectFastPath

In the component illustrated in Figure 6.7, both the clock and the output enable are causing a '1' to be scheduled on pin Q. However, the propagation delay for the output enable, OE, is longer than the delay for the clock. If RejectFastPath is FALSE, Q will transition from 'Z' to '1' after a delay of 4. If RejectFastPath is TRUE, Q will transition from 'Z' to '1' after a delay of 8. In most cases, 8 more closely models the actual component. It is important to note that RejectFast-Path does not work with simultaneous inputs.

OutputMap

The output type of a VPD is std_logic. The part you are modeling may not be capable of driving the full range of values supported by std_logic. For example, the component being modeled may have an open collector output. In such a case the OutputMap parameter can be used to modify an output value before scheduling.

```
OutputMap => "UX0ZZWLZ-";
```

This statement will cause any '1' or 'H' computed result to be transformed to 'Z'. This is done by a simple positional mapping between the VitalDefault-OutputMap and the user-defined map:

```
UX01ZWLH- --VitalDefaultOutputMap
UX0ZZWLZ- --user defined map
```

Note that the OutputMap parameter is available only in the VitalPathDelay01Z procedure. Also note that because the transformation takes place before

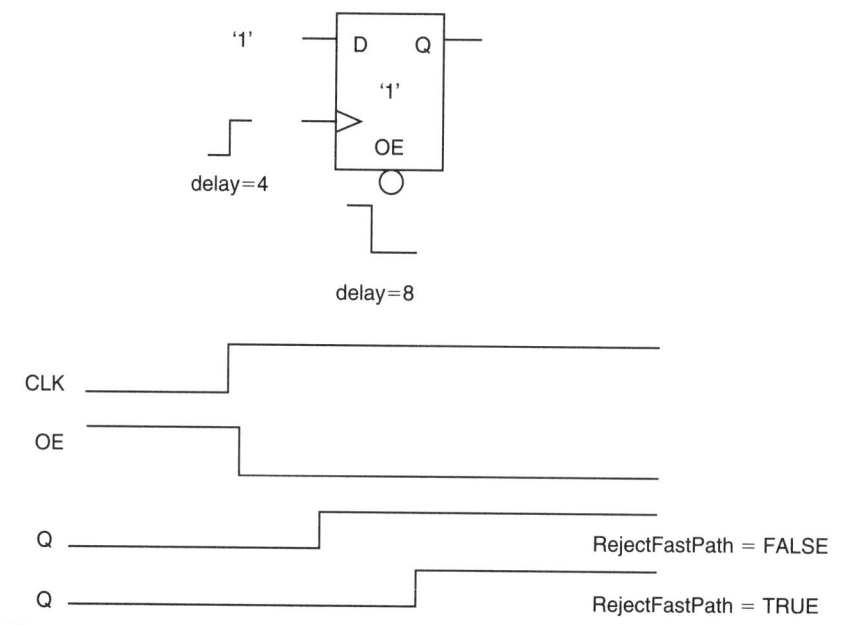

Figure 6.7   Fast path

scheduling, the delay chosen is based on the new transition values, in this case
0 –> Z rather than 0 –> 1.

6.5   Using VPDs

The `VitalPathDelay` procedures are usually called from within the behavioral
process that is computing the output values. Before they can be used, some variables must be declared. These are the temporary or zero delay result variable and
the glitch variable.

```
-- Functionality Results Variables
VARIABLE ECLKOUT_zd  : std_ulogic;

-- Output Glitch Detection Variables
VARIABLE ECLKOUT_GlitchData   : VitalGlitchDataType;
```

The result variable must be of type `std_ulogic`. The glitch variable is of type
`VitalGlitchDataType`. The glitch variable has a local scope, and although it must
always be declared, it must never be used outside the VPD call.

In the functionality section, results are assigned to the temporary variable.

```
    ECLKOUT_zd := ECLKIN;
```

The delay section is at the end of the process.

```
    ----------------------------------------------------------------------
    -- Path Delay Section
    ----------------------------------------------------------------------
        VitalPathDelay01 (
            OutSignal        => ECLK_int,
            OutSignalName    => "ECLKOUT",
            OutTemp          => ECLKOUT_zd,
            GlitchData       => ECLKOUT_GlitchData,
            XOn              => XOn,
            MsgOn            => MsgOn,
            Paths            => (
                0 => (InputChangeTime => ECLKIN'LAST_EVENT,
                        PathDelay      => tpd_ECLKIN_ECLKOUT,
                        PathCondition  => TRUE)
            )
        );
    END PROCESS ECLK;
```

In most cases, each output port will have its own VPD call. However, for components with duplicated outputs, such as clock drivers, this may not be necessary. Instead, a single VPD can be used to drive an internal signal and VITAL concurrent procedure calls or simple signal assignments used to drive all the ports (see the fct807 model in Figure 6.8).

Note in Figure 6.8 that the one VPD updates an internal signal. That delayed internal signal then triggers and provides the value to 10 concurrent procedure calls.

6.6   Generates and VPDs

Having the VPDs in the same process that is computing the output values is simplest and most flexible. In the case of working with large output buses, writing a separate VPD for each bit of the bus takes too long and makes the model too verbose. At the expense of `Vital_Level1` compliance, a VPD can be placed in its own process inside a generate statement. A VitalPathDelay nested inside a generate is shown in Figure 6.9. Because this technique requires interprocess communications, the zero delay output must be a signal rather than a variable.

6.7   Device Delays

Sometimes there is a need to backannotate into a model a timing value that is internal to the model and not associated with any port. An example is the refresh

```
-------------------------------------------------------------------------------
--  File Name: fct807.vhd
-------------------------------------------------------------------------------
--  Copyright (C) 2001, 2003 Free Model Foundry; http://eda.org/fmf/
--
--  This program is free software; you can redistribute it and/or modify
--  it under the terms of the GNU General Public License version 2 as
--  published by the Free Software Foundation.
--
--  MODIFICATION HISTORY:
--
--  version | author  | mod date | changes made
--    V1.0  R. Munden  01 FEB 16  Initial release
--    V2.0  R. Munden  03 JAN 22  Flattened model to match customer usage
-------------------------------------------------------------------------------
--  PART DESCRIPTION:
--
--  Library:    CLOCK
--  Technology: CMOS
--  Part:       FCT807
--
--  Description:  1 to 10 Clock Driver
-------------------------------------------------------------------------------
LIBRARY IEEE;     USE IEEE.std_logic_1164.ALL;
                  USE IEEE.VITAL_timing.ALL;
                  USE IEEE.VITAL_primitives.ALL;
LIBRARY FMF;      USE FMF.gen_utils.ALL;


-------------------------------------------------------------------------------
-- ENTITY DECLARATION
-------------------------------------------------------------------------------
ENTITY fct807 IS
    GENERIC (
        -- tipd delays: interconnect path delays
        tipd_A              : VitalDelayType01 := VitalZeroDelay01;
        -- tpd delays
        tpd_A_Y1            : VitalDelayType01 := UnitDelay01;
        -- generic control parameters
        MsgOn               : BOOLEAN   := DefaultMsgOn;
        XOn                 : Boolean   := DefaultXOn;
        InstancePath        : STRING    := DefaultInstancePath;
        -- For FMF SDF techonology file usage
        TimingModel         : STRING    := DefaultTimingModel
    );
    PORT (
        A           : IN    std_ulogic := 'U';
        Y1          : OUT   std_ulogic := 'U';
        Y2          : OUT   std_ulogic := 'U';
        Y3          : OUT   std_ulogic := 'U';
        Y4          : OUT   std_ulogic := 'U';
        Y5          : OUT   std_ulogic := 'U';
        Y6          : OUT   std_ulogic := 'U';
        Y7          : OUT   std_ulogic := 'U';
        Y8          : OUT   std_ulogic := 'U';
        Y9          : OUT   std_ulogic := 'U';
        Y10         : OUT   std_ulogic := 'U'
    );
    ATTRIBUTE VITAL_LEVEL0 of fct807 : ENTITY IS TRUE;
END fct807;


-------------------------------------------------------------------------------
-- ARCHITECTURE DECLARATION
-------------------------------------------------------------------------------
ARCHITECTURE vhdl_behavioral of fct807 IS
    ATTRIBUTE VITAL_LEVEL1 of vhdl_behavioral : ARCHITECTURE IS TRUE;

    CONSTANT partID     : STRING := "FCT807";

    SIGNAL A_ipd        : std_ulogic := 'U';
    SIGNAL Y            : std_logic  := 'U';
```

Figure 6.8 Model with duplicated outputs

```
BEGIN
    ----------------------------------------------------------------------------
    -- Wire Delays
    ----------------------------------------------------------------------------
    WireDelay : BLOCK
    BEGIN

        w_1: VitalWireDelay (A_ipd, A, tipd_A);

    END BLOCK;


    ----------------------------------------------------------------------------
    -- Concurrent procedure calls
    ----------------------------------------------------------------------------
    a_1 : VitalBuf (q => Y1, a => Y);
    a_2 : VitalBuf (q => Y2, a => Y);
    a_3 : VitalBuf (q => Y3, a => Y);
    a_4 : VitalBuf (q => Y4, a => Y);
    a_5 : VitalBuf (q => Y5, a => Y);
    a_6 : VitalBuf (q => Y6, a => Y);
    a_7 : VitalBuf (q => Y7, a => Y);
    a_8 : VitalBuf (q => Y8, a => Y);
    a_9 : VitalBuf (q => Y9, a => Y);
    a_10 : VitalBuf (q => Y10, a => Y);


        ------------------------------------------------------------------------
        -- VITALBehavior Process
        ------------------------------------------------------------------------
        VITALBehavior : PROCESS(A_ipd)

        -- Functionality Results Variables
        VARIABLE Y_zd        : std_ulogic := 'U';

        -- Output Glitch Detection Variables
        VARIABLE Y_GlitchData : VitalGlitchDataType;

    BEGIN

        --------------------------------------------------------------------
        -- Functionality Section
        --------------------------------------------------------------------
            Y_zd := VitalBUF(data=> A_ipd);


        --------------------------------------------------------------------
        -- Path Delay Section
        --------------------------------------------------------------------
        VitalPathDelay01 (
            OutSignal        =>  Y,
            OutSignalName    =>  "Y",
            OutTemp          =>  Y_zd,
            XOn              => XOn,
            MsgOn            => MsgOn,
            Paths            => (
                0 => (InputChangeTime    => A_ipd'LAST_EVENT,
                      PathDelay          => tpd_A_Y1,
                      PathCondition      => TRUE ) ),
            GlitchData       => Y_GlitchData );

    END PROCESS;

END vhdl_behavioral;
```

Figure 6.8 Model with duplicated outputs *(continued)*

```
        SIGNAL D_zd     : std_logic_vector(HiDbit DOWNTO 0);
...
        D_zd <= DataDrive;
...
        -----------------------------------------------------------------------
        -- Path Delay Processes generated as a function of data width
        -----------------------------------------------------------------------
        DataOut_Width : FOR i IN HiDbit DOWNTO 0 GENERATE
            DataOut_Delay : PROCESS (D_zd(i))
                VARIABLE D_GlitchData:VitalGlitchDataArrayType(HiDbit Downto 0);
            BEGIN
                VitalPathDelay01Z (
                    OutSignal       => DataOut(i),
                    OutSignalName   => "Data",
                    OutTemp         => D_zd(i),
                    Mode            => OnEvent,
                    GlitchData      => D_GlitchData(i),
                    Paths           => (
                        0 => (InputChangeTime => OENeg_ipd'LAST_EVENT,
                              PathDelay       => tpd_OENeg_DO,
                              PathCondition   => TRUE),
                        1 => (InputChangeTime => CENeg_ipd'LAST_EVENT,
                              PathDelay       => tpd_CENeg_DO,
                              PathCondition   => TRUE),
                        2 => (InputChangeTime => AddressIn'LAST_EVENT,
                              PathDelay => VitalExtendToFillDelay(tpd_A0_D0),
                              PathCondition   => TRUE)
                    )
                );
            END PROCESS;
        END GENERATE;
```

Figure 6.9   VPD inside a generate statement

interval for a dynamic memory. In such cases device delays can be used. The generic
prefix for this type of delay is tdevice.

```
    -- tdevice values: values for internal delays
    tdevice_REF              : VitalDelayType    := 15_625 ns;
```

A reasonable default value should be assigned. Otherwise the model might be
unusable should the user decide not to backannotate timing. If the value is constant across vendors and speed grades of this component, it may be left out of the
SDF file.

Because the suffix of the tdevice generic is expected to match the label of a
VITAL primitive concurrent procedure call, the generic must be used with a VITAL
primitive.

```
    -----------------------------------------------------------------------
    -- Internal Delays
    -----------------------------------------------------------------------
    - Artificial VITAL primitives to incorporate internal delays
    REF : VitalBuf (refreshed_out, refreshed_in, (UnitDelay, tdevice_REF));
```

Please note, however, that although the generic must be used with a VITAL primitive, it may also be used elsewhere in the model.

```
IF (NOW > Next_Ref AND PoweredUp AND Ref_Cnt > 0) THEN
  Ref_Cnt := Ref_Cnt – 1;
  Next_Ref := NOW + tdevice_REF;
END IF;
```

In other cases, the tdevice generic may be used to control the pulse width of an
internal signal. In the following example, a flash memory is designed to time out
if an erase command sequence is not completed in a certain amount of time after
it is initiated.

First, the generic is declared and given a default value:

```
        --sector erase command sequence timeout
tdevice_CTMOUT          : VitalDelayType    := 50 us;
```

Input and output signals are declared:

```
SIGNAL CTMOUT            : std_ulogic := '0'; –Sector Erase TimeOut
SIGNAL CTMOUT_in         : std_ulogic := '0';
```

Than a `VitalBuf` is instantiated using the signals and the generic:

```
TCTMOUT :VitalBuf(CTMOUT, CTMOUT_in, (tdevice_CTMOUT ,UnitDelay));
```

Now, whenever CTMOUT_in is driven high, CTMOUT will follow after 50 us. In the
model's state machine, in a process sensitive to CTMOUT,

```
ELSIF DataLo=16#30# THEN
      --put selected sector to sec. ers. queue
      --start timeout
      Ers_Queue  <= (OTHERS => '0');
      Ers_Queue(SecAddr) <= '1';
      CTMOUT_in <= '1';
```

Then CTMOUT is monitored:

```
WHEN SERS             =>
     IF CTMOUT = '1' THEN
          CTMOUT_in <= '0';
          START_T1_in <= '0';
          ESTART <= '1', '0' AFTER 1 ns;
          ESUSP  <= '0';
          ERES   <= '0';
```

If CTMOUT goes high before the state machine moves to the next state, the operation is aborted and CTMOUT_in is reset to '0'.

There are many other way the tdevice generic might be used. They are limited only by your need and your imagination.

6.8   Backannotating Path Delays

The VITAL standard is strict regarding how timing generics are named but more relaxed about how they are used. Taking advantage of this can result in some economies in certain types of models and provide utility in others.

Most components can have multiple delay paths described by a single timing specification. The data sheet for a 2-input nand gate will usually state the same delay from either input to the output. There is no need to write separate generics for each input and then backannotate identical values to both of them. A single generic will do. This is all the more true for a model with a single clock and a 16-bit output bus. If all bits have the same delay specification, a single generic will do.

A more interesting case is when a component exhibits different delays based on an internal state. This frequently is true for synchronous DRAMs, which can be programmed for different CAS latencies. A CAS latency of 3 provides an output with a shorter delay relative to the clock than a CAS latency of 2, but with a three-clock-cycle latency. A CAS latency of 2 means there will be a longer delay relative to the clock but only a two-clock-cycle latency.

In the km432s2030 model, the two sets of delays are brought in using two tpd generics:

```
-- tpd delays
tpd_CLK_DQ2              : VitalDelayType01Z := UnitDelay01Z;
tpd_CLK_DQ3              : VitalDelayType01Z := UnitDelay01Z;
```

They are backannotated from two SDF IOPATH statements:

```
(DELAY (ABSOLUTE
  (IOPATH CLK DQ2  (3:6:8) (3:6:8) (3:6:8) (3:6:8) (3:6:8) (3:6:8))
  (IOPATH CLK DQ3  (2:4:6) (2:4:6) (2:4:6) (2:4:6) (2:4:6) (2:4:6))
))
```

Any of the DQ ports could have been used in the generics and SDF, but it is easy to remember DQ2 and DQ3 are for CAS latencies of 2 and 3.

The model incorporates a signal called CAS_Lat that gets its value from a programmable register. The user (of the component or the model) configures the memory by programming this and other registers. In the model, the VPD uses CAS_Lat as a condition in selecting the path delays for the output bus.

A path delay procedure call that uses the value of an internal register to select a path is shown in Figure 6.10.

If you are trying to write compatible VHDL and Verilog models of components, be warned that these tricks will not translate to Verilog. The specification of multiple path delay values in Verilog requires the use of conditional path delay

```
------------------------------------------------------------------------
-- Path Delay Process
------------------------------------------------------------------------
DataOutBlk : FOR i IN 31 DOWNTO 0 GENERATE
    DataOut_Delay : PROCESS (D_zd(i))
        VARIABLE D_GlitchData:VitalGlitchDataArrayType(31 Downto 0);
    BEGIN
        VitalPathDelay01Z (
            OutSignal        => DataOut(i),
            OutSignalName    => "Data",
            OutTemp          => D_zd(i),
            Mode             => OnEvent,
            GlitchData       => D_GlitchData(i),
            Paths            => (
                1 => (InputChangeTime => CLKIn'LAST_EVENT,
                        PathDelay => tpd_CLK_DQ2,
                        PathCondition   => CAS_Lat = 2),
                2 => (InputChangeTime => CLKIn'LAST_EVENT,
                        PathDelay => tpd_CLK_DQ3,
                        PathCondition   => CAS_Lat = 3)
            )
        );

    END PROCESS;
END GENERATE;
```

Figure 6.10   Selecting a path delay with an internal register

generics (described in Chapter 10). However, those may not depend on internal states so they will not work for this example.

6.9   Interconnect Delays

Interconnect delays represent the time it takes a signal transition to propagate from one component to another through the copper traces on a printed circuit board or through other media. As components get faster and clock cycle times get shorter, these interconnect delays become more important to the correct timing of our designs.

On a PCB, the delay is dependent on the length of the trace between the driving and receiving pins, the construction of the board, the materials from which it is built, and the signaling system employed. As a board is designed, progressively more accurate delay values can be extracted. Before component placement, zero delays are assumed. After component placement, delays can estimated based on manhattan distances. After routing the board, delays can be estimated based on the actual length of the interconnect. For the most accurate analysis, a signal integrity analyzer in employed. It uses analog models of the drivers and receivers as well as any terminators to determine the time from when a driver begins to change state to when the receiver switches. It can take into account capacitive loads, transmission line reflections, and series and parallel terminations.

Any good PCB layout or signal integrity tool should be capable of extracting interconnect delays and writing them to an SDF file. In the SDF file they will be expressed in an INTERCONNECT statement. In VITAL it is mapped to a `tipd` generic. As seen in Chapter 5, the delay is applied to an input pin. It must be applied to an input because VHDL (or Verilog for that matter) has no way to simulate a wire. A `WireDelay` block delays the input signal by the value in the `tipd` generic and updates a new internal (_ipd) signal for use throughout the model.

As long as there is only one driver on a net, there can be any number of receivers, and interconnect delays remain straightforward. However, when there is more than one driver on a net, the situation becomes more complex. This is called a multisource interconnect delay. SDF has no difficulty expressing independent delays from multiple drivers to multiple receivers.

VITAL defines two mechanisms for dealing with multisource delays. In one scenario, the simulator's SDF annotator picks one of the delay values. The simulator may allow the user to select the minimum, maximum, or last delay in the list. In the other scenario, the simulator accepts all the delays and puts them into an array. Each time an event is detected at an input port, the simulator determines the source driver. It then applies the correct delay for that driver through the `WireDelay` block.

Which capabilities you get depends on which simulator you buy. Read your manuals for instructions on how to control backannotation of multisource interconnect delays.

6.10   Summary

There are two methods of delay modeling in VITAL. They are distributed delays and pin-to-pin delays. For most models, pin-to-pin delays are preferred because they are easier, more flexible, and allow greater control over glitch handling and path selection.

The VITAL path delay procedures are used to implement output delays. They are usually called from within the process that computes the output values whenever practical. In most cases, using VPDs is preferred over modeling with distributed delays. For modeling components with buses they may be embedded in a generate statement. Multiple VPDs may read the same path delay generic, reducing the size of the timing and SDF files required.

Signal assignments or concurrent VITAL procedure calls may be used to drive multiple identical outputs from a single `VitalPathDelay` procedure call. VITAL models may use an internal signal or variable in the selection of a path delay.

Interconnect delays are the physical delay associated with a printed circuit board (or other) implementation. Most PCB tools provide some means of generating an SDF file for backannotation.

Whenever a model requires a timing value that is not directly associated with a pin pair, a device delay may be used. They are very flexible and can be fit to almost any need.

7

VITAL Tables

One of the strengths of Verilog is its user-defined primitives. The VITAL standard brings similar capability to VHDL. VITAL tables can be used to model a variety of combinatorial and state-dependent behaviors.

In the early 1990s, the VITAL team was formed and tasked with finding a way to improve the gate-level simulation performance of VHDL so it could compete with Verilog. In performing this mandate, they were not shy about borrowing from Verilog's strengths. One of the features they borrowed was Verilog's UDPs. The UDP has been a useful tool for Verilog model writers since its inception because it gives the writer the ability to concisely define the behavior of small digital circuits such as gates, multiplexers, decoders, and counters. The result was VITAL truth tables and VITAL state tables. Truth tables and state tables are defined in the VITAL_Primitives package.

In the VITAL2000 revision, the use of tables has been extended to include modeling static memories.

7.1   Advantages of Truth and State Tables

The difference between truth tables and state tables is truth tables work with a current stimulus and state tables can store a previous condition. Truth tables and state tables offer a concise means of representing complex combinatorial and sequential logic. Commonly used operations, such as decoders and flip-flops, can be written as truth and state tables. Because truth and state tables are either functions or procedures, it is convenient to put them in packages where they can be used by many models. The Free Model Foundry ff_package contains 37 varieties of latch and flip-flop state tables that can be instantiated in your models or used as examples for creating something new.

Because VITAL tables are accelerated, they may execute faster than other methods of encoding low-level functions. Their table format also makes them easy to read and comprehend.

7.2   Truth Tables

Truth tables are used for modeling combinatorial logic such as decoders. Inputs consist entirely of asynchronous signals. They have no states.

7.2.1   Truth Table Construction

Truth tables are constructed of a set of input patterns and an associated set of output values. Wildcard characters can be used in the input patterns to reduce the table size and improve readability. Figure 7.1 is a truth table for the function (A OR B) and not (C). VITAL truth tables are two-dimensional arrays or type Vital-TruthTableType. The first dimension (rows) is the number of patterns. The second dimension (columns) is the total number of inputs and outputs. "–" means don't care.

The VITAL truth table for a 74139, 2 to 4 decoder, is shown in Figure 7.2. In this case the array is of size (0 to 4, 0 to 6).

7.2.2   VITAL Table Symbols

There are 23 symbols available for constructing VITAL truth tables and state tables. Figure 7.3 presents the declaration of the VitalTableSymbolType taken from the VITAL_Timing package. The table symbols are enumeration literals. This means they are case sensitive. The levels and edge transitions corresponding to the elements of VitalTableSymbolType are shown in Table 7.1

A	B	C	Q	QN
-	-	1	0	1
0	0	-	0	1
1	-	0	1	0
-	1	0	1	0

Figure 7.1 Example truth table

```
CONSTANT std139_tab : VitalTruthTableType := (
  ------------------------------------
  -----INPUTS---|-------OUTPUTS------
  --G    B     A | Y0    Y1    Y2    Y3
  ------------------------------------
  ('1', '-', '-', '1', '1', '1', '1'),
  ('0', '0', '0', '0', '1', '1', '1'),
  ('0', '0', '1', '1', '0', '1', '1'),
  ('0', '1', '0', '1', '1', '0', '1'),
  ('0', '1', '1', '1', '1', '1', '0')
);
```

Figure 7.2 74139 decoder truth table

```
TYPE VitalTableSymbolType IS (
   '/',        -- 0 -> 1
   '\',        -- 1 -> 0
   'P',        -- Union of '/' and '^' (any edge to 1)
   'N',        -- Union of '" and 'v' (any edge to 0)
   'r',        -- 0 -> X
   'f',        -- 1 -> X
   'p',        -- Union of '/' and 'r' (any edge from 0)
   'n',        -- Union of '" and 'f' (any edge from 1)
   'R',        -- Union of '^' and 'p' (any possible rising edge)
   'F',        -- Union of 'v' and 'n' (any possible falling edge)
   '^',        -- X -> 1
   'v',        -- X -> 0
   'E',        -- Union of 'v' and '^' (any edge from X)
   'A',        -- Union of 'r' and '^' (rising edge to or from 'X')
   'D',        -- Union of 'f' and 'v' (falling edge to or from 'X')
   '*',        -- Union of 'R' and 'F' (any edge)
   'X',        -- Unknown level
   '0',        -- low level
   '1',        -- high level
   '-',        -- don't care
   'B',        -- 0 or 1
   'Z',        -- High Impedance
   'S'         -- steady value
);
```

Figure 7.3 VitalTableSymbolType

VITAL state tables utilize the entire symbol set. VITAL truth tables use a subset of VitalTableSymbolType in the range of 'X' to 'Z'. A truth table is divided into Input Pattern and Response sections. Valid symbols for the stimulus section of a VITAL truth table are limited to 'X', '0', '1', '-', and 'B'. Valid symbols for the results section of a truth table are 'X', '0', '1', and 'Z'. The symbol 'B' matches any non-'X' value. The symbol '-' is a "don't care" symbol that matches any value.

State tables use more complex transition symbols. For example, an 'R' is used to indicate "all rising transitions," including 0 to X, X to 1, and 0 to 1.

Symbol Matching

During truth table and state table processing, input data are automatically converted using the std_logic function To_X01. The resulting values are then compared to the stimulus portion of the table using the matching rules in Table 7.2.

7.2.3   Truth Table Usage

Truth tables are usually collected in a package but may be included in a model. The function version of the VitalTruthTable is used inside a process. The procedure version is used in a concurrent procedure call. The function version is more commonly employed in component models.

Using the FMF std138 model as an example of how to use a truth table, we see the first step is declaring the table as a constant:

Table 7.1 Truth table and state table symbol semantics

	$0 \to 0$	$1 \to 0$	$X \to 0$	$1 \to 1$	$0 \to 1$	$X \to 1$	$0 \to X$	$1 \to X$	$X \to X$
'/'					*				
'\'		*							
'P'					*	*			
'N'		*	*						
'r'							*		
'f'								*	
'p'					*		*		
'n'		*						*	
'R'					*	*	*		
'F'		*	*						
'∧'						*			
'v'			*						
'E'			*			*			
'A'						*	*		
'D'			*					*	
'*'		*	*		*	*	*	*	
'X'							*	*	*
'0'	*	*	*						
'1'				*	*	*			
'-'	*	*	*	*	*	*	*	*	*
'B'	*	*	*	*	*	*			
'Z'									
'S'	*			*					

Table 7.2 Symbol matching rules

Table Stimulus Portion	DataInX01 := To_X01 (DataIn)	Result of Comparison
'X'	'X'	'X' only matches with 'X'
'0'	'0'	'0' only matches with '0'
'1'	'1'	'1' only matches with '1'
'-'	'X', '0', '1'	'-' matches with any value of DataInX01
'B'	'0', '1'	'B' only matches with '0' or '1'

```
-------------------------------------------------------------------------
-- Decode Process
-------------------------------------------------------------------------
Decode : PROCESS (G1_ipd, G2int, C_ipd, B_ipd, A_ipd)

        CONSTANT std138_tab : VitalTruthTableType := (
        -----------------------------------------------------------------
        ---------INPUTS--------- | ----------------OUTPUTS----------------
        --G1  G2   C    B    A | Y0   Y1   Y2   Y3   Y4   Y5   Y6   Y7
        -----------------------------------------------------------------
        ('-', '1', '-', '-', '-', '1', '1', '1', '1', '1', '1', '1', '1'),
        ('0', '-', '-', '-', '-', '1', '1', '1', '1', '1', '1', '1', '1'),
        ('1', '0', '0', '0', '0', '0', '1', '1', '1', '1', '1', '1', '1'),
        ('1', '0', '0', '0', '1', '1', '0', '1', '1', '1', '1', '1', '1'),
        ('1', '0', '0', '1', '0', '1', '1', '0', '1', '1', '1', '1', '1'),
        ('1', '0', '0', '1', '1', '1', '1', '1', '0', '1', '1', '1', '1'),
        ('1', '0', '1', '0', '0', '1', '1', '1', '1', '0', '1', '1', '1'),
        ('1', '0', '1', '0', '1', '1', '1', '1', '1', '1', '0', '1', '1'),
        ('1', '0', '1', '1', '0', '1', '1', '1', '1', '1', '1', '0', '1'),
        ('1', '0', '1', '1', '1', '1', '1', '1', '1', '1', '1', '1', '0')
    );
```

This table functions as a 3 to 8 decoder. It has two enable inputs, one active high, the other active low. The outputs are active low. When the table is searched, the enables will be checked first. If both are active, the input to be decoded is selected. In the first two rows, if either enable is inactive, it does not matter what values are on the other inputs. That is why '-'s are used.

When a call is made to a VitalTruthTable primitive, the stimulus, DataIn, is compared with the input pattern of each row, starting with the top row. The comparison stops with the first matching entry. The outputs of the matching row are converted to std_logic X01Z. If no match is found, 'X's are returned.

Because tables are searched from top to bottom and the search stops at the match, the order of the rows can be important. Although this example does not illustrate it, some tables will produce incorrect results if the rows are not ordered correctly. This is particularly true for state tables.

Next, the required variables are declared:

```
    -- Functionality Results Variables
    VARIABLE YData          : std_logic_vector(0 to 7);
    ALIAS Y0_zd             : std_ulogic IS YData(0);
    ALIAS Y1_zd             : std_ulogic IS YData(1);
    ALIAS Y2_zd             : std_ulogic IS YData(2);
    ALIAS Y3_zd             : std_ulogic IS YData(3);
    ALIAS Y4_zd             : std_ulogic IS YData(4);
    ALIAS Y5_zd             : std_ulogic IS YData(5);
    ALIAS Y6_zd             : std_ulogic IS YData(6);
    ALIAS Y7_zd             : std_ulogic IS YData(7);
```

```
           -- Output Glitch Detection Variables
           VARIABLE Y0_GlitchData    : VitalGlitchDataType;
           VARIABLE Y1_GlitchData    : VitalGlitchDataType;
           VARIABLE Y2_GlitchData    : VitalGlitchDataType;
           VARIABLE Y3_GlitchData    : VitalGlitchDataType;
           VARIABLE Y4_GlitchData    : VitalGlitchDataType;
           VARIABLE Y5_GlitchData    : VitalGlitchDataType;
           VARIABLE Y6_GlitchData    : VitalGlitchDataType;
           VARIABLE Y7_GlitchData    : VitalGlitchDataType;
```

The first variable, YData, will get the results of the truth table call. The aliases will be used to transfer the results to the path delay procedures. The GlitchData variables are also for use in the path delay procedures.

The call to the VitalTruthTable function is in the functionality section of the process (by FMF convention):

```
           BEGIN
               ------------------------------------------------------------------
               -- Functionality Section
               ------------------------------------------------------------------
               Ydata := VitalTruthTable (
                       TruthTable  => std138_tab,
                       DataIn      => (G1_ipd, G2int, C_ipd, B_ipd, A_ipd)
                   ) ;
```

It is a simple function call, with YData getting the results. The two arguments to the call are the name of the truth table and a list of the inputs. The inputs make up an array and must be listed in the same order in which they appear in the table.

Finally, the results are sent to a VPD so they can be delayed by an appropriate time before being assigned to the output ports.

```
               ------------------------------------------------------------------
               -- Path Delay Section
               ------------------------------------------------------------------
               VitalPathDelay01 (
                   OutSignal       => Y0Neg,
                   OutSignalName   => "Y0Neg",
                   OutTemp         => Y0_zd,
                   GlitchData      => Y0_GlitchData,
                   XOn             => XOn,
                   MsgOn           => MsgOn,
                   Paths           => (
                     0 => (
                           InputChangeTime => G1_ipd'LAST_EVENT,
                           PathDelay       => tpd_G1_Y0Neg,
                           PathCondition   => TRUE),
```

```
    1 => (
        InputChangeTime    => G2int'LAST_EVENT,
        PathDelay          => tpd_G2ANeg_Y0Neg,
        PathCondition      => TRUE),
    2 => (
        InputChangeTime    => A_ipd'LAST_EVENT,
        PathDelay          => tpd_A_Y0Neg,
        PathCondition      => ((G1_ipd = '1') AND G2int = '1')),
    3 => (
        InputChangeTime    => B_ipd'LAST_EVENT,
        PathDelay          => tpd_A_Y0Neg,
        PathCondition      => ((G1_ipd = '1') AND G2int = '1')),
    4 => (
        InputChangeTime    => C_ipd'LAST_EVENT,
        PathDelay          => tpd_A_Y0Neg,
        PathCondition      => ((G1_ipd = '1') AND G2int = '1')))
);
```

Because there were five inputs to the function, we need five paths to analyze to determine the correct delay. Only one VPD is shown here. There are seven more like it in the model.

7.3   State Tables

State tables are similar to truth tables but are used for describing behavior that includes an internal state. A state table defines a transition from the present state and present inputs to the next state and its outputs. Unlike the `VitalTruthTable`, the `VitalStateTable` has only versions that are used as procedures. One version accepts variable inputs for use in sequential processes and the other version accepts signal inputs for use as a concurrent statement outside of any process. For component modeling, the sequential version will most often be used.

7.3.1   State Table Symbols

State tables can use the full set of symbols shown in Figure 7.3 including all the edge transition symbols. Indeed, it is the use of edge transition symbols that makes state tables suitable for modeling synchronous behavior.

7.3.2   State Table Construction

State tables are a bit more complicated to construct than truth tables. Although they can still be divided into stimulus and response sections, the stimulus section consists of columns for all the inputs followed by columns for each of the previous outputs. The response section will have a column for each output. The ordering

```
CONSTANT DFF_tab : VitalStateTableType  := (

----INPUTS--|PREV|-OUTPUT--
-- CLK    D  | QI | Q'     --
----------------|-----|---------
( '/', '0', '-', '0'), -- active clock edge
( '/', '1', '-', '1'), -- active clock edge
( '-', '-', '-', 'S')  -- default

); -- end of VitalStateTableType definition
```

Figure 7.4   An oversimplified D flip-flop

of the columns for the previous outputs must match the ordering of the subsequent columns for outputs.

A state table for an oversimplified D flip-flop is presented in Figure 7.4. In the table, when there is a rising edge (0 to 1) on CLK, 'Q' takes the value of D. At all other times, the output is stable.

7.3.3   State Table Usage

To use a state table we must first declare some variables. The code in this example is taken from the FMF std273 model:

```
-----------------------------------------------------------------------
-- Main Behavior Process
-----------------------------------------------------------------------

VitalBehavior : PROCESS (CLK_ipd, D_ipd, CLRint)

    -- Timing Check Variables
...

    VARIABLE Violation      : X01 := '0';

    -- Functionality Results Variables
    VARIABLE Q_zd           : std_ulogic;
    VARIABLE PrevData       : std_logic_vector (0 to 3);
```

Here, Q_zd is the temporary output variable. PrevData is an array for storing the previous state of the state table. The range of this array should begin at zero. The variable Violation takes the combined output of the timing checks (not shown). If there is a timing violation, its value will be 'X'. Otherwise, its value will be '0'.

The VitalStateTable procedure call is in the functionality section of the process:

```
VitalStateTable     (
    StateTable      => DFFR_tab,
    DataIn          => (Violation, CLK_ipd, D_ipd, CLRint),
    Result          => Q_zd,
    PreviousDataIn  => PrevData
) ;
```

Where

StateTable takes the name of a previously defined state table. In this model, the state table definition resides in a separate package. The state table is

```
-------------------------------------------------------------------------
-- D-flip/flop with Reset active high
-------------------------------------------------------------------------

CONSTANT DFFR_tab : VitalStateTableType := (

        ------- INPUTS -------|PREV-|-OUTPUT--
        -- Viol CLK   D     R  | QI | Q'     --
        --------------------|-----|---------
    ( 'X', '-', '-', '-', '-', 'X'), -- timing violation
    ( '-', 'B', '-', 'X', '0', '0'), -- reset unknown
    ( '-', '/', '0', 'X', '0', '0'), -- reset unknown
    ( '-', '-', '-', 'X', '-', 'X'), -- reset unknown
    ( '-', '-', '-', '1', '-', '0'), -- reset asserted
    ( '-', 'X', '0', '0', '0', '0'), -- clk unknown
    ( '-', 'X', '1', '0', '1', '1'), -- clk unknown
    ( '-', 'X', '-', '0', '-', 'X'), -- clk unknown
    ( '-', '/', '0', '0', '-', '0'), -- active clock edge
    ( '-', '/', '1', '0', '-', '1'), -- active clock edge
    ( '-', '/', '-', '0', '-', 'X'), -- active clock edge
    ( '-', '-', '-', '-', '-', 'S')  -- default

    ) ; -- end of VitalStateTableType definition
```

DataIn takes an array of input values. These are the present inputs. Result is of mode INOUT. It returns the output values and is read as the previous state.

PreviousDataIn is of mode INOUT. It holds the previous DataIn values and saves the present inputs for use in the table's next call.

The Q_zd output is used to supply an input value to a VitalPathDelay procedure.

7.3.4   State Table Algorithm

When the VitalStateTable procedure is called, the state table is searched, row by row, until an entry matching the current input transitions (edges and steady values) and state values is found. If a match is found, the output and next state values corresponding to the match are returned. If no match is found, the result is set to all 'X' values. The PreviousDataIn vector is updated automatically.

It is important to remember that the row-by-row traversal algorithm results in a priority equivalent to an IF / ELSIF / ELSE expression in a program. It also means that asynchronous inputs should be handled by placing their entries near the top of the table. Also each table should have at least one row with an 'S' or a '-' in the clock column to account for the table being entered when there is no clock transition.

7.4   Reducing Pessimism

The construction of a state table should be carefully thought out. Although it may not be difficult to design a table to match the way a component is specified to work under proper conditions, it takes more consideration to envision how it will work under all possible conditions.

The state table in Figure 7.4 describes a D flip-flop the way a data sheet might. But how will the circuit behave if D is unknown? Figure 7.5 adds a row to cover that condition. This is an improvement over our first try.

Next, consider how the circuit responds to an unknown clock input. Figure 7.6 covers the basic case of an unknown clock but is needlessly pessimistic. If D has the same value as Q, then the output will remain stable regardless of the clock input. This is reflected in the next version of the state table in Figure 7.7. If having an unknown clock is an event that should cause user notification, that could be done using an assertion statement, as described in Chapter 16.

A model that is overly pessimistic will output 'X's under conditions in which the actual component performs satisfactorily. This will either cause the designer to waste time trying to fix nonexistent problems or make the simulation output more laborious to interpret. The state table in Figure 7.7 accounts for all the possible inputs. However, because a flip-flop is a synchronous circuit, it should also consider the effect of timing violations.

```
CONSTANT DFF_tab : VitalStateTableType  := (

    ----INPUTS--|PREV-|-OUTPUT--
    -- CLK    D  | QI  | Q'      --
    ----------------|-----|---------
    ( '/', '0', '-', '0'), -- active clock edge
    ( '/', '1', '-', '1'), -- active clock edge
    ( '/', '-', '-', 'X'), -- active clock edge
    ( '-', '-', '-', 'S')  -- default

    ); -- end of VitalStateTableType definition
```

Figure 7.5   D flip-flop with unknown D added

```
CONSTANT DFF_tab : VitalStateTableType  := (

    ----INPUTS--|PREV-|-OUTPUT--
    -- CLK    D  | QI  | Q'      --
    ----------------|-----|---------
    ( 'X', '-', '-', 'X'), -- clk unknown
    ( '/', '0', '-', '0'), -- active clock edge
    ( '/', '1', '-', '1'), -- active clock edge
    ( '/', '-', '-', 'X'), -- active clock edge
    ( '-', '-', '-', 'S')  -- default

    ); -- end of VitalStateTableType definition
```

Figure 7.6   D flip-flop with unknown CLK added

```
CONSTANT DFF_tab : VitalStateTableType  := (

        ----INPUTS--|PREV-|-OUTPUT--
        -- CLK   D | QI | Q'    --
        ---------------|-----|---------
      ( 'X', '0', '0', '0'), -- clk unknown
      ( 'X', '1', '1', '1'), -- clk unknown
      ( 'X', '-', '-', 'X'), -- clk unknown
      ( '/', '0', '-', '0'), -- active clock edge
      ( '/', '1', '-', '1'), -- active clock edge
      ( '/', '-', '-', 'X'), -- active clock edge
      ( '-', '-', '-', 'S')  -- default

   ); -- end of VitalStateTableType definition
```

Figure 7.7   D flip-flop with pessimism removed

```
CONSTANT DFF_tab : VitalStateTableType  := (

        ----INPUTS-------|PREV-|-OUTPUT--
        -- Viol CLK   D | QI | Q'    --
        ------------------|-----|---------
      ( 'X', '-', '-', '-', 'X'), -- timing violation
      ( '-', 'X', '0', '0', '0'), -- clk unknown
      ( '-', 'X', '1', '1', '1'), -- clk unknown
      ( '-', 'X', '-', '-', 'X'), -- clk unknown
      ( '-', '/', '0', '-', '0'), -- active clock edge
      ( '-', '/', '1', '-', '1'), -- active clock edge
      ( '-', '/', '-', '-', 'X'), -- active clock edge
      ( '-', '-', '-', '-', 'S')  -- default

   ); -- end of VitalStateTableType definition
```

Figure 7.8   D flip-flop with timing violation added

Because in the event of a timing violation circuit behavior is unknown, the table will specify an 'X' in that situation, as shown in Figure 7.8. Now the table is complete. Although the data sheet may have described the part with a table of 3 or 4 rows, the state table for our model has 8 rows in order to correctly cover all the nonspecified conditions and to avoid excess pessimism.

7.5   Memory Tables

The 2000 revision of the VITAL standard added the VITALMemory package, which includes the VitalMemoryTable procedure. It can be thought of as an extension of the VitalStateTable that has been specialized for describing static memories. VITAL memory tables can be used to model memories in much the same manner as VITAL state tables are used to model flip-flops and latches.

7.5.1   Memory Table Symbols

The VitalMemorySymbolType enumerates the symbols that may be used in a memory table. They are listed with explanation in Figure 7.9.

```
TYPE VitalMemorySymbolType IS (
    '/',        -- 0 -> 1
    '"',        -- 1 -> 0
    'P',        -- Union of '/' and '^' (any edge to 1)
    'N',        -- Union of '"' and 'v' (any edge to 0)
    'r',        -- 0 -> X
    'f',        -- 1 -> X
    'p',        -- Union of '/' and 'r' (any edge from 0)
    'n',        -- Union of '"' and 'f' (any edge from 1)
    'R',        -- Union of '^' and 'p' (any possible rising edge)
    'F',        -- Union of 'v' and 'n' (any possible falling edge)
    '^',        -- X -> 1
    'v',        -- X -> 0
    'E',        -- Union of 'v' and '^' (any edge from X)
    'A',        -- Union of 'r' and '^' (rising edge to or from 'X')
    'D',        -- Union of 'f' and 'v' (falling edge to or from 'X')
    '*',        -- Union of 'R' and 'F' (any edge)
    'X',        -- Unknown level
    '0',        -- low level
    '1',        -- high level
    '-',        -- don't care
    'B',        -- 0 or 1
    'Z',        -- High Impedance
    'S',        -- steady value

    'g',        -- Good address or data (no transition)
    'u',        -- Unknown address or data (no transition)
    'i',        -- Invalid address or data (no transition)
    'G',        -- Good address or data (with transition)
    'U',        -- Unknown address or data (with transition)
    'I',        -- Invalid address or data (with transition)

    'w',        -- Write data to memory
    's',        -- Retain previous memory contents
    'c',        -- Corrupt entire memory with 'X'
    'l',        -- Corrupt a word in memory with 'X'
    'd',        -- Corrupt a single bit in memory with 'X'
    'e',        -- Corrupt a word with 'X' based on data in
    'C',        -- Corrupt a sub-word entire memory with 'X'
    'L',        -- Corrupt a sub-word in memory with 'X'
    'M',        -- Implicit read data from memory
    'm',        -- Read data from memory
    't'         -- Immediate assign/transfer data in
);
```

Figure 7.9 VITALMemory table symbols

The symbols shown in Figure 7.9 are divided into three groups. The first group is identical to the VitalTableSymbolType shown in Figure 7.3. These symbols retain the semantics given in Table 7.1. Their use, as before, is limited to scalar signals. The next two groups are unique to the VitalMemorySymbolType. Symbols 'g' through 'I' are for use with address and data vectored inputs. Symbols 'w' through 't' define memory table results.

7.5.2   Memory Table Construction

VITAL memory tables are intended for modeling the functionality of static memories. They work in a manner similar to VITAL truth tables but at a somewhat higher level of complexity.

Table 7.3 Allowed symbols for memory tables

Section of Table	Allowed Symbols
Direct Inputs	'/', '\', 'P', 'N', 'r', 'f', 'p', 'n', 'R', 'F', '^', 'v', 'E', 'A', 'D', '*', 'X', '0', '1', '-', 'B', 'S'
EnableBus Interpreted Input	'/', '\', 'P', 'N', 'r', 'f', 'p', 'n', 'R', 'F', '^', 'v', 'E', 'A', 'D', '*', 'X', '0', '1', '-', 'B', 'S'
AddressBus Interpreted Input	'g', 'u', 'i', 'G', 'U', 'I', '*', '-', 'S'
DataInBus Interpreted Input	'g', 'u', 'G', 'U', '*', '-', 'S'
Memory Action	'0', '1', 'w', 's', 'c', 'l', 'd', 'e', 'C', 'D', 'E', 'L'
Output Action	'0', '1', 'Z', 'l', 'd', 'e', 'C', 'D', 'E', 'L', 'M', 'm', 't', 'S'

Memory tables can be broken down into three sections: direct input columns, interpreted input columns, and result columns. The direct inputs are scalar signals such as clocks and enables. The interpreted inputs are vectored signals that correspond to `EnableBus`, `AddressBus`, and `DataInBus`. Results include the action to perform on the memory array and the output bus. The `EnableBus` column is used only when modeling memories that are subword addressable. The three sets of columns must be written in the order listed.

Each section of the table may use only a subset of the symbols enumerated in the `VitalMemorySymbolType`. Which symbols are allowed in which columns is shown in Table 7.3. A memory table for a simple SRAM is shown in Figure 7.10.

7.5.3   Memory Table Usage

As with state tables, using memory tables requires the declaration of some variables. The code in the following examples is from the FMF cy7c185 model.

```
-- VITAL Memory Declaration
VARIABLE Memdat : VitalMemoryDataType :=
    VitalDeclareMemory (
        NoOfWords               => TotalLOC,
        NoOfBitsPerWord         => DataWidth,
        NoOfBitsPerSubWord      => DataWidth,
        MemoryLoadFile          => MemLoadFileName,
        BinaryLoadFile          => FALSE
    ) ;
```

The first declaration creates the memory array. VITAL memory tables work only with memory arrays of `VitalMemoryDataType`.

```
-------------------------------------------------------------------------
-- Asynchronous SRAM with low chip enable and write enable
-------------------------------------------------------------------------
CONSTANT Table_2_cntrl_sram : VitalMemoryTableType := (

-- -------------------------------------------------------------
-- CEN, WEN, Addr, DI, act, DO
-- -------------------------------------------------------------
-- Address initiated read
  ( '0', '1', 'G', '-', 's', 'm' ),
  ( '0', '1', 'U', '-', 's', 'l' ),

-- CEN initiated read
  ( 'N', '1', 'g', '-', 's', 'm' ),
  ( 'N', '1', 'u', '-', 's', 'l' ),

-- Write Enable initiated Write
  ( '0', 'P', 'g', '-', 'w', 'm' ),
  ( '0', 'N', '-', '-', 's', 'Z' ),

-- CEN initiated Write
  ( 'P', '0', 'g', '-', 'w', 'Z' ),
  ( 'N', '0', '-', '-', 's', 'Z' ),

-- Address change during write
  ( '0', '0', '*', '-', 'c', 'Z' ),
  ( '0', 'X', '*', '-', 'c', 'Z' ),

-- if WEN is X
  ( '0', 'X', 'g', '*', 'e', 'e' ),
  ( '0', 'X', 'u', '*', 'c', 'l' ),

-- CEN is unasserted
  ( 'X', '0', 'G', '-', 'e', 'Z' ),
  ( 'X', '0', 'u', '-', 'c', 'Z' ),
  ( 'X', '1', '-', '-', 's', 'l' ),
  ( '1', '-', '-', '-', 's', 'Z' )

); -- end of VitalMemoryTableType definition
```

Figure 7.10 Memory table for simple SRAM

The following variables are required for exclusive use by the `VitalMemoryTable` procedure:

```
VARIABLE Prevcntls      : std_logic_vector(0 to 3);
VARIABLE PrevData       : std_logic_vector(HiDbit downto 0);
VARIABLE Prevaddr       : std_logic_vector(HiAbit downto 0);
VARIABLE PFlag          : VitalPortFlagVectorType(0 downto 0);
VARIABLE Addrvalue      : VitalAddressValueType;
```

They are described in Table 7.4.

The `VitalMemoryTable` procedure call from the cy7c185 model is

```
VitalMemoryTable (
    DataOutBus      => D_zd,
    MemoryData      => Memdat,
```

```
              PrevControls      => Prevcntls,
              PrevDataInBus     => Prevdata,
              PrevAddressBus    => Prevaddr,
              PortFlag          => PFlag,
              Controls          => (CE2In, CE1NegIn, OENegIn, WENegIn),
              DataInBus         => DataIn,
              AddressBus        => AddressIn,
              AddressValue      => Addrvalue,
              MemoryTable       => Table_generic_sram
    );
```

All the possible arguments for the procedure call and their descriptions are listed in Table 7.4.

Table 7.4   Arguments for `VitalMemoryTable`

Argument Name	Type	Description
DataOutBus	STD_LOGIC_VECTOR	Variable for functional output data
MemoryData	VitalMemoryDataType	Pointer to VITAL memory data object
PrevControls	STD_LOGIC_VECTOR	Previous state values of Controls parameter
PrevEnableBus	STD_LOGIC_VECTOR	Previous state values of PrevEnableBus parameter
PrevDataInBus	STD_LOGIC_VECTOR	Previous DataInBus for edge detection
PrevAddressBus	STD_LOGIC_VECTOR	Previous address bus for edge detection
PortFlag	VitalPortFlagType	Indicates operating mode of the port in a single process execution. Possible values are READ, WRITE, CORRUPT, NOCHANGE
PortFlagArray	VitalPortFlagVectorType	Vector form of PortFlag for subword addressable memories. Used in overloaded version
Controls	STD_LOGIC_VECTOR	Aggregate of scalar memory control inputs
EnableBus	STD_LOGIC_VECTOR	Concatenation of vector control inputs
DataInBus	STD_LOGIC_VECTOR	Memory data in bus inputs
AddressBus	STD_LOGIC_VECTOR	Memory address bus inputs
AddressValue	VitalAddressValueType	Decoded integer value of the AddressBus
MemoryTable	VitalMemoryTableType	Memory function table
PortType	VitalPortType	The base type of port (one of READ, WRITE, RDNWR)
PortName	STRING	Port name string for messages
HeaderMsg	STRING	Header string for messages
MsgOn	BOOLEAN	Reporting control of message generation
MsgSeverity	SEVERITY_LEVEL	Severity control of message generation

7.6   Summary

VITAL truth tables may be used to describe combinatorial logic. VITAL state tables may be used to describe synchronous logic. Many complex combinatorial and sequential functions can be written more easily using truth or state tables than by other means.

Tables are constants written as two-dimensional arrays of type `Vital-TruthTableType` or `VitalStateTableType`. Tables are searched row by row, thus earlier rows have a higher priority than later rows. When writing state tables it is important to consider the effect of unknown values on each input and the effect of timing violations.

VITAL2000 introduced memory tables. These can be thought of as a specialized extension of state tables. They are designed for modeling the functionality of static memories. They are more complex than state tables and use an extended set of symbols.

8   Timing Constraints

It has been said in more than one profession that "timing is everything." This is particularly true when you are trying to interface your new ASIC or FPGA design with the other components on the board.

Most modern digital designs are synchronous. The use of synchronous design methods requires that data be stable before and after the clock, resulting in timing constraints. An important part of digital simulation is checking that those timing constraints are met. The `VITAL_Timing` package provides a number of routines for use in performing those checks. This chapter explains what they are and how to use them.

8.1   The Purpose of Timing Constraint Checks

The synchronous circuits that compose most digital designs require that certain timing constraints be met in order to guarantee correct operation. Registers require that data be present and stable for a period of time before and after the active transition of the control input (clock). There are further requirements for pulses to be of at least some minimum duration. When these requirements are not met, circuit operation becomes unpredictable.

A significant part of the effort that goes into the design of a new digital product is determining the timing requirements of all the circuits and then ensuring those requirements are met. Some engineers still analyze a design for timing by building spreadsheets. Such spreadsheets were once an effective tool, but many of today's components have 20 or more timing parameters. Having to deal with so many parameters makes the spreadsheet approach impractical for many designs. By including timing constraint checks in simulation models, we allow for verification that timing requirements are met during logic simulation, at least for the stimulus provided. Dynamic simulation cannot practically cover all cases. For that, static timing is the correct solution.

8.2   Using Timing Constraint Checks in VITAL Models

Adding timing constraint checks to VITAL models is a straightforward process. It begins with adding the timing generics that are needed for the particular model to the entity. Generic names are based on the formulae given in Chapter 5.

Timing checks are always called from within a process. Any signal referenced by a timing check must appear in the sensitivity list of that process. Most compilers will check this for you, but if somehow a signal is missed, incorrect results are likely to be given. Within the process certain variables must be declared, depending on which timing checks are being used. Those variables will be described later, along with the timing check procedure calls.

Timing checks have their own section within a VITAL process. They come after the declaration of variables but before any functional description. Each process may have its own timing check section. This section exists entirely within an IF clause:

```
IF (TimingChecksOn) THEN
...
-- All timing check code
...
END IF;
```

The execution of all the timing checks may be controlled through the `TimingChecksOn` generic. Its value can, if desired, be set on an instance-by-instance basis for a design being simulated. This could be useful if you are concentrating on a section of your design and would like to reduce the number of messages from the simulator regarding other sections.

In Free Model Foundry models, the `TimingChecksOn` generic defaults to FALSE. This is because most designs are first simulated without timing to detect basic errors in logic, then simulated with timing to look for more subtle errors. Defaulting to FALSE means the value must be changed once during the design cycle rather than twice. Each timing check procedure performs its specified check and returns a parameter value indicating whether a constraint violation occurred. These values are the only output of the timing section. It is common practice to OR the various outputs together into a single term.

```
VARIABLE Violation      : X01 := '0';
...
Violation := Tviol_D_CLK OR Pviol_CLK OR Pviol_CLRint;
```

The violation variable may then be used elsewhere with the model.

There are four types of timing constraint checks in VITAL2000: setup/hold, period/pulsewidth, recovery/removal, and skew checks.

8.2.1   Setup/Hold Checks

The `VitalSetupHoldCheck` procedure checks that its `TestSignal` (data) is stable during a specified period before and after the active transition of its `RefSignal`

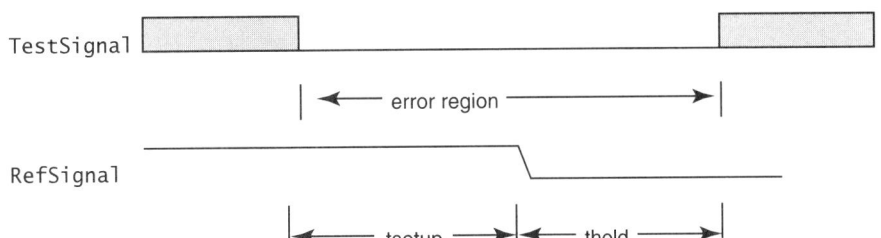

Figure 8.1 Setup/hold check

```
VitalSetupHoldCheck (
    TestSignal      => D_ipd,
    TestSignalName  => "D_ipd",
    RefSignal       => CLK_ipd,
    RefSignalName   => "CLK_ipd",
    SetupHigh       => tsetup_D_CLK,
    SetupLow        => tsetup_D_CLK,
    HoldHigh        => thold_D_CLK,
    HoldLow         => thold_D_CLK,
    CheckEnabled    => TRUE,
    RefTransition   => '/',
    HeaderMsg       => InstancePath & "/std534",
    TimingData      => TD_D_CLK,
    XOn             => XOn,
    MsgOn           => MsgOn,
    Violation       => Tviol_D_CLK
);
```

Figure 8.2 Example of a VitalSetupHoldCheck call

(clock), as illustrated in Figure 8.1. The timing constraint values are specified through generics. This check is usually required for any component model with registered or latched inputs. Real components can produce unpredictable outputs when setup and hold requirements are not met. A good simulation model will warn the user if a hazard exists.

Use of the VitalSetupHoldCheck procedure requires the declaration of two variables for each time it is referenced. The variables are used for timing data storage and for the violation flag output.

```
-- Timing Check Variables
VARIABLE Tviol_D_CLK    : X01 := '0';
VARIABLE TD_D_CLK       : VitalTimingDataType;
```

A VitalSetupHoldCheck procedure call appears in Figure 8.2.

The procedure call in Figure 8.2 will test that D_ipd, the delayed data signal, is stable for the period tsetup_D_CLK before the rising edge of CLK_ipd, the delayed clock signal, and for the period thold_D_CLK after the rising edge of CLK_ipd. In the example,

```
CheckEnabled    => TRUE,
```

causes this check to always be enabled. Otherwise, an expression could be used to cause execution of the check to depend on some dynamic condition. One common condition under which timing checks are disabled is an inactive chip enable.

The parameters to the `VitalSetupHoldCheck` procedure are as follows:

Parameters of Mode IN

`TestSignal`, the value of the test signal. The procedure is overloaded for `TestSignal` to be either `std_logic` or `std_logic_vector`. It should be a delayed input signal.

`TestSignalName`, the name of the test signal. It is of type `STRING` and will be used in any messages generated by the procedure. You should supply a name the user will recognize.

`TestDelay`, not shown. This is the model's internal delay associated with `TestSignal`. It is used only in models having negative timing constraints. It is of type `TIME`. If a value is not provided, it defaults to zero.

`RefSignal`, the value of the reference signal. It is of type `std_logic`. It should be a delayed input signal.

`RefSignalName`, the name of the reference signal. It is of type `STRING` and will be used in any messages generated by the procedure. You should supply a name the user will recognize.

`RefDelay`, not shown. This is the model's internal delay associated with `RefSignal`. It is used only in models having negative timing constraints. It is of type `TIME`. If a value is not provided, it defaults to zero.

`SetupHigh`, the minimum time duration before the active transition of `RefSignal` for which transitions of `TestSignal` are allowed to proceed to the "1" state without causing a setup violation. It is of type `TIME` and usually gets its value from a `tsetup` generic. If a value is not provided, it defaults to zero.

`SetupLow`, the minimum time duration before the active transition of `RefSignal` for which transitions of `TestSignal` are allowed to proceed to the "0" state without causing a setup violation. It is of type `TIME` and usually gets its value from a `tsetup` generic. If a value is not provided, it defaults to zero. Because some components may have asymterical setup constraints, separate `SetupHigh` and `SetupLow` parameters are provided.

`HoldHigh`, the minimum time duration after the active transition of `RefSignal` for which transitions of `TestSignal` are allowed to proceed to the "1" state without causing a setup violation. It is of type `TIME` and usually gets its value from a `thold` generic. If a value is not provided, it defaults to zero.

`HoldLow`, the minimum time duration after the active transition of `RefSignal` for which transitions of `TestSignal` are allowed to proceed to the "0" state without causing a setup violation. It is of type TIME and usually gets its value from a thold generic. If a value is not provided, it defaults to zero. Because some components may have asymterical hold constraints, separate HoldHigh and HoldLow parameters are provided.

CheckEnabled, an expression of type BOOLEAN. A check is performed if TRUE. If a value is not provided, it defaults to TRUE. This parameter enables or disables the entire procedure call. Expressions may be used to make execution of the procedure dependent on the state of one or more pins, internal registers, or states.

RefTransition, the active transition of RefSignal. It is of type VitalEdge-SymbolType.

HeaderMsg, text that will accompany any assertion messages produced. It is of type STRING. It should, at a minimum, help the user determine the origin of the message. Additional information may be added.

XOn, a BOOLEAN that controls the violation output parameter. If TRUE, the output parameter is set to 'X' in the event of a violation. Otherwise, violation is always '0'. If a value is not provided, it defaults to TRUE. This parameter could be used to allow execution of the procedure while disabling the Violation output.

MsgOn, a BOOLEAN that controls the emission of violation messages. If TRUE, setup and hold violation messages will be generated. Otherwise no messages are generated, even upon violations. If a value is not provided, it defaults to TRUE.

MsgSeverity, not shown. Severity level for the assertion. It is of type SEVER-ITY_LEVEL. It can be used to control message display and simulation execution. If a value is not provided, it defaults to WARNING. The simulator may allow masking of low-severity messages or pausing or aborting simulation in the event of a high-severity message.

EnableSetupOnTest, not shown. If FALSE at the time that the TestSignal signal changes, no setup check will be performed. It is of type BOOLEAN. If a value is not provided, it defaults to TRUE.

EnableSetupOnRef, not shown. If FALSE at the time that the RefSignal signal changes, no setup check will be performed. It is of type BOOLEAN. If a value is not provided, it defaults to TRUE.

EnableHoldOnTest, not shown. If FALSE at the time that the TestSignal signal changes, no hold check will be performed. It is of type BOOLEAN. If a value is not provided, it defaults to TRUE.

EnableHoldOnRef, not shown. If FALSE at the time that the RefSignal signal changes, no hold check will be performed. It is of type BOOLEAN. If a value is not provided, it defaults to TRUE.

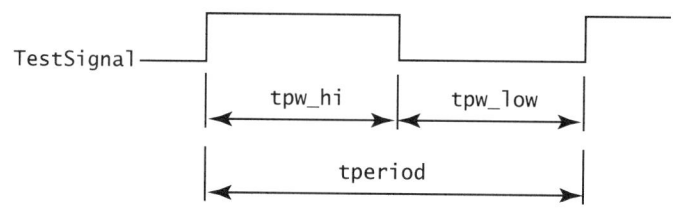

Figure 8.3   Period/pulsewidth check

Parameters of Mode INOUT

TimingData, an information storage area for the procedure. It is used internally to detect reference edges and record the time of the last edge. It is of type VitalTimingDataType. It must be declared but must not be used outside the procedure.

Parameters of Mode OUT

Violation, the violation flag returned. It is of type X01.

In this list, note that CheckEnabled enables or disables the entire procedure. In VITAL2000, the parameters EnableSetupOnTest, EnableSetupOnRef, EnableHoldOnRef, and EnableHoldOnTest were added to allow more precise control over which checks are performed and when. They can all take their values from expressions based on control signals, state, or register values.

8.2.2   Period/Pulsewidth Checks

The VitalPeriodPulseCheck procedure has two functions. It tests the TestSignal for maximum periodicity (1/frequency) and for minimum pulse width for '0' and '1' values. It is illustrated in Figure 8.3.

Use of the VitalPeriodPulseCheck procedure requires the declaration of two variables for each use of the procedure. The variables are used for timing data storage and for the violation flag output.

```
VARIABLE Pviol_CLK        : X01 := '0';
VARIABLE PD_CLK           : VitalPeriodDataType := VitalPeriodDataInit;
```

A VitalPeriodPulseCheck procedure call appears in Figure 8.4.

The procedure call in Figure 8.4 will test that CLK_ipd has a minimum period of tperiod_CLK_posedge, during which it is high for at least tpw_CLK_posedge and low for at least tpw_CLK_negedge. This check as shown is always enabled.

This procedure uses several of the same parameter names and type as the VitalSetupHoldCheck procedure, plus some new ones. Together, they are as follows:

Parameters of Mode IN

TestSignal, the value of the test signal. For this procedure TestSignal must be std_logic. It should be a delayed input signal.

TestSignalName, the name of the test signal. It is of type STRING and will be used in any messages generated by the procedure. You should supply a name the user will recognize.

TestDelay, not shown. This is the model's internal delay associated with TestSignal. It is used only in models having negative timing constraints. It is of type TIME. If a value is not provided, it defaults to zero.

Period, the minimum period allowed between consecutive rising ('P') or falling ('F') transitions. It is of type TIME. If a value is not provided, it defaults to zero.

PulseWidthHigh, the minimum time allowed for a high ('1' or 'H') pulse. It is of type TIME. If a value is not provided, it defaults to zero.

PulseWidthLow, the minimum time allowed for a low ('0' or 'L') pulse. It is of type TIME. If a value is not provided, it defaults to zero.

CheckEnabled, an expression of type BOOLEAN. A check is performed if TRUE. If a value is not provided, it defaults to TRUE. This parameter enables or disables the entire procedure call. Expressions may be used to make execution of the procedure dependent on the state of one or more pins, internal registers, or states.

HeaderMsg, text that will accompany any assertion messages produced. It is of type STRING. It should, at a minimum, help the user determine the origin of the messsage. Additional information may be supplied.

XOn, a BOOLEAN that controls the violation output parameter. If TRUE, the output parameter is set to 'X' in the event of a violation. Otherwise, violation is always '0'. If a value is not provided, it defaults to TRUE. This parameter could be used to allow execution of the procedure while disabling the Violation output.

```
VitalPeriodPulseCheck (
    TestSignal       => CLK_ipd,
    TestSignalName   => "CLK_ipd",
    Period           => tperiod_CLK_posedge,
    PulseWidthHigh   => tpw_CLK_posedge,
    PulseWidthLow    => tpw_CLK_negedge,
    CheckEnabled     => TRUE,
    HeaderMsg        => InstancePath & "/std534",
    PeriodData       => PD_CLK,
    XOn              => XOn,
    MsgOn            => MsgOn,
    Violation        => Pviol_CLK
);
```

Figure 8.4   Example VitalPeriodPulseCheck call

MsgOn, a BOOLEAN that controls the emission of violation messages. If TRUE, setup and hold violation messages will be generated. Otherwise no messages are generated, even upon detection of violations. If a value is not provided, it defaults to TRUE.

MsgSeverity, not shown. It is the severity level for the assertion. It is of type SEVERITY_LEVEL. It can be used to control message display and simulation execution. If a value is not provided, it defaults to WARNING. The simulator may allow masking of low-severity messages or pausing or aborting simulation in the event of a high-severity message.

Parameters of Mode INOUT

PeriodData, an information storage area for the procedure. It is used internally to detect reference edges and record the pulse and period times. It is of type VitalPeriodDataType. It must be declared and initialized to VitalPeriodDataInit but must not be used outside this procedure.

Parameters of Mode OUT

Violation, the violation flag returned. It is of type X01.

Period is the minimum allowed time for a full cycle of the TestSignal. It corresponds to tperiod in Figure 8.2 and is used when TestSignal is a clock or other periodic signal. For nonperiodic signals, such as reset, Period is omitted.

PulseWidthHigh and PulseWidthLow correspond to tpw_hi and tpw_low in Figure 8.3. They represent the minimum allowed time for a high or low pulse, respectively.

8.2.3   Recovery/Removal Checks

The VitalRecoveryRemovalCheck procedure is used to test for the presence of a recovery or removal violation on the TestSignal with respect to the correspond-

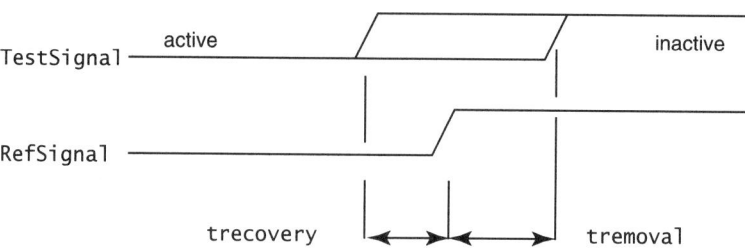

Figure 8.5   Recovery/removal check

```
VitalRecoveryRemovalCheck (
    TestSignal        => CLRint,
    TestSignalName    => "CLRint",
    RefSignal         => CLK_ipd,
    RefSignalName     => "CLK_ipd",
    Recovery          => trecovery_CLRNeg_CLK,
    ActiveLow         => TRUE,
    CheckEnabled      => TRUE,
    RefTransition     => '\',
    HeaderMsg         => InstancePath & "/std174",
    TimingData        => TD_CLRNeg_CLK,
    XOn               => XOn,
    MsgOn             => MsgOn,
    Violation         => Rviol_CLRNeg_CLK
);
```

Figure 8.6 Example `VitalRecoveryRemovalCheck` call

ing `RefSignal`, as illustrated in Figure 8.5. The most common use for this is to check for the timely deassertion of the Clear signal prior to arrival of a Clock edge on a resetable flip-flop.

As always, the use of the `VitalRecoveryRemovalCheck` procedure requires the declaration of two variables for each reference. The variables are used for timing data storage and for the violation flag output.

```
VARIABLE Rviol_CLRNeg_CLK    : X01 := '0';
VARIABLE TD_CLRNeg_CLK       : VitalTimingDataType;
```

A `VitalRecoveryRemovalCheck` procedure call appears in Figure 8.6.

The procedure call in Figure 8.6 will test that `CLRint` is inactive for the period `trecovery_CLRNeg_CLK` before the active edge of `CLK_ipd`. This check, as written, is always enabled.

The parameters to the `VitalRecoveryRemovalCheck` procedure are as follows:

Parameters of Mode IN

`TestSignal`, the value of the test signal. The procedure is overloaded for `Test-Signal` to be either `std_logic` or `std_logic_vector`. It should be a delayed input signal (`_ipd`).

`TestSignalName`, the name of the test signal. It is of type `STRING` and will be used in any messages generated by the procedure. You should supply a name the user will recognize.

`TestDelay`, not shown. This is the model's internal delay associated with TestSignal. It is only used in models having negative timing constraints. It is of type `TIME`. If a value is not provided, it defaults to zero.

`RefSignal`, the value of the reference signal. It is of type `std_logic`. It should be a delayed input signal.

RefSignalName, the name of the reference signal. It is of type STRING and will be used in any messages generated by the procedure. You should supply a name the user will recognize.

RefDelay, not shown. This is the model's internal delay associated with RefSignal. It is used only in models having negative timing constraints. It is of type TIME. If a value is not provided, it defaults to zero.

Recovery, the minimum time the asynchronous TestSignal must be not asserted prior to a reference edge on RefSignal. It is of type TIME and if not provided, defaults to zero.

Removal, not shown. This is the minimum time the asynchronous TestSignal, if already asserted, must remain asserted after a reference edge on RefSignal. It is of type TIME and if not provided, defaults to zero.

ActiveLow, a flag to indicate if TestSignal is asserted when low ('0'). FALSE indicates that TestSignal is asserted when it has a value '1'. It is of type BOOLEAN. If a value is not provided, it defaults to TRUE.

CheckEnabled, an expression of type BOOLEAN. A check is performed if TRUE. If a value is not provided, it defaults to TRUE. This parameter enables or disables the entire procedure call. Expressions may be used to make execution of the procedure dependent on the state of one or more pins, internal registers, or states.

RefTransition, the active transition of RefSignal. It is of type VitalEdge-SymbolType.

HeaderMsg, text that will accompany any assertion messages produced. It is of type STRING. It should, at a minimum, help the user determine the origin of the messsage. Additional information may be supplied.

XOn, a BOOLEAN that controls the violation output parameter. If TRUE, the output parameter is set to 'X' in the event of a violation. Otherwise, violation is always '0'. If a value is not provided, it defaults to TRUE.

MsgOn, a BOOLEAN that controls the emission of violation messages. If TRUE, setup and hold violation messages will be generated. Otherwise no messages are generated, even upon violations. If a value is not provided, it defaults to TRUE. This parameter could be used to allow execution of the procedure while disabling the Violation output.

MsgSeverity, not shown. It is the severity level for the assertion. It is of type SEVERITY_LEVEL. MsgSeverity can be used to control message display and simulation execution. If a value is not provided, it defaults to WARNING. The simulator may allow masking of low-severity messages or pausing or aborting simulation in the event of a high-severity message.

EnableSetupOnTest, not shown. If FALSE at the time that the TestSignal signal changes, no setup check will be performed. It is of type BOOLEAN. If a value is not provided, it defaults to TRUE.

EnableSetupOnRef, not shown. If FALSE at the time that the RefSignal signal changes, no setup check will be performed. It is of type BOOLEAN. If a value is not provided, it defaults to TRUE.

EnableHoldOnTest, not shown. If FALSE at the time that the TestSignal signal changes, no hold check will be performed. It is of type BOOLEAN. If a value is not provided, it defaults to TRUE.

EnableHoldOnRef, not shown. If FALSE at the time that the RefSignal signal changes, no hold check will be performed. It is of type BOOLEAN. If a value is not provided, it defaults to TRUE.

Parameters of Mode OUT

Violation, the violation flag returned. It is of type X01.

To utilize the Removal check, you must use SDF 3.0 or above. The REMOVAL key word is not present in earlier versions.

8.2.4   Skew Checks

The VITAL skew timing checks detect a skew violation between two signals, Signal1 and Signal2, in any direction. There are two skew check procedures. The VitalInPhaseSkewCheck procedure is used for testing in-phase signals, as shown in Figure 8.7.

The VitalOutPhaseSkewCheck procedure is used for testing out-of-phase signals, as illustrated in Figure 8.8.

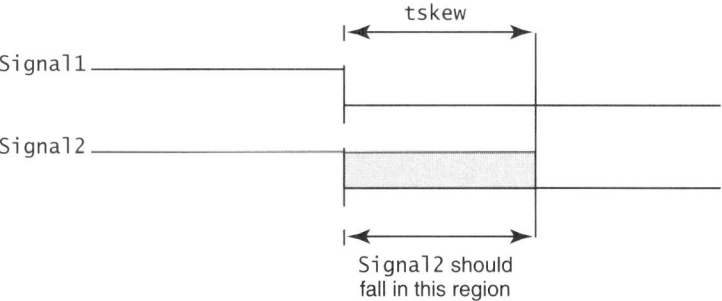

Figure 8.7   In-phase skew check

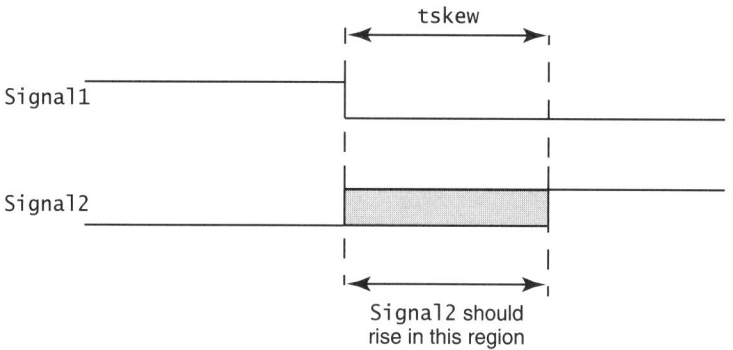

Figure 8.8   Out-of-phase skew check

```
VitalOutPhaseSkewCheck (
        Signal1          => CLKIn,
        Signal1Name      =>  "CLK",
        Signal2          => CLKNegIn,
        Signal2Name      =>  "CLKNeg",
        SkewS1S2RiseFall => tskew_CLK_CLKNeg,
        SkewS2S1RiseFall => tskew_CLK_CLKNeg,
        SkewS1S2FallRise => tskew_CLK_CLKNeg,
        SkewS2S1FallRise => tskew_CLK_CLKNeg,
        CheckEnabled     => TRUE,
        HeaderMsg        => InstancePath & PartID,
        SkewData         => SD_CLK_CLKNeg,
        Trigger          => CKSKWtrg,
        XOn              => XOn,
        MsgOn            => MsgOn,
        Violation        => Sviol_CLK_CLKNeg );
```

Figure 8.9   Example `VitalOutPhaseSkewCheck` call

Using either skew check procedure requires the declaration of a trigger signal and two variables for each reference. The trigger signal must be declared outside the process and must be in the process's sensitivity list.

```
SIGNAL CKSKWtrg              : std_ulogic := '0';
...
DLL: PROCESS(CLKcomb, CLKIn, CLKNegIn, CKSKWtrg)
```

The variables are used for timing data storage and for the violation flag output.

```
VARIABLE Sviol_CLK_CLKNeg  : X01 := '0';
VARIABLE SD_CLK_CLKNeg     : VitalSkewDataType := VitalSkewDataInit;
```

A `VitalOutPhaseSkewCheck` procedure call appears in Figure 8.9.

The procedure call in Figure 8.9 will test that `CLKIn` and `CLKNegIn` remain out of phase and any transition on one of them will be followed by a transition to the opposite state on the other within the period `tskew_CLK_CLKNeg`.

The parameters for `VitalInPhaseSkewCheck` and `VitalOutPhaseSkewCheck` are almost the same. For `VitalInPhaseSkewCheck` they are as follows:

Parameters of Mode IN

`Signal1`, the value of the first signal. It is of type `std_ulogic`. It should be a delayed input signal.

`Signal1Name`, the name of the first signal. It is of type `STRING` and will be used in any messages generated by the procedure.

`Signal1Delay`, not shown. This is the model's internal delay associated with `Signal1`. It is used only in models having negative timing constraints. It is of type `TIME`. If a value is not provided, it defaults to zero.

`Signal2`, the value of the second signal. It is of type `std_ulogic`. It should be a delayed input signal.

`Signal2Name`, the name of the second signal. It is of type `STRING` and will be used in any messages generated by the procedure.

`Signal2Delay`, not shown. This is the model's internal delay associated with `Signal2`. It is used only in models having negative timing constraints. It is of type `TIME`. If a value is not provided, it defaults to zero.

`SkewS1S2RiseRise`, the maximum time duration `Signal2` can remain at 0 after `Signal1` goes to the 1 state, without causing a skew violation. It is of type `TIME`. If a value is not provided, it defaults to `TIME'HIGH` (the end of time).

`SkewS2S1RiseRise`, the maximum time duration `Signal1` can remain at 0 after `Signal2` goes to the 1 state, without causing a skew violation. It is of type `TIME`. If not provided it defaults to `TIME'HIGH` (the end of time).

`SkewS1S2FallFall`, the maximum time duration `Signal2` can remain at 1 after `Signal1` goes to the 0 state, without causing a skew violation. It is of type `TIME`. If a value is not provided, it defaults to `TIME'HIGH` (the end of time).

`SkewS2S1FallFall`, the maximum time duration `Signal1` can remain at 1 after `Signal2` goes to the 0 state, without causing a skew violation. It is of type `TIME`. If a value is not provided, it defaults to `TIME'HIGH` (the end of time).

`CheckEnabled`, an expression of type `BOOLEAN`. A check is performed if `TRUE`. If a value is not provided, it defaults to `TRUE`. This parameter enables or disables the entire procedure call. Expressions may be used to make execution of the procedure dependent on the state of one or more pins, internal registers, or states.

`HeaderMsg`, the text that will accompany any assertion messages produced. It is of type `STRING`. It should, at a minimum, help the user determine the origin of the messsage. Additional information may be supplied.

XOn, a BOOLEAN that controls the violation output parameter. If TRUE, the output parameter is set to 'X' in the event of a violation. Otherwise, violation is always '0'. If a value is not provided, it defaults to TRUE. This parameter could be used to allow execution of the procedure while disabling the Violation output.

MsgOn, a BOOLEAN that controls the emission of violation messages. If TRUE, setup and hold violation messages will be generated. Otherwise no messages are generated, even upon detection of violations. If a value is not provided, it defaults to TRUE.

MsgSeverity, not shown. This is the severity level for the assertion. It is of type SEVERITY_LEVEL. MsgSeverity can be used to control message display and simulation execution. If a value is not provided, it defaults to WARNING. The simulator may allow masking of low-severity messages or pausing or aborting simulation in the event of a high-severity message.

Parameters of Mode INOUT

SkewData, an information storage area for the procedure. It is used internally to detect signal edges and record the time of the last edge. It is of type VitalSkewDataType. It must be declared and initialized to VitalSkew-DataInit but must not be used outside this procedure.

Trigger, a signal used to trigger the process in which the timing check occurs upon expiration of the skew interval. It must be declared and must appear in the process sensitivity list but must not be written to outside this procedure.

Parameters of Mode OUT

Violation, the violation flag returned. It is of type X01.

The VitalOutPhaseSkewCheck procedure has the same parameters except for the four timing parameters. They are replaced by the following:

SkewS1S2RiseFall, the maximum time duration Signal2 can remain at 1 after Signal1 goes to the 1 state, without causing a skew violation. It is of type TIME. If a value is not provided, it defaults to TIME'HIGH (the end of time).

SkewS2S1RiseFall, the maximum time duration Signal1 can remain at 1 after Signal2 goes to the 1 state, without causing a skew violation. It is of type TIME. If a value is not provided, it defaults to TIME'HIGH (the end of time).

SkewS1S2FallRise, the maximum time duration Signal2 can remain at 0 after Signal1 goes to the 0 state, without causing a skew violation. It is of type TIME. If a value is not provided, it defaults to TIME'HIGH (the end of time).

SkewS2S1FallRise, the maximum time duration Signal1 can remain at 0 after Signal2 goes to the 0 state, without causing a skew violation. It is of type TIME. If a value is not provided, it defaults to TIME'HIGH (the end of time).

The VITAL skew check procedures would seem to be a requirement for any model with differential inputs. However, in practice their use is hampered by the lack of any skew specification in most component data sheets. A more likely use is in models with multiple clock inputs, such as multiport memories and FIFOs.

8.3 Violations

Violation flags take the value 'X' when a violation occurs and '0' at all other times. When a timing violation is detected, the user will be notified through the simulator interface or log file. In many cases, some additional action should be taken. What action is appropriate depends on the component being modeled and, possibly, the nature of the violation. Sometimes the violation flags are ORed together and the result used as input to a state table.

```
Violation := Tviol_D_CLK OR Pviol_CLK;

VitalStateTable (
      StateTable       => DFFQN_tab,
      DataIn           => (Violation, CLK_ipd, D_ipd),
      Result           => Qint,
      PreviousDataIn   => PrevData
);
```

This can be used to cause the state table to output 'X's or modify its behavior in some other way. Other models have used the violation flag to warn the user there is a problem.

```
Violation := Pviol_WENeg OR Tviol_DO_WENeg OR Tviol_DO_CENeg;

ASSERT Violation = '0'
      REPORT InstancePath & partID & ": simulation may be" &
            "inaccurate due to timing violations"
      SEVERITY SeverityMode;
```

Some memory models use the violation flag to corrupt the memory location being written when the violation occurred.

```
IF Violation = 'X' THEN
      MemDataA(MemAddr1) := -1;
ELSE
      MemDataA(MemAddr1) := to_nat(DatAIn);
END IF;
```

When you write a component model think about what the model should do in the event of a timing violation.

8.4   Summary

Meeting timing constraints is an important part of synchronous digital design. Embedding timing checks into component models can help you find timing violations in your design. The execution of timing checks can be controlled through generics enabling or disabling checks for a specific instance or for the entire design.

VITAL2000 provides procedures for four type of timing checks: setup/hold, period/pulsewidth, recovery/removal, and skew checks. Using any of these procedures requires the declaration of variables specific to each procedure type for each procedure call. In addition, the skew check procedure requires the declaration of a trigger signal and its inclusion in the process sensitivity list.

The violation flag outputs of the timing checks may be used in a number of ways depending on the component being modeled.

III

Modeling Basics

Parts I and II covered the construction of basic models of simple components with a focus on using various packages as resources. In Part III we build on this foundation by demonstrating how to write complete models for small- and medium-scale integrated circuits. Components with registers are the primary focus: With registers come additional constraint requirements, such as conditional timings and negative timing constraints. The structure and syntax of timing files are also covered.

Chapter 9 explores the requirements for modeling devices containing latches and registers. New generics and their associated timing checks are introduced. The use of state tables is illustrated.

Chapter 10 discusses how to work with conditional delays and conditional constraints. Alternative methods are examined. The mapping of conditional statements from SDF to VITAL is also covered.

Chapter 11 explains how negative timing constraints arise and how they are modeled. It discusses how VITAL compliant simulators handle negative constraints and the implications for model functionality.

Chapter 12 covers the structure and syntax of timing files and an available tool to aid in their creation. It discusses how timing files are used to generate SDF files. How to apply SDF files to backannotate hierarchical designs is demonstrated.

9

Modeling Components with Registers

In the world of synchronous design, most components have registers. Modeling registers is a basic skill for anyone creating models for board-level simulation. This chapter, by pulling together the techniques and concepts covered in earlier chapters, enables us to move to the next level of complexity: modeling simple latches and registers.

We start by modeling a component containing simple flip-flops. We examine the new generics required in the model's entity and how they are used in the model's timing check section. The flip-flops are modeled using VITAL state tables, and the outputs are described in `VitalPathDelay` procedures. Similar treatment is given to modeling a component containing latches.

9.1   Anatomy of a Flip-Flop

The flip-flop example that follows is a model of a 74xx952 registered transceiver with 3-state outputs. It is a component that might be used to control bidirectional data flow between two buses. Each section of the component contains two D flip-flops for temporary storage of data flowing in either direction.

The Texas Instruments data sheet for the 74ABT16952 [3] offers a schematic similar to Figure 9.1. The flip-flops are active on the rising edge of their respective clocks. They each have a clock enable. The A and B pins are 3-state and bidirectional. This part is available as a 74ABT16952, a 74ACT16952, and possibly others. The component is modeled in a technology-independent fashion, so a single model is sufficient to cover any technology.

9.1.1   The Entity

We first examine the model entity in three sections. The top model has the copyright, revision data, and description.

```
--------------------------------------------------------------------
-- File Name: std952.vhd
--------------------------------------------------------------------
-- Copyright (C) 1997, 2003 Free Model Foundry; http://eda.org/fmf/
--
```

```
--  This program is free software; you can redistribute it and/or modify
--  it under the terms of the GNU General Public License version 2 as
--  published by the Free Software Foundation.
--
--  MODIFICATION HISTORY:
--
--  version: |   author:  | mod date: | changes made:
--    V1.0     R. Munden    97 AUG 14   Conformed to style guide
--    V1.1     R. Munden    03 MAR 25   Updated style
--    V1.2     R. Munden    03 OCT 11   Fix bug in clock enable
--
--------------------------------------------------------------------------------
--  PART DESCRIPTION:
--
--  Library:    STD
--  Technology: 54/74XXXX
--  Part:       STD952
--
--  Desciption: Registered Transceiver with 3-State Output
--------------------------------------------------------------------------------
LIBRARY IEEE;   USE IEEE.std_logic_1164.ALL;                        --  1
                USE IEEE.VITAL_timing.ALL;                          --  2
                USE IEEE.VITAL_primitives.ALL;                      --  3
LIBRARY FMF;    USE FMF.gen_utils.ALL;                              --  4
                USE FMF.ff_package.ALL;                             --  5
```

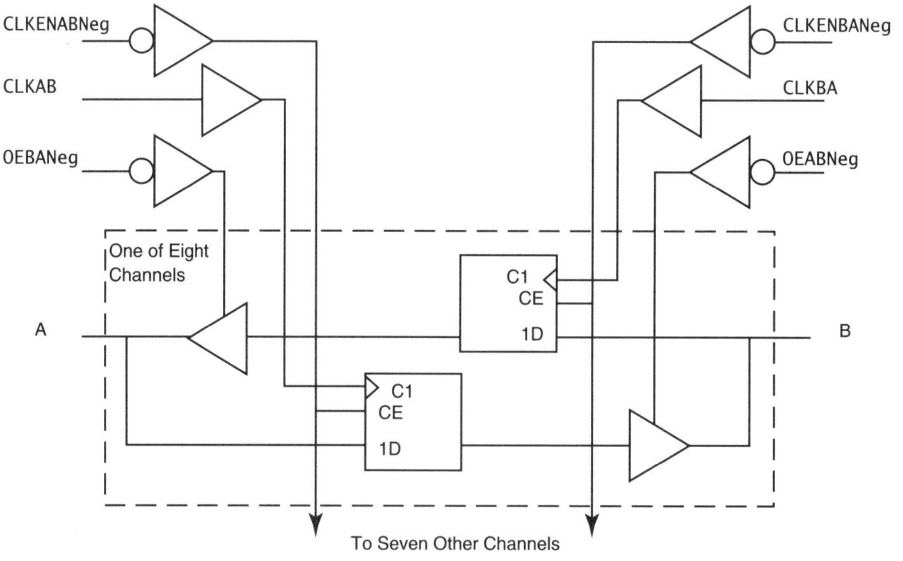

Figure 9.1 74xx952 schematic

Following that, the first five lines of code declare the libraries and packages to be used. This is a VITAL model so the VITAL timing and primitives packages are needed along with `std_logic_1164`. From the FMF library this model will use the `gen_utils` package for a number of predeclared constants and the `ff_package` for the state tables that define the component's flip-flops.

Lines 7 through 33 declare the generics:

```
------------------------------------------------------------------------------
-- ENTITY DECLARATION
------------------------------------------------------------------------------
ENTITY std952 IS                                                    --  6
  GENERIC (                                                         --  7
    -- tipd delays: interconnect path delays
    tipd_CLKENABNeg        : VitalDelayType01 := VitalZeroDelay01;  --  8
    tipd_CLKAB             : VitalDelayType01 := VitalZeroDelay01;  --  9
    tipd_OEABNeg           : VitalDelayType01 := VitalZeroDelay01;  -- 10
    tipd_A                 : VitalDelayType01 := VitalZeroDelay01;  -- 11
    tipd_CLKENBANeg        : VitalDelayType01 := VitalZeroDelay01;  -- 12
    tipd_CLKBA             : VitalDelayType01 := VitalZeroDelay01;  -- 13
    tipd_OEBANeg           : VitalDelayType01 := VitalZeroDelay01;  -- 14
    tipd_B                 : VitalDelayType01 := VitalZeroDelay01;  -- 15
    -- tpd delays
    tpd_CLKAB_B            : VitalDelayType01  := UnitDelay01;      -- 16
    tpd_OEBANeg_A          : VitalDelayType01Z := UnitDelay01Z;    -- 17
    -- tsetup values: setup times
    tsetup_A_CLKAB         : VitalDelayType := UnitDelay;          -- 18
    tsetup_B_CLKBA         : VitalDelayType := UnitDelay;          -- 19
    tsetup_CLKENABNeg_CLKAB : VitalDelayType := UnitDelay;         -- 20
    tsetup_CLKENBANeg_CLKBA : VitalDelayType := UnitDelay;         -- 21
    -- thold values: hold times
    thold_A_CLKAB          : VitalDelayType := UnitDelay;          -- 22
    thold_B_CLKBA          : VitalDelayType := UnitDelay;          -- 23
    thold_CLKENABNeg_CLKAB : VitalDelayType := UnitDelay;          -- 24
    thold_CLKENBANeg_CLKBA : VitalDelayType := UnitDelay;          -- 25
    -- tpw values: pulse widths
    tpw_CLKAB_posedge      : VitalDelayType := UnitDelay;          -- 26
    tpw_CLKBA_posedge      : VitalDelayType := UnitDelay;          -- 27
    -- generic control parameters
    InstancePath    : STRING   := DefaultInstancePath;             -- 28
    TimingChecksOn  : BOOLEAN  := DefaultTimingChecks;             -- 29
    MsgOn           : BOOLEAN  := DefaultMsgOn;                     -- 30
    XOn             : BOOLEAN  := DefaultXon;                       -- 31
    -- For FMF SDF technology file usage
    TimingModel     : STRING   := DefaultTimingModel               -- 32
  );                                                               -- 33
```

Lines 8 through 15 give the interconnect path delay generics (`tipd_`). A `tipd` generic is declared for every port of mode IN or INOUT. They are all of type `VitalDelayType01` and are given default values of `VitalZeroDelay01`. This way, if there are no values backannotated, the delays will be zero.

Lines 16 and 17 give the path `delay` generics. This component is symmetric in that it has the same delays on each side, so it is not necessary to have separate generics for the A and B ports. Line 16 has the clock to output generic `tpd_CLKAB_B`. The clock can cause the output to go either high or low, so the generic is of type `VitalDelayType01`. Line 17 has the generic for the output enables `tpd_OEBANeg_A`. The enables can cause the output to switch between high impedance and low impedance, so its type is `VitalDelayType01 Z`. The two generics get default values of `UnitDelay01` and `UnitDelay01Z`. These two constants are defined as 1 nanosecond in the FMF `gen_utils` package. If no values are backannotated, the delays will each be 1 nanosecond.

Lines 18 through 25 have the setup and hold generics. Here separate generics are declared for each flip-flop. Because, as previously stated, the part has symmetric timing, these could have been reduced to a single set of generics and shared by the two flip-flops. Each flip-flop has timing constraints for data to clock setup and hold and clock enable to clock setup and hold.

Lines 26 and 27 hold the values for the minimum pulse width constraints for the clocks. There are separate generics for the two flip-flops but the same generic will be used for both high and low pulses.

Lines 28 through 31 hold the control generics. They are all given default values defined in the `FMF.gen_utils` package. `InstancePath` provides a string for use in messages from the timing constraint procedures. The default value given is `*`. `TimingChecksOn` controls the execution of those procedures. It is given a default value of `FALSE`. `MsgOn` controls the emission of messages from the timing checks should a violation be detected. Its default value is `TRUE`. The value of `XOn` determines whether or not the violation flags are driven to `'X'` if a violation is detected. The default value from the `gen_utils` package is `TRUE`. All of these generics may be set on a per-instance basis for the entire design by providing values at a higher level or through a configuration specification.

The entity ends with the port declaration and VITAL attribute:

```
PORT (                                                          -- 34
     CLKENABNeg      : IN    std_ulogic := 'U';                 -- 35
     CLKAB           : IN    std_ulogic := 'U';                 -- 36
     OEABNeg         : IN    std_ulogic := 'U';                 -- 37
     A               : INOUT std_ulogic := 'U';                 -- 38
     CLKENBANeg      : IN    std_ulogic := 'U';                 -- 39
     CLKBA           : IN    std_ulogic := 'U';                 -- 40
     OEBANeg         : IN    std_ulogic := 'U';                 -- 41
     B               : INOUT std_ulogic := 'U'                  -- 42
);                                                              -- 43
     ATTRIBUTE VITAL_LEVEL0 of std952 : ENTITY IS TRUE;         -- 44
END std952;                                                     -- 45
```

The ports on this model are all scalar and are of modes IN and INOUT. They are all of type std_ulogic. This is the preferred type for model ports. The ports will all connect to signals of type std_logic in the netlist. In this model, all ports are initialized to 'U'. Should we need to leave any port unconnected in the netlist (perhaps not a good idea in this case), we would be unable to do so without providing an explicit default value. VHDL requires that unconnected inputs be initialized if they are to be associated with the key word OPEN.

This is a VITAL model so the VITAL_LEVEL0 attribute is required for compiler checking and code optimization.

This model simulates one slice or channel of a std952-type component. A 74ABT952 has eight channels. A 74ACT16952 has two separate circuits with eight channels each. We rely on the user's schematic capture system and VHDL netlister to instantiate as many channels as required to simulate the number of specific components in the design. There is an example of this in Chapter 15. This strategy is useful for most small- and medium-size components.

9.1.2   The Architecture

The architecture unit of the model begins with the VITAL_LEVEL1 attribute:

```
-------------------------------------------------------------------------------
-- ARCHITECTURE DECLARATION
-------------------------------------------------------------------------------
ARCHITECTURE vhdl_behavioral of std952 IS                              --  46
    ATTRIBUTE VITAL_LEVEL1 of vhdl_behavioral : ARCHITECTURE IS TRUE;  --  47
```

This is a level 1 model and its level of abstraction is gate level. By supplying the VITAL_LEVEL1 attribute we allow the compiler to further optimize its output for better memory usage and faster execution.

The first part of a VHDL architecture is signal declarations. Here we declare the delayed input signals and any internal signals that will be needed in the model:

```
SIGNAL CLKENABNeg_ipd        : std_ulogic := 'U';                     --  48
SIGNAL CLKAB_ipd             : std_ulogic := 'U';                     --  49
SIGNAL OEABNeg_ipd           : std_ulogic := 'U';                     --  50
SIGNAL CLKENBANeg_ipd        : std_ulogic := 'U';                     --  51
SIGNAL CLKBA_ipd             : std_ulogic := 'U';                     --  52
SIGNAL OEBANeg_ipd           : std_ulogic := 'U';                     --  53
SIGNAL A_ipd                 : std_ulogic := 'U';                     --  54
SIGNAL B_ipd                 : std_ulogic := 'U';                     --  55
SIGNAL Aint                  : std_ulogic := 'U';                     --  56
SIGNAL Bint                  : std_ulogic := 'U';                     --  57
```

In this case all the signals are of type std_ulogic.

That job out of the way, the model can begin. VITAL models always begin with a WireDelay block. The block must carry the label WireDelay. This is where the interconnect delays are applied to the input signals:

```
BEGIN                                                                     -- 58
    ----------------------------------------------------------------------
    -- Wire Delays
    ----------------------------------------------------------------------
WireDelay : BLOCK                                                         -- 59
    BEGIN                                                                 -- 60

        w_1 : VitalWireDelay (CLKENABNeg_ipd, CLKENABNeg, tipd_CLKENABNeg); -- 61
        w_2 : VitalWireDelay (CLKAB_ipd, CLKAB, tipd_CLKAB);             -- 62
        w_3 : VitalWireDelay (OEABNeg_ipd, OEABNeg, tipd_OEABNeg);       -- 63
        w_4 : VitalWireDelay (B_ipd, B, tipd_B);                         -- 64
        w_5 : VitalWireDelay (CLKENBANeg_ipd, CLKENBANeg, tipd_CLKENBANeg); -- 65
        w_6 : VitalWireDelay (CLKBA_ipd, CLKBA, tipd_CLKBA);             -- 66
        w_7 : VitalWireDelay (OEBANeg_ipd, OEBANeg, tipd_OEBANeg);       -- 67
        w_8 : VitalWireDelay (A_ipd, A, tipd_A);                         -- 68

    END BLOCK;                                                           -- 69
```

Although not required, it is good practice to label each `VitalWireDelay` procedure call. In general labels are an aid to debugging either the model or the design the model is in. When running a simulation interactively, the simulator's graphical interface will usually make labels visible, improving the accessibility of the objects to which they are attached.

Following the `WireDelay` block is a concurrent procedure calls section. In this section are two VITAL primitive procedure calls:

```
    ----------------------------------------------------------------------
    -- Concurrent procedure calls
    ----------------------------------------------------------------------
a_1: VitalBUFIF0 (                                                       -- 70
            q            => A,                                           -- 71
            data         => Arint,                                       -- 72
            Enable       => OEBANeg_ipd,                                 -- 73
            tpd_enable_q => tpd_OEBANeg_A);                              -- 74

a_2: VitalBUFIF0 (                                                       -- 75
            q            => B,                                           -- 76
            data         => Bint,                                        -- 77
            Enable       => OEABNeg_ipd,                                 -- 78
            tpd_enable_q => tpd_OEBANeg_A);                              -- 79
```

In this model, `VitalBUFIF0` primitives are employed to model the 3-state behavior of the two data ports. Each primitive will drive an output port with a signal delayed by `tpd_OEBANeg_A`. The input signals `Aint` and `Bint` are already delayed by path delay procedures. However, the generic `tpd_OEBANeg_A` has null values for the $0 \rightarrow 1$ and $1 \rightarrow 0$ transitions. Thus, the delays will not be additive.

This model could be written without these VITAL primitive calls. The 3-state control could be included in the processes that follow. In this case, taking 3-state control out of the processes and making it separate might make the model easier to understand. You be the judge.

9.1.3   A VITAL Process

Most of the behavior of this model is described in two VITAL processes. One models the data flow from the A port to the B port, the other from B to A. Each has its own timing checks and path delay sections. We will call this process the "A" side, and the other the "B" side of the component.

The process begins with a process statement. It is good practice to label each process. The sensitivity list must include every signal that is read inside the process. The first part of any process is the declarative section:

```
-------------------------------------------------------------------------------
-- Main Behavior Process
-------------------------------------------------------------------------------

VitalBehavior1 : PROCESS (CLKAB_ipd, CLKENABNeg_ipd, A_ipd)            -- 80

    -- Timing Check Variables
    VARIABLE Tviol_A_CLKAB          : X01 := '0';                      -- 81
    VARIABLE TD_A_CLKAB             : VitalTimingDataType;             -- 82

    VARIABLE Tviol_CLKENABNeg_CLKAB    : X01 := '0';                   -- 83
    VARIABLE TD_CLKENABNeg_CLKAB       : VitalTimingDataType;          -- 84

    VARIABLE Pviol_CLKAB        : X01 := '0';                          -- 85
    VARIABLE PD_CLKAB           : VitalPeriodDataType := VitalPeriodDataInit;

    VARIABLE Violation          : X01 := '0';                          -- 87

    -- Functionality Results Variables
    VARIABLE Q_zd               : std_ulogic;                          -- 88
    VARIABLE PrevData           : std_logic_vector(0 to 3);            -- 89

    -- Output Glitch Detection Variables
    VARIABLE Q_GlitchData       : VitalGlitchDataType;                 -- 90
```

Timing check variables are declared for each timing check in the process. The functionality result variables that are declared include the zero delay result variable, Q_zd, and the PrevData array that stores the previous state for the state table that describes the D flip-flop in this process. There is also a declaration for the glitch detection variable, Q_GlitchData that is used in the path delay procedure.

The next section in the process is the timing check section. The timing checks are all inside an IF statement controlled by the TimingChecksOn generic. There are three timing checks. The first is a VitalSetupHoldCheck to check the clock enable:

```
    BEGIN                                                      --  91
    ---------------------------------------------------------------
    -- Timing Check Section
    ---------------------------------------------------------------

    IF (TimingChecksOn) THEN                                   --  92

        VitalSetupHoldCheck (                                  --  93
            TestSignal        => CLKENABNeg_ipd,               --  94
            TestSignalName    => "CLKENABNeg_ipd",             --  95
            RefSignal         => CLKAB_ipd,                    --  96
            RefSignalName     => "CLKAB_ipd",                  --  97
            SetupHigh         => tsetup_CLKENABNeg_CLKAB,      --  98
            SetupLow          => tsetup_CLKENABNeg_CLKAB,      --  99
            HoldHigh          => thold_CLKENABNeg_CLKAB,       -- 100
            HoldLow           => thold_CLKENABNeg_CLKAB,       -- 101
            CheckEnabled      => TRUE,                         -- 102
            RefTransition     => '/',                          -- 103
            HeaderMsg         => InstancePath & "/std952",     -- 104
            TimingData        => TD_CLKENABNeg_CLKAB,          -- 105
            XOn               => XOn,                          -- 106
            MsgOn             => MsgOn,                         -- 107
            Violation         => Tviol_CLKENABNeg_CLKAB        -- 108
        );                                                     -- 109
```

All the signals read by the procedure are delayed (_ipd) signals. The check is always enabled on line 102, although the procedure will not be executed if TimingChecksOn is FALSE. Line 103 says we are performing the check against the time of the rising edge ('/') of the clock, CLKAB_ipd.

The second constraint check checks the setup and hold times of the "A" port relative to its clock, CLKAB_ipd:

```
        VitalSetupHoldCheck (                                  -- 110
            TestSignal        => A_ipd,                        -- 111
            TestSignalName    => "A_ipd",                      -- 112
            RefSignal         => CLKAB_ipd,                    -- 113
            RefSignalName     => "CLKAB_ipd",                  -- 114
            SetupHigh         => tsetup_A_CLKAB,               -- 115
            SetupLow          => tsetup_A_CLKAB,               -- 116
            HoldHigh          => thold_A_CLKAB,                -- 117
            HoldLow           => thold_A_CLKAB,                -- 118
            CheckEnabled      => CLKENABNeg_ipd = '0',         -- 119
            RefTransition     => '/',                          -- 120
            HeaderMsg         => InstancePath & "/std952",     -- 121
            TimingData        => TD_A_CLKAB,                   -- 122
            XOn               => XOn,                          -- 123
            MsgOn             => MsgOn,                         -- 124
            Violation         => Tviol_A_CLKAB                 -- 125
        );                                                     -- 126
```

In this procedure call and the next, the check is enabled only when CLKENAB–Neg_ipd is low.

The third timing check in this process is VitalPeriodPulseCheck:

```
VitalPeriodPulseCheck (                                     -- 127
    TestSignal        => CLKAB_ipd,                         -- 128
    TestSignalName    => "CLKAB_ipd",                       -- 129
    PulseWidthHigh    => tpw_CLKAB_posedge,                 -- 130
    PulseWidthLow     => tpw_CLKAB_posedge,                 -- 131
    CheckEnabled      => CLKENABNeg_ipd = '0',              -- 132
    HeaderMsg         => InstancePath & "/std952",          -- 133
    PeriodData        => PD_CLKAB,                          -- 134
    XOn               => XOn,                               -- 135
    MsgOn             => MsgOn,                             -- 136
    Violation         => Pviol_CLKAB                        -- 137
);                                                          -- 138

END IF;                                                     -- 139
```

It checks the pulse width, high and low, of the clock.

Finally, after the end of the IF statement, the violation flags are ORed together and the timing check section is closed with:

```
Violation := Tviol_A_CLKAB OR Tviol_CLKENABNeg_CLKAB OR Pviol_CLKAB;
```

Although it would seem to make sense to combine the violation flags inside the IF statement, a strict VITAL compiler would consider it an error.

9.1.4   Functionality Section

The functionality section holds the functional behavior of the process. In this process it consists entirely of a call to a VITAL state table.

```
---------------------------------------------------------------------------
-- Functionality Section
---------------------------------------------------------------------------
VitalStateTable (                                           -- 141
    StateTable      => DFFCEN_tab,                          -- 142
    DataIn          => (Violation,CLKENABNeg_ipd,           -- 143
                        CLKAB_ipd,                           -- 144
                        A_ipd),                              -- 145
    Result          => Q_zd,                                -- 146
    PreviousDataIn  => PrevData                             -- 147
);                                                          -- 148
```

Line 142 provides the name of the state table. The table itself is shown in Figure 9.2. Lines 143, 144, and 145 give the four elements of the input array. The state table models a D flip-flop with active-low clock enable. The output of the state table is assigned to Q_zd. The array PrevData stores the inputs for use with the next call to the state table.

```
-----------------------------------------------------------------
-- D-flip/flop with active low clock enable
-----------------------------------------------------------------
CONSTANT DFFCEN_tab : VitalStateTableType := (

    ----INPUTS-----------|PREV-|-OUTPUT--
    -- Viol CEN  CLK   D  | QI  | Q'     --
    --------------------|-----|---------
    ( 'X', '-', '-', '-', '-', 'X'), -- timing violation
    ( '-', '-', 'X', '0', '0', '0'), -- clk unknown
    ( '-', '-', 'X', '1', '1', '1'), -- clk unknown
    ( '-', '-', 'X', '-', '-', 'X'), -- clk unknown
    ( '-', 'X', '-', '0', '0', '0'), -- clken unknown
    ( '-', 'X', '-', '1', '1', '1'), -- clken unknown
    ( '-', 'X', '-', '-', '-', 'X'), -- clken unknown
    ( '-', '0', '/', '0', '-', '0'), -- active clock edge
    ( '-', '0', '/', '1', '-', '1'), -- active clock edge
    ( '-', '0', '/', '-', '-', 'X'), -- active clock edge
    ( '-', '-', '-', '-', '-', 'S')  -- default

); -- end of VitalStateTableType definition
```

Figure 9.2   DFFCEN state table

9.1.5   Path Delay

This process ends with a path delay section containing a single call to the Vital-
PathDelay01 procedure:

```
-------------------------------------------------------------------------------
-- Path Delay Section
-------------------------------------------------------------------------------
VitalPathDelay01 (                                               -- 149
    OutSignal        => Bint,                                    -- 150
    OutSignalName    => "Bint",                                  -- 151
    OutTemp          => Q_zd,                                    -- 152
    GlitchData       => Q_GlitchData,                            -- 153
    XOn              => XOn,                                     -- 154
    MsgOn            => MsgOn,                                   -- 155
    Paths            => (                                        -- 156
        0 => (InputChangeTime => CLKAB_ipd'LAST_EVENT,           -- 157
              PathDelay        => tpd_CLKAB_B,                   -- 158
              PathCondition    => CLKENABNeg_ipd = '0')          -- 159
        )                                                        -- 160
    );                                                           -- 161

END PROCESS;                                                     -- 162
```

The output of the procedure is written to an internal signal rather than directly to
a port. Only one path is listed but it has the path condition CLKENABNeg_ipd =
'0' on line 159. This condition would seem unnecessary but harmless. However,
when writing a component model the author should consider all the conditions

the model might encounter. In this case consider what would happen if the user tied the clock enable to ground through a resistor. The signal CLKENABNeg_ipd would have a constant value of 'L' rather than '0'. The state table would function correctly but the only path in the path delay procedure would be disqualified. If the user was simulating a 74ACT16952 part with maximum delays, the tpd_CLKAB_B would be 10.7 nanoseconds, but with no qualified path the VitalPathDelay01 procedure would use the default delay. Unless otherwise specified, the default delay is 0 nanoseconds. This would likely cause timing errors elsewhere in the simulation without an obvious indication of the actual source of the error. So in this case, the path condition may not add any value but can cause serious but subtle errors. It would be better if the PathCondition was:

```
PathCondition        => (CLKENABNeg_ipd = '0' OR CLKENABNeg_ipd = 'L'))
```

9.1.6   The "B" Side

The other half of the model, modeling the port B to A data flow, differs from what has been shown only in its signal names:

```
VitalBehavior2 : PROCESS (CLKBA_ipd, CLKENBANeg_ipd, B_ipd)

    -- Timing Check Variables
    VARIABLE Tviol_B_CLKBA        : X01 := '0';
    VARIABLE TD_B_CLKBA           : VitalTimingDataType;

    VARIABLE Tviol_CLKENBANeg_CLKBA    : X01 := '0';
    VARIABLE TD_CLKENBANeg_CLKBA       : VitalTimingDataType;

    VARIABLE Pviol_CLKBA          : X01 := '0';
    VARIABLE PD_CLKBA             : VitalPeriodDataType := VitalPeriodDataInit;

    VARIABLE Violation            : X01 := '0';

    -- Functionality Results Variables
    VARIABLE Q_zd                 : std_ulogic;
    VARIABLE PrevData             : std_logic_vector(0 to 3);

    -- Output Glitch Detection Variables
    VARIABLE Q_GlitchData         : VitalGlitchDataType;

BEGIN
    -------------------------------------------------------------------------
    -- Timing Check Section
    -------------------------------------------------------------------------
    IF (TimingChecksOn) THEN

        VitalSetupHoldCheck (
            TestSignal        => CLKENBANeg_ipd,
            TestSignalName    => "CLKENBANeg_ipd",
```

```
        RefSignal           => CLKBA_ipd,
        RefSignalName       => "CLKBA_ipd",
        SetupHigh           => tsetup_CLKENBANeg_CLKBA,
        SetupLow            => tsetup_CLKENBANeg_CLKBA,
        HoldHigh            => thold_CLKENBANeg_CLKBA,
        HoldLow             => thold_CLKENBANeg_CLKBA,
        CheckEnabled        => TRUE,
        RefTransition       => '/',
        HeaderMsg           => InstancePath & "/std952",
        TimingData          => TD_CLKENBANeg_CLKBA,
        XOn                 => XOn,
        MsgOn               => MsgOn,
        Violation           => Tviol_CLKENBANeg_CLKBA
    );

    VitalSetupHoldCheck (
        TestSignal          => B_ipd,
        TestSignalName      => "B_ipd",
        RefSignal           => CLKBA_ipd,
        RefSignalName       => "CLKBA_ipd",
        SetupHigh           => tsetup_B_CLKBA,
        SetupLow            => tsetup_B_CLKBA,
        HoldHigh            => thold_B_CLKBA,
        HoldLow             => thold_B_CLKBA,
        CheckEnabled        => CLKENBANeg_ipd = '0',
        RefTransition       => '/',
        HeaderMsg           => InstancePath & "/std952",
        TimingData          => TD_B_CLKBA,
        XOn                 => XOn,
        MsgOn               => MsgOn,
        Violation           => Tviol_B_CLKBA
    );

    VitalPeriodPulseCheck (
        TestSignal          => CLKBA_ipd,
        TestSignalName      => "CLKBA_ipd",
        PulseWidthHigh      => tpw_CLKBA_posedge,
        CheckEnabled        => CLKENBANeg_ipd = '0',
        HeaderMsg           => InstancePath & "/std952",
        PeriodData          => PD_CLKBA,
        XOn                 => XOn,
        MsgOn               => MsgOn,
        Violation           => Pviol_CLKBA
    );

END IF;
```

```
        Violation := Tviol_B_CLKBA OR Tviol_CLKENBANeg_CLKBA OR Pviol_CLKBA;

        ------------------------------------------------------------------------
        -- Functionality Section
        ------------------------------------------------------------------------
        VitalStateTable (
            StateTable        => DFFCEN_tab,
            DataIn            => (Violation,CLKENBANeg_ipd,
                                  CLKBA_ipd,
                                  B_ipd),
            Result            => Q_zd,
            PreviousDataIn    => PrevData
        );

        ------------------------------------------------------------------------
        -- Path Delay Section
        ------------------------------------------------------------------------
        VitalPathDelay01 (
            OutSignal         => Aint,
            OutSignalName     => "Aint",
            OutTemp           => Q_zd,
            GlitchData        => Q_GlitchData,
            XOn               => XOn,
            MsgOn             => MsgOn,
            Paths             => (
                0 => (InputChangeTime   => CLKBA_ipd'LAST_EVENT,
                      PathDelay         => tpd_CLKAB_B,
                      PathCondition     => CLKENBANeg_ipd = '0')
                )
            );

    END PROCESS;

END vhdl_behavioral;
```

Although this model has been used for several years without negative comment, it has some shortcomings. It was in use for six years before a flaw regarding the clock enable was noticed and the state table changed to correct it. The path conditions still need to be changed to accommodate for weak inputs. All this says that although the model is imperfect, it has still been useful.

9.2   Anatomy of a Latch

This is a model of a 74xx16334 16-bit universal bus driver with 3-state outputs. As in the previous model, this model is of a single bit. This component is unidirectional. It operates according to the function table in Table 9.1, which is based on the Texas Instruments SN74ALVC162334 data sheet [4].

When the latch enable pin, LENeg, is low, the component is in transparent mode. Any input to A appears on Y, assuming the output enable pin, OENeg, is low.

Table 9.1 Function table

INPUTS				OUTPUT
OENeg	LENeg	CLK	A	Y
H	X	X	X	Z
L	L	X	L	L
L	L	X	H	H
L	H	/	L	L
L	H	/	H	H
L	H	L or H	X	Y0

When LENeg is high, the component acts as a D flip-flop and data are clocked from A to Y on the rising edge of CLK.

9.2.1   The Entity

The entity begins with the copyright, revision history, and description:

```
-------------------------------------------------------------------------------
-- File Name: std16334.vhd
-------------------------------------------------------------------------------
-- Copyright (C) 2002 Free Model Foundry; http://eda.org/fmf/
--
-- This program is free software; you can redistribute it and/or modify
-- it under the terms of the GNU General Public License version 2 as
-- published by the Free Software Foundation.
--
-- MODIFICATION HISTORY:
--
-- version: |   author:   | mod date: | changes made:
--    V1.0     R. Munden    02 OCT 11    Initial release
--
-------------------------------------------------------------------------------
-- PART DESCRIPTION:
--
-- Library:    STND
-- Technology: 74XXXX
-- Part:       std16334
--
-- Description:   16-BIT UNIVERSAL BUS DRIVER WITH 3-STATE OUTPUTS
-------------------------------------------------------------------------------
LIBRARY IEEE;   USE IEEE.std_logic_1164.ALL;
                USE IEEE.VITAL_timing.ALL;
                USE IEEE.VITAL_primitives.ALL;
```

```
LIBRARY FMF;    USE FMF.gen_utils.ALL;
                USE FMF.ff_package.ALL;
```

The library clauses are the same as for the previous model. The `ff_package` is used again, so the model can utilize a state table defined in that package.

The generics section has some notable items:

```
---------------------------------------------------------------------------
-- ENTITY DECLARATION
---------------------------------------------------------------------------
ENTITY std16334 IS
  GENERIC (
    -- tipd delays: interconnect path delays
    tipd_CLK                  : VitalDelayType01 := VitalZeroDelay01;
    tipd_OENeg                : VitalDelayType01 := VitalZeroDelay01;
    tipd_LENeg                : VitalDelayType01 := VitalZeroDelay01;
    tipd_A                    : VitalDelayType01 := VitalZeroDelay01;
    -- tpd delays
    tpd_A_Y         : VitalDelayType01  := UnitDelay01;    --tPLH, tPHL
    tpd_LENeg_      : VitalDelayType01  := UnitDelay01;    --tPLH, tPHL
    tpd_CLK_Y       : VitalDelayType01  := UnitDelay01;    --tPLH, tPHL
    tpd_OENeg_Y     : VitalDelayType01Z := UnitDelay01Z;   --tPZH, tPZL,
    -- tsetup values: setup times
    tsetup_A_CLK              : VitalDelayType := UnitDelay; --tSU
    tsetup_A_LENeg_CLK_EQ_1 : VitalDelayType := UnitDelay; --tSU,
                                                           -- CLK-HIGH

    tsetup_A_LENeg_CLK_EQ_0 : VitalDelayType := UnitDelay; --tSU,
                                                           -- CLK-LOW

    -- thold values: hold times
    thold_A_CLK              : VitalDelayType := UnitDelay; --tSU
    thold_A_LENeg            : VitalDelayType := UnitDelay; --tSU,
                                                     --CLK-HIGH i CLK-LOW

    -- tpw values: pulse widths
    tpw_LENeg_negedge        : VitalDelayType := UnitDelay; --tW Low
    tpw_CLK_posedge          : VitalDelayType := UnitDelay; --tW(H),
                                                           --tW(L)

    -- tperiod_min: minimum clock period = 1/max freq
    tperiod_CLK_posedge      : VitalDelayType := UnitDelay;
    -- generic control parameters
    InstancePath             : STRING  := DefaultInstancePath;
    TimingChecksOn           : BOOLEAN := DefaultTimingChecks;
    MsgOn                    : BOOLEAN := DefaultMsgOn;
    XOn                      : BOOLEAN := DefaultXon;
    --For FMF SDF technology file usage
    TimingModel              : STRING  := DefaultTimingModel
  );
```

The path delay and timing constraint generics are commented with the parameter names from the data sheet. This facilitates the addition of new component timings in the timing file by stating which data sheet parameter is mapped to each timing generic. This is particularly helpful if the new timings are being added by someone other than the original author of the model. Timing files are discussed in Chapter 12.

It is of note that the data sheet defines different setup times for A relative to LENeg for CLK high and low. Therefore, there are two generics for this constraint: tsetup_A_LENeg_CLK_EQ_1 and tsetup_A_LENeg_CLK_EQ_0. These are conditional generics, as discussed in Chapters 4 and 10.

The entity ends with the port list and VITAL_LEVEL0 attribute.

```
PORT (
    A              : IN std_ulogic := U';
    OENeg          : IN std_ulogic := 'U';
    CLK            : IN std_ulogic := 'U';
    LENeg          : IN std_ulogic := 'U';
    Y              : OUT std_ulogic := 'U'
    );
    ATTRIBUTE VITAL_LEVEL0 of std16334 : ENTITY IS TRUE;
END std16334;
```

This model has ports of modes IN and OUT.

9.2.2   The Architecture

The architecture begins with a VITAL attribute. This is a VITAL_LEVEL0 model. It will not be optimized as much by the compiler, but is written at a slightly higher level of abstraction that may make up for that. It is important to note that declaring VITAL_LEVEL1 FALSE may not give the same results as VITAL_LEVEL0 TRUE. This is because declaring VITAL_LEVEL1 FALSE should (but is not guaranteed to) have the same interpretation as no declaration at all.

```
-------------------------------------------------------------------------
-- ARCHITECTURE DECLARATION
-------------------------------------------------------------------------
ARCHITECTURE vhdl_behavioral of std16334 IS
    ATTRIBUTE VITAL_LEVEL0 of vhdl_behavioral : ARCHITECTURE IS TRUE;
    CONSTANT partID        : STRING := "std16334";

    SIGNAL OENeg_ipd       : std_ulogic := 'U';
    SIGNAL CLK_ipd         : std_ulogic := 'U';
    SIGNAL LENeg_ipd       : std_ulogic := 'U';
    SIGNAL A_ipd           : std_ulogic := 'U';

BEGIN
```

```
-------------------------------------------------------------------
-- Wire Delays
-------------------------------------------------------------------
WireDelay : BLOCK
BEGIN
      w_1 : VitalWireDelay (OENeg_ipd, OENeg, tipd_OENeg);
      w_2 : VitalWireDelay (CLK_ipd, CLK, tipd_CLK);
      w_3 : VitalWireDelay (LENeg_ipd, LENeg, tipd_LENeg);
      w_4 : VitalWireDelay (A_ipd, A, tipd_A);

END BLOCK;
```

There is a new constant declared, `partID`. It is assigned the string that will be used for the `HeaderMsg` parameter in the timing checks. Using a constant makes it easier to write new models by cutting and pasting from old ones. The signal declarations and `WireDelay` block contain nothing we haven't seen before. The behavior process does have a few things that deserve closer examination:

```
-------------------------------------------------------------------
-- Behavior Process
-------------------------------------------------------------------
VitalBehavior : PROCESS (OENeg_IPD, CLK_ipd, LENeg_ipd, A_ipd)

    -- Timing Check Variables
        VARIABLE Tviol_A_CLK       : X01 := '0';
        VARIABLE TD_A_CLK          : VitalTimingDataType;

        VARIABLE Tviol_A_LENeg_CLKhigh       : X01 := '0';
        VARIABLE TD_A_LENeg_CLKhigh          : VitalTimingDataType;

        VARIABLE Tviol_A_LENeg_CLKlow        : X01 := '0';
        VARIABLE TD_A_LENeg_CLKlow           : VitalTimingDataType;

        VARIABLE PD_CLK      : VitalPeriodDataType := VitalPeriodDataInit;
        VARIABLE Pviol_CLK   : X01 := '0';

        VARIABLE PD_LENeg    : VitalPeriodDataType := VitalPeriodDataInit;
        VARIABLE Pviol_LENeg : X01 := '0';

        VARIABLE Violation        : X01 := '0';

        -- Functionality Results Variables
        VARIABLE Y_zd    : std_ulogic;

        VARIABLE LENeg_inv : std_ulogic;

        -- Temporary Variables for tri state out
        VARIABLE QA_int       : std_ulogic := 'U';

        --       Prevdata for LATNDFF tab
        VARIABLE PrevData0    : std_logic_vector(0 to 3);
```

```
            -- Output Glitch Detection Variables
            VARIABLE Y_GlitchData        : VitalGlitchDataType;

            -- No Weak Values Variables
            VARIABLE CLK_nwv         : UX01 := 'X';
            VARIABLE OENeg_nwv       : UX01 := 'X';
            VARIABLE LENeg_nwv       : UX01 := 'X';

        BEGIN

        CLK_nwv       := To_UX01 (s => CLK_ipd);
        OENeg_nwv     := To_UX01 (s => OENeg_ipd);
        LENeg_nwv     := To_UX01 (s => LENeg_ipd);
```

The new items in this section are additional variables for internal nets, LENeg_inv, and QA_int, and three "no weak values" (_nwv) variables, CLK_nwv, OENeg_nwv, and LENeg_nwv. Each of these_nwv variables is assigned a version of its named signal through a To_UX01 transformation. The purpose of the transformation is to change 'L's to '0's and 'H's to '1's. That allows us to simplify IF statements and other tests. Instead of writing

```
    IF (OENeg_ipd = 'H' OR OENeg_ipd = '1') THEN
```

We can write

```
    IF (OENeg_nwv = '1') THEN
```

This is more efficient and easier to write if a signal is going to be tested in many places in the process. The first use of a no weak value variable occurs in the timing section. Placing the To_UX01 conversions ahead of the timing checks is what prevents this model from being compiled as VITAL_LEVEL1.

```
------------------------------------------------------------------------------
-- Timing Check Section
------------------------------------------------------------------------------

IF (TimingChecksOn) THEN

    VitalSetupHoldCheck (
        TestSignal      => A_ipd,
        TestSignalName  => "A_ipd",
        RefSignal       => CLK_ipd,
        RefSignalName   => "CLK_ipd",
        SetupHigh       => tsetup_A_CLK,
        SetupLow        => tsetup_A_CLK,
        HoldHigh        => thold_A_CLK,
        HoldLow         => thold_A_CLK,
        CheckEnabled    => TRUE,
        RefTransition   => '/',
        HeaderMsg       => InstancePath & partID,
        TimingData      => TD_A_CLK,
```

```
        XOn              => XOn,
        MsgOn            => MsgOn,
        Violation        => Tviol_A_CLK
);

VitalSetupHoldCheck (
        TestSignal       => A_ipd,
        TestSignalName   => "A_ipd",
        RefSignal        => LENeg_ipd,
        RefSignalName    => "LENeg_ipd",
        SetupHigh        => tsetup_A_LENeg_CLK_EQ_0,
        SetupLow         => tsetup_A_LENeg_CLK_EQ_0,
        HoldHigh         => thold_A_LENeg,
        HoldLow          => thold_A_LENeg,
        CheckEnabled     => CLK_nwv = '0',
        RefTransition    => '/',
        HeaderMsg        => InstancePath & partID,
        TimingData       => TD_A_LENeg_CLKLow,
        XOn              => XOn,
        MsgOn            => MsgOn,
        Violation        => Tviol_A_LENeg_CLKLow
);

VitalSetupHoldCheck (
        TestSignal       => A_ipd,
        TestSignalName   => "A_ipd",
        RefSignal        => LENeg_ipd,
        RefSignalName    => "LENeg_ipd",
        SetupHigh        => tsetup_A_LENeg_CLK_EQ_1,
        SetupLow         => tsetup_A_LENeg_CLK_EQ_1,
        HoldHigh         => thold_A_LENeg,
        HoldLow          => thold_A_LENeg,
        CheckEnabled     => CLK_nwv = '1',
        RefTransition    => '/',
        HeaderMsg        => InstancePath & partID,
        TimingData       => TD_A_LENeg_CLKhigh,
        XOn              => XOn,
        MsgOn            => MsgOn,
        Violation        => Tviol_A_LENeg_CLKhigh
);

VitalPeriodPulseCheck (
        TestSignal       => CLK_ipd,
        TestSignalName   => "CLK_ipd",
        Period           => tperiod_CLK_posedge,
        PulseWidthHigh   => tpw_CLK_posedge,
```

```
              PulseWidthlow      => tpw_CLK_posedge,
              CheckEnabled       => TRUE,
              HeaderMsg          => InstancePath & partID,
              PeriodData         => PD_CLK,
              XOn                => XOn,
              MsgOn              => MsgOn,
              Violation          => Pviol_CLK
         );

     VitalPeriodPulseCheck (
              TestSignal         => LENeg_ipd,
              TestSignalName     => "LENeg_ipd",
              PulseWidthlow      => tpw_LENeg_negedge,
              CheckEnabled       => TRUE,
              HeaderMsg          => InstancePath & partID,
              PeriodData         => PD_LENeg,
              XOn                => XOn,
              MsgOn              => MsgOn,
              Violation          => Pviol_LENeg
         );

    END IF;
```

Note that there are separate `VitalSetupHoldCheck` procedure calls for checking `A_ipd` relative to `LENeg_ipd`. One or the other will be enabled depending on the value of `CLK_nwv`.

In this model, the violation flags are combined in the functionality section:

```
--------------------------------------------------------------------------------
-- Functionality Section
--------------------------------------------------------------------------------
Violation   := Tviol_A_CLK   OR
     Tviol_A_LENeg_CLKHigh OR
     Tviol_A_LENeg_CLKLow  OR
     Pviol_LENeg           OR
     Pviol_CLK;

LENeg_inv := VitalINV (Data => LENeg_ipd);

VitalStateTable (
     StateTable       => LATNDFF_tab,
     DataIn           => (Violation, LENeg_inv, CLK_ipd, A_ipd),
     Result           => QA_int,
     PreviousDataIn   => PrevData0
);

Y_zd := VitalBUFIF0 (data => QA_int,
                     nable => OENeg_ipd);
```

It might seem be more logical to do that in the timing checks section. If timing checks are disabled by `TimingChecksOn` being set `FALSE`, there is no point in combining the

```
------------------------------------------------------------------
-- Latch/D-flip/flop with LEN transparent high and active high clock
------------------------------------------------------------------
CONSTANT LATNDFF_tab : VitalStateTableType  := (

    ----INPUTS----------|PREV-|-OUTPUT--
    -- Viol LEN  CLK  D  | QI  | Q'      --
    --------------------|-----|---------
    ( 'X', '-', '-', '-', '-', 'X'), -- timing violation
    ( '-', 'X', 'X', '0', '0', '0'), -- len and clk unknown
    ( '-', 'X', 'X', '1', '1', '1'), -- len and clk unknown
    ( '-', 'X', '-', '-', '-', 'X'), -- len unknown
    ( '-', '0', 'X', '-', '-', 'X'), -- clk unknown
    ( '-', '1', '-', '0', '-', '0'), -- Latch transparent
    ( '-', '1', '-', '1', '-', '1'), -- Latch transparent
    ( '-', '1', '-', '-', '-', 'X'), -- Latch transparent unknown D
    ( '-', '0', 'X', '0', '0', '0'), -- clk unknown
    ( '-', '0', 'X', '1', '1', '1'), -- clk unknown
    ( '-', '0', 'X', '-', '-', 'X'), -- clk unknown
    ( '-', '0', '/', '0', '-', '0'), -- ff active clock edge
    ( '-', '0', '/', '1', '-', '1'), -- ff active clock edge
    ( '-', '0', '/', '-', '-', 'X'), -- ff active clock edge unknown D
    ( '-', '-', '-', '-', '-', 'S')  -- default

); -- end of VitalStateTableType definition
```

Figure 9.3   LATNDFF state table

violation flags. However, the VITAL specification prohibits anything other than calls to the timing check procedures from being placed in the timing check section.

The behavior of this component is modeled using two VITAL primitive function calls as well as a `VitalStateTable` procedure call. The first function call inverts the value of the latch enable, LENeg, so it will have the correct polarity for the state table. The second function call drives the output to high impedance if the output enable, OENeg, is high. The state table LATNDFF_tab in the FMF ff_package describes the behavior of the latch itself, as shown in Figure 9.3.

The model ends with the path delay section:

```
------------------------------------------------------------------------
-- Path Delay Section
------------------------------------------------------------------------
VitalPathDelay01Z (
    OutSignal       => Y,
    OutSignalName   => "Y",
    OutTemp         => Y_zd,
    GlitchData      => Y_GlitchData,
    XOn             => XOn,
    MsgOn           => MsgOn,
    Paths           => (
        0 => (InputChangeTime    => CLK_ipd'LAST_EVENT,
              PathDelay          => VitalExtendToFillDelay(tpd_CLK_Y),
              PathCondition      => (OENeg_nwv = '0'
                                     AND LENeg_nwv = '1' )),
```

```
    1 => (InputChangeTime    => OENeg_ipd'LAST_EVENT,
          PathDelay          => tpd_OENeg_Y,
          PathCondition      => TRUE),
    2 => (InputChangeTime    => LENeg_ipd'LAST_EVENT,
          PathDelay          => VitalExtendToFillDelay(tpd_LENeg_Y),
          PathCondition      => OENeg_nwv = '0'),
    3 => (InputChangeTime    => A_ipd'LAST_EVENT,
          PathDelay          => VitalExtendToFillDelay(tpd_A_Y),
          PathCondition      => (LENeg_nwv = '0' AND
                                   OENeg_nwv = '0'))
                 )
      );
  END PROCESS;
END vhdl_behavioral;
```

Because the output can be driven to 'Z', the model uses the VitalPathDelay01Z path delay procedure. There are four paths defined in this procedure call. Three of them have expressions for the PathCondition parameters. In this model, the path delay procedure call drives the output port directly.

The PathDelay parameter in a VitalPathDelay01Z procedure call must have six values. A generic of type VitalDelayType01 Z has six values, but a generic of type VitalDelayType01 has only two. This can be remedied by using the function VitalExtendToFillDelay. This function will assign the two supplied values to the four missing elements and return an array of type VitalDelayType01Z. The four padded elements will never be needed during simulation but are required to satisfy the procedure call.

9.3   Summary

Registers, latches, and flip-flops are common in glue logic and as elements in more complex components. The definitions of latches and flip-flops can be stored in packages as VITAL state tables and reused by other models.

Whenever there is a register in a model, there are setup/hold and period/ pulsewidth timing checks. The timing values for these timing constraint checks, as well as for the path delays, should not be hard coded into the model, but should be annotated through SDF and timing generics.

All ports in a model should have explicit initialization so if some ports are left unconnected a legal netlist may still be generated.

When testing port values, it should be considered that an input port might be assigned a weak value, 'L' or 'H'. Otherwise, incorrect behavior could result under some conditions. Creating a no weak value variable or signal and using it in tests can help.

10 Conditional Delays and Timing Constraints

Delay and constraint timing values can be selected based on conditions. However, to make use of this capability you must know how conditional statements are mapped between SDF and VITAL.

We have seen in previous chapters the selection of a delay from multiple candidates based on path, but sometimes the propagation delay from an input to an output depends on the state of one or more other pins. We call that a conditional delay. It usually requires two or more path delay generics. Part of the data sheet for a TI SN74GTL1655 (Figure 10.1) illustrates a conditional delay. Both SDF and VITAL have sufficient syntax to describe conditional timing and there is a prescribed mapping between them. This chapter discusses the implementation of conditional timing in SDF and in VITAL.

10.1 Conditional Delays in VITAL

VITAL path delay generics are constructed by a formula:

```
<tpd_generic> [condition]
```

For example, the generic for the path delay from port A to port Y would be

```
tpd_A_Y
```

Now let us suppose the delay is dependent on the value of a third port S. Two generics are required:

```
tpd_A_Y_S_EQ_0
```

for when S is low, and

```
tpd_A_Y_S_EQ_1
```

for when S is high.

PARAMETER	FROM (INPUT)	TO (OUTPUT)	SN54GTL1655		SN74GTL1655		UNIT
			MIN	MAX	MIN	MAX	
t_{max}			160		160		MHz
t_{PLH}	A $V_{ERC}=V_{CC}$	B	3.1	5.2	3.1	5.2	ns
t_{PHL}			2.6	6.2	2.6	6.2	
t_{PLH}	CLK $V_{ERC}=V_{CC}$	B	3.4	5.5	3.4	5.5	ns
t_{PHL}			2.4	5.8	2.4	5.8	
t_{PLH}	LEAB $V_{ERC}=V_{CC}$	B	3.5	5.8	3.5	5.8	ns
t_{PHL}			2.6	6.4	2.6	6.4	
t_{en}	$\overline{OEAB}$ or $\overline{OE}$ $V_{ERC}=V_{CC}$	B	3.3	5.4	3.3	5.4	ns
t_{dls}			2.7	5.9	2.7	5.9	
t_{PLH}	A $V_{ERC}=GND$	B	2.3	4.3	2.3	4.3	ns
t_{PHL}			1.9	4.3	1.9	4.3	
t_{PLH}	CLK $V_{ERC}=GND$	B	2.7	4.8	2.7	4.8	ns
t_{PHL}			1.8	4.3	1.8	4.3	
t_{PLH}	LEAB $V_{ERC}=GND$	B	2.8	4.9	2.8	4.9	ns
t_{PHL}			2	4.8	2	4.8	
t_{en}	$\overline{OEAB}$ or $\overline{OE}$ $V_{ERC}=GND$	B	2.5	4.5	2.5	4.5	ns
t_{dls}			2	4.2	2	4.2	

Figure 10.1 Timing table for part with conditional delays

Conditional delays can be more complex. They may involve multiple ports. If there is one delay when both S and M are high and another if either is low, the generics could be written as

```
tpd_A_Y_S_EQ_1_AN_M_EQ_1
```

for the first case, and

```
tpd_A_Y_S_EQ_0_OR_M_EQ_0
```

for the second.

Although the generics seem to have their own logic, for VHDL models they are really just useful mnemonics. Their purpose is to allow additional unique generics to which more complex timing can be backannotated. Although the generic may describe the conditions, the VitalPathDelay procedure call must specify the conditions in the PathCondition parameter.

The FMF stdh1655 model has a number of conditional delays. Let's look at the port A0 to port B0 delays:

```
-- tpd delays
tpd_A0_B0_VERC_EQ_0         : VitalDelayType01 := UnitDelay01;
tpd_A0_B0_VERC_EQ_1         : VitalDelayType01 := UnitDelay01;
```

In the Paths section of the `VitalPathDelay01Z` procedure call, the selection is made in the following two paths:

```
Paths =>    (
     0 => (InputChangeTime    => B0int'LAST_EVENT,
          PathDelay           =>
          VitalExtendToFillDelay(tpd_A0_B0_VERC_EQ_0),
          PathCondition       => (Benable = '1' AND VERC_ipd = '0')),
     1 => (InputChangeTime    => B0int'LAST_EVENT,
          PathDelay           =>
              VitalExtendToFillDelay(tpd_A0_B0_VERC_EQ_1),
          PathCondition       => (Benable = '1' AND VERC_ipd = '1')),
```

Again, note that it is the `PathCondition` and not the generic name that selects the correct delay. Using the long generic name is still good practice because it documents the purpose of each generic. This code also includes another example of using the `VitalExtendToFillDelay` function to expand a generic of type `VitalDelayType01` so it can be used where a generic of type `VitalDelayType01Z` is expected.

10.2   Conditional Delays in SDF

In SDF, a conditional path delay consists of a condition applied to a path delay (`iopath_def`). To review, an `iopath_def` is defined as

```
iopath_def ::=
  (IOPATH port_spec port_instance { retain_def } deval_list )
retain_def ::=
  (RETAIN retval_list )
```

where

IOPATH is the key word.

port_spec is the input port.

port_instance is the output port.

retain_def will be discussed shortly.

delval_list is the delay data.

As shown in Chapter 4, the formal syntax for conditional path delay is

```
cond_def ::=
     ( COND [ qstring ] conditional_port_expr iopath_def )
```

where

COND is the key word.

qstring is an optional symbolic name. Its mapping in VITAL is not well documented, so we shall avoid using it.

conditional_port_expr is the description of the state dependency of the path delay. A particular conditional path delay will be used only if the condition is TRUE. Only expressions using ports are legal.

iopath_def has the same meaning as described earlier.

The `conditional_port_expr` is an expression constructed of port names, the constants 0 and 1, and operators. For the conditional path delay generics shown earlier, the corresponding SDF expression is

```
(COND      VERC == 0 (IOPATH A0 B0 (1.5:3.8:4.5) (1.5:3.8:4.5)))
(COND      VERC == 1 (IOPATH A0 B0 (1.5:4.5:5.5) (1.5:4.5:5.5)))
```

This provides one set of delays when VERC is low and another when VERC is high.

Table 10.1 shows the operators available for constructing conditional expressions in SDF for both delays and timing checks. Although the range of possible expressions is vast, simple expressions are usually sufficient.

10.3   Conditional Delay Alternatives

It is not uncommon for a delay value to be based on an internal condition rather than the state of a port. In such a case a different approach must be used. In the following VPD call, the delay selection is based on the value on an internal signal:

```
-- tpd delays
tpd_CLK_DQ2             : VitalDelayType01Z := UnitDelay01Z;
tpd_CLK_DQ3             : VitalDelayType01Z := UnitDelay01Z;
    VitalPathDelay01Z (
            OutSignal        => DataOut(i),
            OutSignalName    => "Data",
            OutTemp          => D_zd(i),
            Mode             => OnEvent,
            GlitchData       => D_GlitchData(i),
            Paths            => (
                1 => (InputChangeTime => CLKIn'LAST_EVENT,
                      PathDelay => tpd_CLK_DQ2,
                      PathCondition => CAS_Lat = 2),
                2 => (InputChangeTime => CLKIn'LAST_EVENT,
                      PathDelay => tpd_CLK_DQ3,
                      PathCondition => CAS_Lat = 3)
            )
        );
```

In this case CAS_Lat is an internal signal signifying the CAS latency of an SDRAM. SDF and VITAL do not provide a means of supplying a discriptive generic name here so we do the best we can while still writing legal VITAL code.

Table 10.1 Operators for expressions in SDF

unary operators	expression
+	arithmetic identity
–	arithmetic negation
!	logical negation
~	bit-wise unary negation
&	reduction unary AND
~&	reduction unary NAND
|	reduction unary OR
~|	reduction unary NOR
^	reduction unary XOR
^~	reduction unary XNOR
~^	reduction unary XNOR

inversion operators	expression
!	logical negation
~	bit-wise unary negation

binary operators	expression
+	arithmetic sum
–	arithmetic difference
*	arithmetic product
/	arithmetic quotient
%	modulus
==	logical equality
!=	logical inequality
===	case equality
!==	case inequality
&&	logical AND
||	logical OR
<	relational
<=	relational
>	relational
>=	relational
&	bit-wise binary AND
|	bit-wise binary inclusive OR
^	bit-wise binary exclusive OR
^~	bit-wise binary equivalence
~^	bit-wise binary equivalence
>>	right shift
<<	left shift

equality operators	expression
==	logical equality
!=	logical inequality
===	case equality
!==	case inequality

It is also possible to calculate a delay value (even in a `VITAL_level1` model). Modeling some parts requires delaying an output by a time that is dependent on the frequency of an external clock or an internal phase locked loop. The following code is from a DSP model:

```
0 => (InputChangeTime    => ECLK_int'LAST_EVENT,
     PathDelay           => tpd_ECLKIN_BUSREQ,
     PathCondition       => RESET_int = '1'),
1 => (InputChangeTime    => RESETNeg'LAST_EVENT,
     PathDelay           => VitalExtendToFillDelay
                               (4 * PERIOD + 3 * EPERIOD),
     PathCondition       => RESETNeg = '0'),
2 => (InputChangeTime    => RESETNeg'LAST_EVENT,
     PathDelay           => VitalExtendToFillDelay
                               (6 * PERIOD + 4 * EPERIOD),
     PathCondition       =>  RESETNeg = '1')
)
```

Paths 1 and 2 have delays based on the signals `PERIOD` and `EPERIOD`. These signals are of type `TIME` and their values are computed by measuring the periods of two input clocks.

10.4   Mapping SDF to VITAL

SDF was originally developed for use with Verilog. When VITAL was developed, it made more sense to reuse SDF than to develop a new language from scratch. However, the reuse of SDF comes with the cost of having to map the SDF representation into VITAL.

The mapping is done per character or per operator and is shown in Table 10.2. Underscores are used to separate elements.

Using Table 10.2, we can see that the SDF entry

```
(COND RESET == 1 && CLK == 1 (IOPATH A Y (10) (20)))
```

maps to the VITAL generic

```
tpd_A_Y_RESET_EQ_1_AN_CLK_EQ_1
```

and

```
(COND PIPER == 0 (IOPATH CLKR IORO (5:10:15) (5:10:15) (1:2:3) (1:2:3)
(1:2:3) (1:2:3)))
```

maps to the VITAL generic:

```
tpd_CLKR_IORO_PIPER_EQ_0
```

Table 10.2   SDF to VITAL symbol mapping

SDF	VITAL	SDF	
		VITAL	
(	OP		
)	CP	&&	AN
{	OB	||	OR
}	CB	<	LT
[	OSB	<=	LE
]	CSB	>	GT
,	CM	>=	GE
?	QM	&	ANB
:	CLN	|	ORB
+	PL	^	XOB
–	MI	^~	XNB
*	MU	~^	XNB
/	DI	>>	RS
%	MOD	<<	LS
==	EQ	!	NT
!=	NE	~	NTB
===	EQ3	~&	NA
!==	NE3	~|	NO

10.5   Conditional Timing Checks in VITAL

Just as delays can be conditional, so can timing checks. The same general rules apply. SDF statements can be built to provide multiple values for timing checks. The mapping described earlier can be used to construct the corresponding VITAL generics. The actual selection of timing values must be done in the VITAL timing check procedure calls.

For example, the IDT709079 is a synchronous pipelined dual-port SRAM. The timing file for this part has the following SDF statements:

```
(WIDTH (COND PIPER_EQ_0_negedge CLKR) (6.5))
(WIDTH (COND PIPER_EQ_0_posedge CLKR) (6.5))
(WIDTH (COND PIPER_EQ_1_negedge CLKR) (4))
(WIDTH (COND PIPER_EQ_1_posedge CLKR) (4))
(PERIOD (COND PIPER_EQ_0_posedge CLKR) (19))
(PERIOD (COND PIPER_EQ_1_posedge CLKR) (10))
```

This has one set of values to apply when PIPER is high and another when it is low. Separate values are given for positive and negative pulse widths of CLKR. This is not necessary for the clock period. In the model for the IDT709079 the generics that map to the earlier SDF are

```
-- tpw values: pulse widths
-- tLC1
tpw_CLKR_PIPER_EQ_0_negedge      : VitalDelayType    := UnitDelay;
-- tHC1
tpw_CLKR_PIPER_EQ_0_posedge      : VitalDelayType    := UnitDelay;
-- tLC2
tpw_CLKR_PIPER_EQ_1_negedge      : VitalDelayType    := UnitDelay;
-- tHC2
tpw_CLKR_PIPER_EQ_1_posedge      : VitalDelayType    := UnitDelay;
-- tperiod_min: minimum clock period = 1/max freq
-- tCYC1
tperiod_CLKR_PIPER_EQ_0_posedge : VitalDelayType := UnitDelay;
-- tCYC2
tperiod_CLKR_PIPER_EQ_1_posedge : VitalDelayType := UnitDelay;
```

Using these two sets of generics require the following two `VitalPeriod-PulseCheck` procedure calls.

```
VitalPeriodPulseCheck (
     TestSignal       => CLKRIn,
     TestSignalName   => "CLKR",
     Period           => tperiod_CLKR_PIPER_EQ_1_posedge,
     PulseWidthLow    => tpw_CLKR_PIPER_EQ_1_negedge,
     PulseWidthHigh   => tpw_CLKR_PIPER_EQ_1_posedge,
     PeriodData       => TD_CLKRIn,
     XOn              => XOn,
     MsgOn            => MsgOn,
     HeaderMsg        => InstancePath & PartID,
     CheckEnabled     => (PIPER_nwv        =        '1'),
     Violation        => Pviol_CLKRIn );

VitalPeriodPulseCheck (
     TestSignal       => CLKRIn,
     TestSignalName   => "CLKR",
     Period           => tperiod_CLKR_PIPER_EQ_0_posedge,
     PulseWidthLow    => tpw_CLKR_PIPER_EQ_0_negedge,
     PulseWidthHigh   => tpw_CLKR_PIPER_EQ_0_posedge,
     PeriodData       => TD_CLKRIn,
     XOn              => XOn,
     MsgOn            => MsgOn,
     HeaderMsg        => InstancePath & PartID,
     CheckEnabled     => (PIPER_nwv        =        '0'),
     Violation        => Pviol_CLKRIn      );
```

The selection of which `VitalPeriodPulseCheck` is executed is determined by the `CheckEnabled` parameters. To simplify the `CheckEnabled` parameters, the

variable PIPER_nwv is used. PIPER_nwv is PIPER stripped of weak values by means
of a To_UX01 function call. As a matter of style, all such variables are recognizable
by the _nwv suffix.

A slightly more complex example comes from the model of a 54AS869 synchronous 8-bit up/down counter. The timing check section of the timing file has
the following SDF code:

```
(TIMINGCHECK
        (SETUP (COND S0 == 0 && S1 == 0 S0) CLK (13:13:13))
        (SETUP (COND S0 == 1 && S1 == 0 S0) CLK (52:52:52))
        (SETUP (COND S0 == 0 && S1 == 1 S0) CLK (13:13:13))
        (SETUP (COND S0 == 1 && S1 == 1 S0) CLK (52:52:52))
```

The corresponding VITAL generics in the model are

```
-- tsetup values: setup times
tsetup_S0_CLK_S0_EQ_0_AN_S1_EQ_0 : VitalDelayType := UnitDelay;
tsetup_S0_CLK_S0_EQ_1_AN_S1_EQ_0 : VitalDelayType := UnitDelay;
tsetup_S0_CLK_S0_EQ_0_AN_S1_EQ_1 : VitalDelayType := UnitDelay;
tsetup_S0_CLK_S0_EQ_1_AN_S1_EQ_1 : VitalDelayType := UnitDelay;
```

The VitalSetupHoldCheck procedure calls are

```
VitalSetupHoldCheck (
        TestSignal      => S0_ipd,1
        TestSignalName  => "S0_ipd",
        RefSignal       => CLK_ipd,
        RefSignalName   => "CLK_ipd",
        SetupHigh       => tsetup_S0_CLK_S0_EQ_1_AN_S1_EQ_0,
        SetupLow        => tsetup_S0_CLK_S0_EQ_0_AN_S1_EQ_0,
        HoldHigh        => thold_S0_CLK,
        HoldLow         => thold_S0_CLK,
        CheckEnabled    => (S1nwv = '0'),
        RefTransition   => '/',
        HeaderMsg       => InstancePath & "/std869",
        TimingData      => TD_S0_CLK_S1_EQ_0,
        XOn             => XOn,
        MsgOn           => MsgOn,
        Violation       => Tviol_S0_CLK_S1_EQ_0 );

VitalSetupHoldCheck (
        TestSignal      => S0_ipd,
        TestSignalName  => "S0_ipd",
        RefSignal       => CLK_ipd,
        RefSignalName   => "CLK_ipd",
        SetupHigh       => tsetup_S0_CLK_S0_EQ_1_AN_S1_EQ_1,
        SetupLow        => tsetup_S0_CLK_S0_EQ_0_AN_S1_EQ_1,
```

```
    HoldHigh            => thold_S0_CLK,
    HoldLow             => thold_S0_CLK,
    CheckEnabled        => (S1nwv = '1'),
    RefTransition       => '/',
    HeaderMsg           => InstancePath & "/std869",
    TimingData          => TD_S0_CLK_S1_EQ_1,
    XOn                 => XOn,
    MsgOn               => MsgOn,
    Violation           => Tviol_S0_CLK_S1_EQ_1 );
```

In this case, are there two levels of selection. The first is controlled by the Check-Enabled parameter. It is based on the value of S1nwv and determines which procedure call is executed. Next, within each procedure call distinct setup constraint values are selected for the case where S0_ipd is high and S0_ipd is low. This is done through the SetupHigh and SetupLow parameters.

10.6   Summary

There are many situations where multiple sets of values for propagation delays and timing constraints are available, each set corresponding to particular conditions or states. Conditional delay statements and conditional timing statements can be constructed in SDF to satisfy such situations. These statements can be mapped to VITAL generics through well-defined transformations with elements separated by underscores. The actual selection of which timing values are used is determined within the VHDL/VITAL model.

11 Negative Timing Constraints

Occasionally you will need to model a component with a negative timing constraint. Perhaps a negative hold time is specified. If this is your first experience designing with a component that has a negative hold time, you may be suspicious about how a signal can be clocked into a part if the clock arrives after the signal has already changed or, in the case of a negative setup time, the clock arrives before the signal is stable. These conditions are illustrated in Figure 11.1.

What causes this condition, which seems like an aberration, is simply internal delay. For components with negative hold times, the `TestSignal` is delayed prior to reaching the register. For parts with negative setup times, it is the clock that is internally delayed. A component can never have both negative setup and negative hold constraints. If a part with negative constraints was modeled without taking those constraints into account, it could display incorrect behavior beyond just reporting incorrect timing violations.

To model a component with negative constraints, we must also model the internal delays. Determining just how much to delay an internal signal could be a problem.

Fortunately, VITAL can do that for you.

11.1   How Negative Constraints Work

Negative constraints are an indication that some inputs to a device have more delay than others. The model must compensate for the imbalanced delays not just to correctly model setup and hold constraints, but usually to get accurate behavior. If a component has a slow data path, changes in the data signal can lead the clock by some amount and the component will still function correctly. In order to see that same behavior in the model, the clock must be delayed to bring the two signals back into alignment, as they are inside the component. The VITAL signal delay block does just that.

Now, you should be wondering, how much delay should be applied to the fast signal and how that value gets into the model. With a correctly written model, the simulator does it for you, but only if you are using a VITAL-compliant simulator.

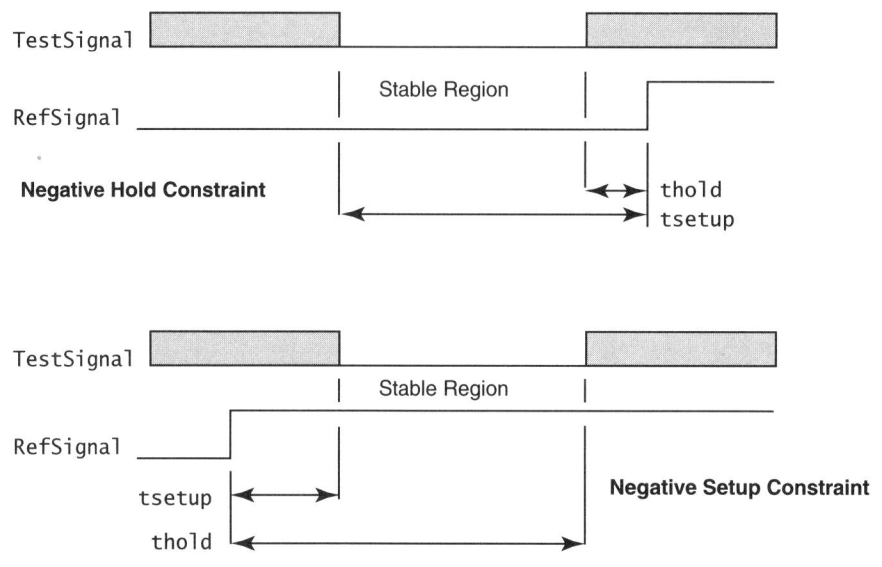

Figure 11.1   Negative setup/hold constraints

When a VITAL model is written for negative constraints, generics are included to hold the delay values. After SDF backannotation, the correct delay values are calculated based on the negative constraint values. The fast signal is delayed. Everywhere in the model the delayed signal is used rather than the actual input signal. This ensures correct operation. The path delay generics are adjusted by the simulator to compensate for the delayed signals. When the output delays are added, the total delay from input to output comes out correct. In the sections that follow, we will see how this is done.

11.2   Modeling Negative Constraints

The first condition for modeling negative constraints is recognizing they exist in a component. Figure 11.2 is an excerpt from the ON semiconductor data sheet for the MC100E445 [5]. For this component, the setup time given for signals SINA and SINB are always negative numbers. This means you must model the component using negative constraints.

The VITALTiming package has a special procedure named VitalSignalDelay. It works something like VitalWireDelay but is specifically designed for delaying signals to accommodate negative timing constraints.

The modeling of negative timing constraints begins with the declaration of some special generics. For each clock (RefSignal) associated with a negative setup or recovery constraint, an internal clock delay generic must be declared. These generics require the prefix ticd:

```
-- ticd values: delayed clocks for negative constraint calculation
ticd_CLK1           : VitalDelayType := VitalZeroDelay;
ticd_CLK2           : VitalDelayType := VitalZeroDelay;
```

For each data signal (`TestSignal`) associated with a negative hold or removal constraint, an internal signal delay generic must be declared. These generics require the prefix `tisd`:

```
-- tisd values: delayed signals for negative constraint calculation
tisd_SEL_CLK1       : VitalDelayType := VitalZeroDelay;
```

If the model contained paths that were dependent upon multiple clocks, it would be necessary to declare a biased propagation delay generic. The prefix for such a generic is `tbpd`. (The author has not yet modeled a component requiring the use of this generic.)

The `ticd` and `tisd` generics are used in a signal delay block, but first the delayed signals must be declared:

```
SIGNAL CLK1_dly       : std_ulogic := 'X';
SIGNAL CLK2_dly       : std_ulogic := 'X';
SIGNAL SEL_dly        : std_ulogic := 'X';
```

Then, after the wire delay block, comes the signal delay block:

```
-------------------------------------------------------------------------
-- Negative Timing Constraint Delays
-------------------------------------------------------------------------

SignalDelay : BLOCK
BEGIN

  s_1: VitalSignalDelay (CLK1_dly, CLK1_ipd, ticd_CLK1);
  s_2: VitalSignalDelay (CLK2_dly, CLK2_ipd, ticd_CLK2);
  s_3: VitalSignalDelay (SEL_dly,  SEL_ipd,  tisd_SEL_CLK1);

END BLOCK;
```

Symbol	Characteristic	0°C			25°C			85°C			Unit
		Min	Typ	Max	Min	Typ	Max	Min	Typ	Max	
t_{MAX}	Maximum Conversion Frequency	2.0			2.0			2.0			Gb/s NRZ
t_{PLH} t_{PHL}	Propagation Delay to Output CLK to Q. Reset to Q CLK to SOUT (Diff) CLK to CL/4 (Diff) CLK to CL/S (Diff)	1500 800 1100 1100	1900 975 1325 1325	2100 1150 1550 1550	1500 800 1100 1100	1900 975 1325 1325	2100 1150 1550 1550	1500 800 1100 1100	1900 975 1325 1325	2100 1150 1550 1550	ps
t_S	Setup Time SINA SINB SEL	-100 0	-250 -200		-100 0	-250 -200		-100 0	-250 -200		ps
t_h	Hold Time SINA, SINB, SEL	450	300		450	300		450	300		ps

Figure 11.2 Negative setup timing characteristics

Only one signal delay block is allowed and it is required to carry the label `Sig-nalDelay`. After this point it is usually necessary to use the delayed (`_dly`) signal versions throughout the model in order to achieve correct behavior. However, using nondelayed signals is not prohibited.

The timing check procedure calls are written as usual, with the following exceptions: The delayed signals are used and the `RefDelay` or `TestDelay` parameters are supplied and associated with the `ticd` or `tisd` generics. In Figure 11.3 the signal `CLKint` has been derived from the signals `CLK1_dly` and `CLK2_dly`.

To put all this in context, let us take a close look at a component that has a negative setup constraint. The component is an MC100E445 ECL 4-bit serial-to-parallel converter. Figure 11.4 is a schematic of the portion of the component that introduces the negative constraint.

The component contains a D flip-flop with a mux in series with its data input. The added delay of the mux relative to the clock causes the negative setup constraint. The full model for this component is given in Figure 11.5. The code pertaining specifically to negative constraints is explained after the model.

Because the data signal is delayed by the mux, causing the negative setup constraint, the simulator must delay the clock to bring the two signals back into alignment. To enable it to do so, internal clock delay (`icd`) generics must be declared:

```
-- ticd values: delayed clock times for negative timing constraints
ticd_CLK            : VitalDelayType := VitalZeroDelay;
ticd_CLKNeg         : VitalDelayType := VitalZeroDelay;
```

The `ticd` generics will be annotated by the simulator during a special calculation phase described in the next section. They are not backannotated from an SDF file.

The annotated delays are applied to the clock signals in the signal delay block:

```
------------------------------------------------------------------------------
-- Negative Timing Constraint Delays
------------------------------------------------------------------------------
SignalDelay : BLOCK
BEGIN

    s_1: VitalSignalDelay (CLK_dly, CLK_ipd, ticd_CLK);
    s_2: VitalSignalDelay (CLKNeg_dly, CLKNeg_ipd, ticd_CLKNeg);

END BLOCK;
```

The delayed differential clock signals are used to generate a delayed single-ended clock:

```
------------------------------------------------------------------------------
-- ECL Clock Process
------------------------------------------------------------------------------
ECLClock : PROCESS (CLK_dly, CLKNeg_dly)
```

This is a two-step process. First, the model determines whether the clock inputs are indeed connected to a differential driver, or if one input is connected to VBB and

```
VitalSetupHoldCheck (
    TestSignal      =>  Dint,
    TestSignalName  => "Dint",
    RefSignal       =>  CLKint,
    RefSignalName   => "CLKint",
    RefDelay        =>  ticd_CLK1,
    SetupHigh       =>  tsetup_D_CLK1,
    SetupLow        =>  tsetup_D_CLK1,
    HoldHigh        =>  thold_D_CLK1,
    HoldLow         =>  thold_D_CLK1,
    CheckEnabled    =>  TRUE,
    RefTransition   =>  '/',
    HeaderMsg       =>  InstancePath & "/eclps143",
    TimingData      =>  TD_D_CLK,
    XOn             => XOn,
    MsgOn           => MsgOn,
    Violation       =>  Tviol_D_CLK
);

VitalSetupHoldCheck (
    TestSignal      =>  SEL_dly,
    TestSignalName  => "SEL_ipd",
    TestDelay       =>  tisd_SEL_CLK1,
    RefSignal       =>  CLKint,
    RefSignalName   => "CLKint",
    RefDelay        =>  ticd_CLK1,
    SetupHigh       =>  tsetup_SEL_CLK1,
    SetupLow        =>  tsetup_SEL_CLK1,
    HoldHigh        =>  thold_SEL_CLK1,
    HoldLow         =>  thold_SEL_CLK1,
    CheckEnabled    =>  TRUE,
    RefTransition   =>  '/',
    HeaderMsg       =>  InstancePath & "/eclps143",
    TimingData      =>  TD_SEL_CLK,
    XOn             => XOn,
    MsgOn           => MsgOn,
    Violation       =>  Tviol_SEL_CLK
);

VitalRecoveryRemovalCheck (
    TestSignal      => MR_ipd,
    TestSignalName  => "MR_ipd",
    RefSignal       => CLKint,
    RefSignalName   => "CLKint",
    RefDelay        => ticd_CLK1,
    Recovery        => trecovery_MR_CLK1,
    ActiveLow       => FALSE,
    CheckEnabled    => TRUE,
    RefTransition   => '/',
    HeaderMsg       => InstancePath & "/eclps143",
    TimingData      => TD_MR_CLK,
    XOn             => XOn,
    MsgOn           => MsgOn,
    Violation       => Rviol_MR_CLK
);
```

Figure 11.3   `SetupHold` and `RecoveryRemoval` checks with negative constraints

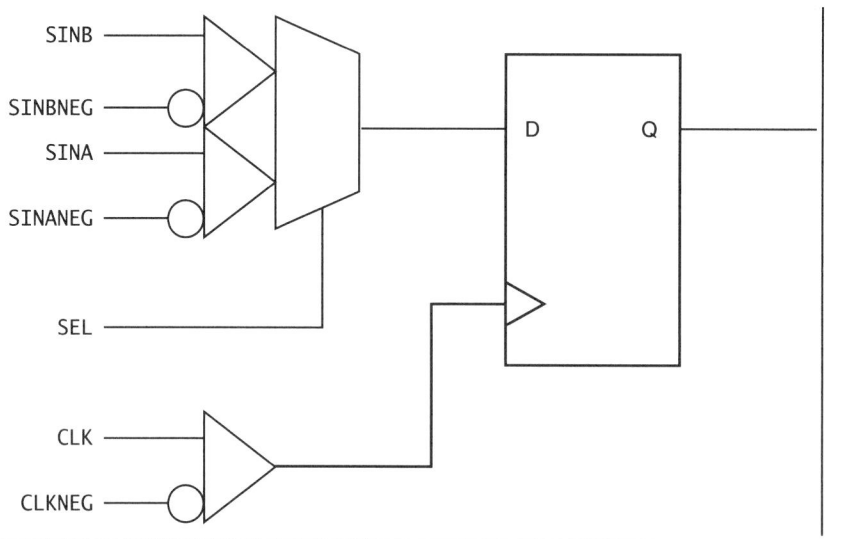

Figure 11.4 Partial schematic of an MC100E445

```
-------------------------------------------------------------------------------
--    File Name : eclps445.vhd
-------------------------------------------------------------------------------
--  Copyright (C) 1997, 2002 Free Model Foundry; http://eda.org/fmf/
--
--  This program is free software; you can redistribute it and/or modify
--  it under the terms og the GNU General Public License version 2 as
--  published by the Free Software Foundation.
--
--  MODIFICATION HISTORY :
--
--  version: |   author:  | mod date: | changes made:
--    V2.0        rev3      96 MAR 22  Conformed to style guide,
--                                     New ecl_utils package with more constants
--    V2.1     R. Munden   96 MAY 19  Changed tpd's for VITAL compliance
--    V2.2     R. Steele   96 SEP 18  Change trelease to trecovery
--    V2.3     R. Steele   96 OCT 11  Updated timing generics, clock process
--    V2.4     R. Munden   97 MAR 01  Changed XGenerationOn to XOn, added MsgOn,
--                                     and updated TimingChecks & PathDelays
--    V2.5     R. Steele   97 JUN 30  Made PathCondition true for Q0
--    V2.6     R. Munden   98 APR 04  Modified for VITAL NTC
--    V2.7     R. Munden   98 OCT 14  Changed from inertial delay to transport
--                                         and added period checks
--    V2.8     R. Munden   02 APR 24   Fixed Dummy VPD
--    V2.9     R. Munden   03 OCT 28   Corrected NTC behavior
-------------------------------------------------------------------------------
--  PART DESCRIPTION :
--
--  Library:        ECLPS
--  Technology:     ECL
--  Part:           ECLPS445
--
```

Figure 11.5 Models of component with negative constraints

```
--  Description:    4-Bit Serial/Parallel Converter
--
-------------------------------------------------------------------------------
LIBRARY IEEE;    USE IEEE.std_logic_1164.ALL;
                 USE IEEE.VITAL_timing.ALL;
                 USE IEEE.VITAL_primitives.ALL;
LIBRARY FMF;     USE FMF.ecl_utils.ALL;
                 USE FMF.ff_package.ALL;


-------------------------------------------------------------------------------
--  ENTITY DECLARATION
-------------------------------------------------------------------------------
ENTITY eclps445 IS
    GENERIC (
        -- tipd delays: interconnect path delays
        tipd_SINA            : VitalDelayType01 := VitalZeroDelay01;
        tipd_SINANeg         : VitalDelayType01 := VitalZeroDelay01;
        tipd_SINB            : VitalDelayType01 := VitalZeroDelay01;
        tipd_SINBNeg         : VitalDelayType01 := VitalZeroDelay01;
        tipd_SEL             : VitalDelayType01 := VitalZeroDelay01;
        tipd_CLK             : VitalDelayType01 := VitalZeroDelay01;
        tipd_CLKNeg          : VitalDelayType01 := VitalZeroDelay01;
        tipd_MODE            : VitalDelayType01 := VitalZeroDelay01;
        tipd_SYNC            : VitalDelayType01 := VitalZeroDelay01;
        tipd_RESET           : VitalDelayType01 := VitalZeroDelay01;
        -- ticd values: delayed clock times for negative timing constraints
        ticd_CLK             : VitalDelayType := VitalZeroDelay;
        ticd_CLKNeg          : VitalDelayType := VitalZeroDelay;
        -- tpd delays: propagation delays
        tpd_CLK_Q0           : VitalDelayType01 := ECLUnitDelay01;
        tpd_CLK_SOUT         : VitalDelayType01 := ECLUnitDelay01;
        tpd_CLK_CL4          : VitalDelayType01 := ECLUnitDelay01;
        tpd_CLK_CL8          : VitalDelayType01 := ECLUnitDelay01;
        tpd_RESET_CL4        : VitalDelayType01 := ECLUnitDelay01;
        -- tsetup values: setup times
        tsetup_SINA_CLK    : VitalDelayType := ECLUnitDelay;
        tsetup_SINA_CLKNeg : VitalDelayType := ECLUnitDelay;
        tsetup_SEL_CLK     : VitalDelayType := ECLUnitDelay;
        -- thold values: hold times
        thold_SINA_CLK     : VitalDelayType := ECLUnitDelay;
        thold_SINA_CLKNeg  : VitalDelayType := ECLUnitDelay;
        thold_SEL_CLK      : VitalDelayType := ECLUnitDelay;
        -- trecovery values: release times
        trecovery_RESET_CLK : VitalDelayType := ECLUnitDelay;
        -- tpw values: pulse widths
        tpw_CLK_posedge    : VitalDelayType := ECLUnitDelay;
        tpw_CLK_negedge    : VitalDelayType := ECLUnitDelay;
        tpw_RESET_posedge  : VitalDelayType := ECLUnitDelay;
        -- tperiod_min: minimum clock period = 1/max freq
        tperiod_CLK_posedge  : VitalDelayType    := ECLUnitDelay;
        -- generic control parameters
        InstancePath         : STRING  := DefaultECLInstancePath;
        TimingChecksOn       : Boolean := DefaultECLTimingChecks;
        MsgOn                : BOOLEAN := DefaultECLMsgOn;
        XOn                  : Boolean := DefaultECLXOn;
        -- For FMF SDF technology file usage
        TimingModel          : STRING  := DefaultECLTimingModel
    );
```

Figure 11.5 Models of component with negative constraints *(continued)*

```
      PORT (
          -- 0 denotes internal pull-down resistor
          SINA                : IN  std_ulogic := '0';
          SINANeg             : IN  std_ulogic := '0';
          SINB                : IN  std_ulogic := '0';
          SINBNeg             : IN  std_ulogic := '0';
          SEL                 : IN  std_ulogic := '0';
          CLK                 : IN  std_ulogic := '0';
          CLKNeg              : IN  std_ulogic := '0';
          MODE                : IN  std_ulogic := '0';
          SYNC                : IN  std_ulogic := '0';
          RESET               : IN  std_ulogic := '0';
          Q0                  : OUT std_ulogic := 'U';
          Q1                  : OUT std_ulogic := 'U';
          Q2                  : OUT std_ulogic := 'U';
          Q3                  : OUT std_ulogic := 'U';
          SOUT                : OUT std_ulogic := 'U';
          SOUTNeg             : OUT std_ulogic := 'U';
          CL4                 : OUT std_ulogic := 'U';
          CL4Neg              : OUT std_ulogic := 'U';
          CL8                 : OUT std_ulogic := 'U';
          CL8Neg              : OUT std_ulogic := 'U';
          VBB                 : OUT std_ulogic := ECLVbbValue
      );
      ATTRIBUTE VITAL_level0 OF eclps445 : ENTITY IS TRUE;
END eclps445;


------------------------------------------------------------------------------
--  ARCHITECTURE DECLARATION
------------------------------------------------------------------------------
ARCHITECTURE vhdl_behavioral OF eclps445 IS
      ATTRIBUTE VITAL_level1 OF vhdl_behavioral : ARCHITECTURE IS TRUE;

      SIGNAL SINA_ipd       : std_ulogic := 'X';
      SIGNAL SINANeg_ipd    : std_ulogic := 'X';
      SIGNAL SINB_ipd       : std_ulogic := 'X';
      SIGNAL SINBNeg_ipd    : std_ulogic := 'X';
      SIGNAL SEL_ipd        : std_ulogic := 'X';
      SIGNAL CLK_ipd        : std_ulogic := 'X';
      SIGNAL CLKNeg_ipd     : std_ulogic := 'X';
      SIGNAL CLK_dly        : std_ulogic := 'X';
      SIGNAL CLKNeg_dly     : std_ulogic := 'X';
      SIGNAL MODE_ipd       : std_ulogic := 'X';
      SIGNAL SYNC_ipd       : std_ulogic := 'X';
      SIGNAL RESET_ipd      : std_ulogic := 'X';
      SIGNAL SINAint        : std_ulogic := 'X';
      SIGNAL SINBint        : std_ulogic := 'X';
      SIGNAL SINint         : std_ulogic := 'X';
      SIGNAL CLKint         : std_ulogic := 'X';
      SIGNAL Q0int          : std_ulogic := 'X';
      SIGNAL Q1int          : std_ulogic := 'X';
      SIGNAL Q2int          : std_ulogic := 'X';
      SIGNAL Q3int          : std_ulogic := 'X';
      SIGNAL SOUTint        : std_ulogic := 'X';
      SIGNAL CL4int         : std_ulogic := 'X';
      SIGNAL CL8int         : std_ulogic := 'X';
```

Figure 11.5 Models of component with negative constraints *(continued)*

```
BEGIN

    --------------------------------------------------------------------------
    -- Wire Delays
    --------------------------------------------------------------------------
    WireDelay : BLOCK
    BEGIN

        w_1: VitalWireDelay (SINA_ipd, SINA, tipd_SINA);
        w_2: VitalWireDelay (SINANeg_ipd, SINANeg, tipd_SINANeg);
        w_3: VitalWireDelay (SINB_ipd, SINB, tipd_SINB);
        w_4: VitalWireDelay (SINBNeg_ipd, SINBNeg, tipd_SINBNeg);
        w_5: VitalWireDelay (SEL_ipd, SEL, tipd_SEL);
        w_6: VitalWireDelay (CLK_ipd, CLK, tipd_CLK);
        w_7: VitalWireDelay (CLKNeg_ipd, CLKNeg, tipd_CLKNeg);
        w_8: VitalWireDelay (MODE_ipd, MODE, tipd_MODE);
        w_9: VitalWireDelay (SYNC_ipd, SYNC, tipd_SYNC);
        w_10: VitalWireDelay (RESET_ipd, RESET, tipd_RESET);

    END BLOCK;

    --------------------------------------------------------------------------
    -- Negative Timing Constraint Delays
    --------------------------------------------------------------------------
    SignalDelay : BLOCK
    BEGIN

        s_1: VitalSignalDelay (CLK_dly, CLK_ipd, ticd_CLK);
        s_2: VitalSignalDelay (CLKNeg_dly, CLKNeg_ipd, ticd_CLKNeg);

    END BLOCK;

    --------------------------------------------------------------------------
    -- Concurrent Procedures
    --------------------------------------------------------------------------
    a_1: VitalMUX2 (q => SINint, d0 => SINBint, d1 => SINAint, dsel => SEL_ipd);
    a_2: VitalBUF (q => SOUT, a => SOUTint, ResultMap => ECL_wired_or_rmap);
    a_3: VitalINV (q => SOUTNeg, a => SOUTint, ResultMap => ECL_wired_or_rmap);
    a_4: VitalBUF (q => CL4, a => CL4int, ResultMap => ECL_wired_or_rmap);
    a_5: VitalINV (q => CL4Neg, a => CL4int, ResultMap => ECL_wired_or_rmap);
    a_6: VitalBUF (q => CL8, a => CL8int, ResultMap => ECL_wired_or_rmap);
    a_7: VitalINV (q => CL8Neg, a => CL8int, ResultMap => ECL_wired_or_rmap);
    a_8: VitalBUF (q => Q0, a => Q0int, ResultMap => ECL_wired_or_rmap);
    a_9: VitalBUF (q => Q1, a => Q1int, ResultMap => ECL_wired_or_rmap);
    a_10: VitalBUF (q => Q2, a => Q2int, ResultMap => ECL_wired_or_rmap);
    a_11: VitalBUF (q => Q3, a => Q3int, ResultMap => ECL_wired_or_rmap);

    --------------------------------------------------------------------------
    -- SINA inputs Process
    --------------------------------------------------------------------------
    SINA_inputs : PROCESS (SINA_ipd, SINANeg_ipd)

        -- Functionality Results Variables
        VARIABLE SINAint_zd    : std_ulogic;

        -- Output Glitch Detection Variables
        VARIABLE SINA_GlitchData : VitalGlitchDataType;
```

Figure 11.5   Models of component with negative constraints *(continued)*

```
BEGIN

    --------------------------------------------------------------------------
    -- Functionality Section
    --------------------------------------------------------------------------
    SINAint_zd := ECL_s_or_d_inputs_tab (SINA_ipd, SINANeg_ipd);

    --------------------------------------------------------------------------
    -- (Dummy) Path Delay Section
    --------------------------------------------------------------------------
    VitalPathDelay (
        OutSignal       => SINAint,
        OutSignalName   => "SINAint",
        OutTemp         => SINAint_zd,
        GlitchData      => SINA_GlitchData,
        XOn             => XOn,
        MsgOn           => MsgOn,
        Paths           => (
            0 => (InputChangeTime  => SINA_ipd'LAST_EVENT,
                  PathDelay        => VitalZeroDelay,
                  PathCondition    => FALSE))
    );

END PROCESS;

--------------------------------------------------------------------------
-- SINB inputs Process
--------------------------------------------------------------------------
SINB_inputs : PROCESS (SINB_ipd, SINBNeg_ipd)

    -- Functionality Results Variables
    VARIABLE SINBint_zd    : std_ulogic;

    -- Output Glitch Detection Variables
    VARIABLE SINB_GlitchData : VitalGlitchDataType;

BEGIN

    --------------------------------------------------------------------------
    -- Functionality Section
    --------------------------------------------------------------------------
    SINBint_zd := ECL_s_or_d_inputs_tab (SINB_ipd, SINBNeg_ipd);

    --------------------------------------------------------------------------
    -- (Dummy) Path Delay Section
    --------------------------------------------------------------------------
    VitalPathDelay (
        OutSignal       => SINBint,
        OutSignalName   => "SINBint",
        OutTemp         => SINBint_zd,
        GlitchData      => SINB_GlitchData,
        XOn             => XOn,
        MsgOn           => MsgOn,
        Paths           => (
            0 => (InputChangeTime  => SINB_ipd'LAST_EVENT,
                  PathDelay        => VitalZeroDelay,
                  PathCondition    => FALSE))
    );

END PROCESS;
```

Figure 11.5 Models of component with negative constraints *(continued)*

```
---------------------------------------------------------------------
-- ECL Clock Process
---------------------------------------------------------------------
ECLClock : PROCESS (CLK_dly, CLKNeg_dly)

    -- Functionality Results Variables
    VARIABLE Mode1          : X01;
    VARIABLE CLKint_zd      : std_ulogic;
    VARIABLE PrevData       : std_logic_vector(0 to 2);

    -- Glitch Detection Variables
    VARIABLE CLK_GlitchData : VitalGlitchDataType;

BEGIN

    ---------------------------------------------------------------------
    -- Functionality Section
    ---------------------------------------------------------------------
    Mode1 := ECL_diff_mode_tab (CLK_dly, CLKNeg_dly);

    VitalStateTable (
        StateTable        => ECL_clk_tab,
        DataIn            => (CLK_dly, CLKNeg_dly, Mode1),
        Result            => CLKint_zd,
        PreviousDataIn    => PrevData
    );

    ---------------------------------------------------------------------
    -- (Dummy) Path Delay Section
    ---------------------------------------------------------------------
    VitalPathDelay (
        OutSignal         => CLKint,
        OutSignalName     => "CLKint",
        OutTemp           => CLKint_zd,
        GlitchData        => CLK_GlitchData,
        XOn               => XOn,
        MsgOn             => MsgOn,
        Paths             => (
            0 => (InputChangeTime    => CLK_dly'LAST_EVENT,
                  PathDelay          => VitalZeroDelay,
                  PathCondition      => FALSE))
    );

END PROCESS;

---------------------------------------------------------------------
-- Main Behavior Process
---------------------------------------------------------------------
VitalBehavior : PROCESS (CLKint, RESET_ipd, SEL_ipd, MODE_ipd,
                         SYNC_ipd, SINint)

    CONSTANT clkdiv_4_tab    : VitalStateTableType := (
```

Figure 11.5 Models of component with negative constraints *(continued)*

```
     -----INPUTS----------|-PREV-------------|--OUTPUTS-----------------
     -- Viol CLK Rst Sync | Sv1 Sv0  Sd  S2d | Sv1' Sv0' Sd' S2d' CL4  --
     --------------------|------------------|--------------------------
     -- Violation   Reset unknown - need reset
     ---------------------------------------------------------------
     ( 'X', '-', '-', '-', '-', '-', '-', '-', 'X', 'X', '0', '0', 'X'),
     ( '-', '-', 'X', '-', '-', '-', '-', '-', 'X', 'X', '0', '0', 'X'),
     ---------------------------------------------------------------
     -- Reset
     ---------------------------------------------------------------
     ( '-', '-', '1', '-', '-', '-', '-', '-', '0', '0', '0', '0', '0'),
     ---------------------------------------------------------------
     -- CLK unknown, unknown states - need reset
     ---------------------------------------------------------------
     ( '-', 'X', '0', '-', '-', '-', '-', '-', 'X', 'X', '0', '0', 'X'),
     ( '-', '/', '0', '-', 'X', '-', '-', '-', 'X', 'X', '0', '0', 'X'),
     ( '-', '/', '0', '-', '-', 'X', '-', '-', 'X', 'X', '0', '0', 'X'),
     ---------------------------------------------------------------
     -- 1st clock: state 0->1 no sync, 1st sync clk, 3rd or more sync clks
     ---------------------------------------------------------------
     ( '-', '/', '0', '0', '0', '0', '-', '-', '0', '1', '0', '0', '1'),
     ( '-', '/', '0', '1', '0', '0', '0', '-', '0', '1', '1', '0', '1'),
     ( '-', '/', '0', '1', '0', '0', '-', '1', '0', '1', '1', '1', '1'),
     ---------------------------------------------------------------
     -- 2nd clock: state 1->2 no sync, 1st sync clk, 3rd or more sync clks
     ---------------------------------------------------------------
     ( '-', '/', '0', '0', '0', '1', '-', '-', '1', '0', '0', '0', '1'),
     ( '-', '/', '0', '1', '0', '1', '0', '-', '1', '0', '1', '0', '1'),
     ( '-', '/', '0', '1', '0', '1', '-', '1', '1', '0', '1', '1', '1'),
     ---------------------------------------------------------------
     -- 3rd clock: state 2->3 no sync, 1st sync clk, 3rd or more sync clks
     ---------------------------------------------------------------
     ( '-', '/', '0', '0', '1', '0', '-', '-', '1', '1', '0', '0', '0'),
     ( '-', '/', '0', '1', '1', '0', '0', '-', '1', '1', '1', '0', '0'),
     ( '-', '/', '0', '1', '1', '0', '-', '1', '1', '1', '1', '1', '0'),
     ---------------------------------------------------------------
     -- 4th clock: state 3->0 no sync, 1st sync clk, 3rd or more sync clks
     ---------------------------------------------------------------
     ( '-', '/', '0', '0', '1', '1', '-', '-', '0', '0', '0', '0', '0'),
     ( '-', '/', '0', '1', '1', '1', '0', '-', '0', '0', '1', '0', '0'),
     ( '-', '/', '0', '1', '1', '1', '-', '1', '0', '0', '1', '1', '0'),
     ---------------------------------------------------------------
     -- 2nd sync clock: present state repeated
     ---------------------------------------------------------------
     ( '-', '/', '0', '1', '0', '0', '1', '0', '0', '0', '1', '1', '0'),
     ( '-', '/', '0', '1', '0', '1', '1', '0', '0', '0', '1', '1', '1'),
     ( '-', '/', '0', '1', '1', '0', '1', '0', '1', '0', '1', '1', '1'),
     ( '-', '/', '0', '1', '1', '1', '1', '0', '1', '0', '1', '1', '0'),
     ---------------------------------------------------------------
     -- default
     ---------------------------------------------------------------
     ( '-', '-', '-', '-', '-', '-', '-', '-', 'S', 'S', 'S', 'S', 'S')

); --end of VitalStateTable definition

-- Timing Check Variables
VARIABLE Tviol_SEL_CLK  : X01 := '0';
VARIABLE TD_SEL_CLK     : VitalTimingDataType;

VARIABLE Tviol_SIN_CLK  : X01 := '0';
VARIABLE TD_SIN_CLK     : VitalTimingDataType;

VARIABLE Rviol_RESET_CLK : X01 := '0';
VARIABLE TD_RESET_CLK    : VitalTimingDataType;
```

Figure 11.5   Models of component with negative constraints *(continued)*

```
        VARIABLE Pviol_RESET      : X01 := '0';
        VARIABLE PD_RESET         : VitalPeriodDataType := VitalPeriodDataInit;
        VARIABLE Pviol_CLK        : X01 := '0';
        VARIABLE PD_CLK           : VitalPeriodDataType := VitalPeriodDataInit;

        VARIABLE ViolationA       : X01 := '0';
        VARIABLE ViolationB       : X01 := '0';

        -- Functionality Results Variables
        VARIABLE PrevData1        : std_logic_vector(0 to 3);
        VARIABLE PrevData2        : std_logic_vector(0 to 2);
        VARIABLE PrevDataQ0a      : std_logic_vector(0 to 2);
        VARIABLE PrevDataQ1a      : std_logic_vector(0 to 2);
        VARIABLE PrevDataQ2a      : std_logic_vector(0 to 2);
        VARIABLE PrevDataQ3a      : std_logic_vector(0 to 2);
        VARIABLE PrevDataQ0b      : std_logic_vector(0 to 2);
        VARIABLE PrevDataQ1b      : std_logic_vector(0 to 2);
        VARIABLE PrevDataQ2b      : std_logic_vector(0 to 2);
        VARIABLE PrevDataQ3b      : std_logic_vector(0 to 2);

        VARIABLE CL4_result       : std_logic_vector(1 to 5);
        ALIAS CL4_zd              : std_ulogic IS CL4_result(5);

        VARIABLE CL8_zd           : std_ulogic;

        VARIABLE Clk_div          : std_ulogic;

        VARIABLE Q0_zd            : std_ulogic;
        VARIABLE Q1_zd            : std_ulogic;
        VARIABLE Q2_zd            : std_ulogic;
        VARIABLE Q3_zd            : std_ulogic;
        VARIABLE Q0int_zd         : std_ulogic;
        VARIABLE Q1int_zd         : std_ulogic;
        VARIABLE Q2int_zd         : std_ulogic;
        VARIABLE Q3int_zd         : std_ulogic;

        VARIABLE BLANK            : std_ulogic := '0';

        -- Output Glitch Detection Variables
        VARIABLE Q0_GlitchData   : VitalGlitchDataType;
        VARIABLE Q1_GlitchData   : VitalGlitchDataType;
        VARIABLE Q2_GlitchData   : VitalGlitchDataType;
        VARIABLE Q3_GlitchData   : VitalGlitchDataType;
        VARIABLE SOUT_GlitchData : VitalGlitchDataType;
        VARIABLE CL4_GlitchData  : VitalGlitchDataType;
        VARIABLE CL8_GlitchData  : VitalGlitchDataType;

    BEGIN

        -------------------------------------------------------------------------
        -- Timing Check Section
        -------------------------------------------------------------------------
        IF (TimingChecksOn) THEN

            VitalSetupHoldCheck (
                TestSignal      -> SEL_ipd,
                TestSignalName  => "SEL_ipd",
                RefSignal       => CLKint,
                RefSignalName   => "CLKint",
                RefDelay        => ticd_CLK,
                SetupHigh       => tsetup_SEL_CLK,
                SetupLow        => tsetup_SEL_CLK,
```

Figure 11.5   Models of component with negative constraints *(continued)*

```
        HoldHigh           => thold_SEL_CLK,
        HoldLow            => thold_SEL_CLK,
        CheckEnabled       => TRUE,
        RefTransition      => '/',
        HeaderMsg          => InstancePath & "/eclps445",
        TimingData         => TD_SEL_CLK,
        XOn                => XOn,
        MsgOn              => MsgOn,
        Violation          => Tviol_SEL_CLK
    );

    VitalSetupHoldCheck (
        TestSignal         => SINint,
        TestSignalName     => "SINint",
        RefSignal          => CLKint,
        RefSignalName      => "CLKint",
        RefDelay           => ticd_CLK,
        SetupHigh          => tsetup_SINA_CLK,
        SetupLow           => tsetup_SINA_CLK,
        HoldHigh           => thold_SINA_CLK,
        HoldLow            => thold_SINA_CLK,
        CheckEnabled       => TRUE,
        RefTransition      => '/',
        HeaderMsg          => InstancePath & "/eclps445",
        TimingData         => TD_SIN_CLK,
        XOn                => XOn,
        MsgOn              => MsgOn,
        Violation          => Tviol_SIN_CLK
    );

    VitalRecoveryRemovalCheck (
        TestSignal         => RESET_ipd,
        TestSignalName     => "RESET_ipd",
        RefSignal          => CLKint,
        RefSignalName      => "CLKint",
        RefDelay           => ticd_CLK,
        Recovery           => trecovery_RESET_CLK,
        ActiveLow          => FALSE,
        CheckEnabled       => TRUE,
        RefTransition      => '/',
        HeaderMsg          => InstancePath & "/eclps445",
        TimingData         => TD_RESET_CLK,
        XOn                => XOn,
        MsgOn              => MsgOn,
        Violation          => Rviol_RESET_CLK
    );

    VitalPeriodPulseCheck (
        TestSignal         => CLKint,
        TestSignalName     => "CLKint",
        Period             => tperiod_CLK_posedge,
        PulseWidthHigh     => tpw_CLK_posedge,
        PulseWidthLow      => tpw_CLK_negedge,
        HeaderMsg          => InstancePath & "/eclps445",
        CheckEnabled       => TRUE,
        PeriodData         => PD_CLK,
        XOn                => XOn,
        MsgOn              => MsgOn,
        Violation          => Pviol_CLK
    );
```

Figure 11.5   Models of component with negative constraints *(continued)*

```
                  VitalPeriodPulseCheck (
                      TestSignal        => RESET_ipd,
                      TestSignalName    => "RESET_ipd",
                      PulseWidthHigh    => tpw_RESET_posedge,
                      HeaderMsg         => InstancePath & "/eclps445",
                      CheckEnabled      => TRUE,
                      PeriodData        => PD_RESET,
                      XOn               => XOn,
                      MsgOn             => MsgOn,
                      Violation         => Pviol_RESET
                  );

          END IF;

          -------------------------------------------------------------------
          -- Functionality Section
          -------------------------------------------------------------------
          ViolationA := Pviol_RESET OR Pviol_CLK OR Rviol_RESET_CLK;

          ViolationB := Tviol_SEL_CLK OR Tviol_SIN_CLK OR Pviol_CLK;

          VitalStateTable (
              StateTable      => clkdiv_4_tab,
              DataIn          => (ViolationA, CLKint, RESET_ipd, Sync_ipd),
              NumStates       => 4,
              Result          => CL4_result, --> CL4_zd is CL4_result(5)
              PreviousDataIn  => PrevData1
          );

          VitalStateTable (
              StateTable      => TFFR_tab,
              DataIn          => (ViolationA, CL4_zd, RESET_ipd),
              Result          => CL8_zd,
              PreviousDataIn  => PrevData2
          );

          CLK_div := VitalMux2 (
                          data0   => CL4_zd,
                          data1   => CL8_zd,
                          dselect => MODE_ipd
                      );

          -- Input flip/flops first (logic diagram is misleading - Q0 and SOUT
          -- switch simultaneously), so reverse normal order of VHDL variable
          -- assignment to make it work according to the Timing Diagram.
          -- This in effect mimics a delay in the clock CLK_div to the output
          -- f/fs implied in the prop delays for CLK -> outputs.

          VitalStateTable (
              StateTable      => DFF_tab,
              DataIn          => (Pviol_CLK, CLKint, Q1int_zd),
              Result          => Q0int_zd,
              PreviousDataIn  => PrevDataQ0a
          );

          VitalStateTable (
              StateTable      => DFF_tab,
              DataIn          => (Pviol_CLK, CLKint, Q2int_zd),
              Result          => Q1int_zd,
              PreviousDataIn  => PrevDataQ1a
          );
```

Figure 11.5 Models of component with negative constraints *(continued)*

```
VitalStateTable (
    StateTable      => DFF_tab,
    DataIn          => (Pviol_CLK, CLKint, Q3int_zd),
    Result          => Q2int_zd,
    PreviousDataIn  => PrevDataQ2a
);

VitalStateTable (
    StateTable      => DFF_tab,
    DataIn          => (ViolationB, CLKint, SINint),
    Result          => Q3int_zd,
    PreviousDataIn  => PrevDataQ3a
);

-- Output flip/flops

VitalStateTable (
    StateTable      => DFF_tab,
    DataIn          => (BLANK, CLK_div, Q3int_zd),
    Result          => Q3_zd,
    PreviousDataIn  => PrevDataQ3b
);

VitalStateTable (
    StateTable      => DFF_tab,
    DataIn          => (BLANK, CLK_div, Q2int_zd),
    Result          => Q2_zd,
    PreviousDataIn  => PrevDataQ2b
);

VitalStateTable (
    StateTable      => DFF_tab,
    DataIn          => (BLANK, CLK_div, Q1int_zd),
    Result          => Q1_zd,
    PreviousDataIn  => PrevDataQ1b
);

VitalStateTable (
    StateTable      => DFF_tab,
    DataIn          => (BLANK, CLK_div, Q0int_zd),
    Result          => Q0_zd,
    PreviousDataIn  => PrevDataQ0b
);

----------------------------------------------------------------------
-- Path Delay Section
----------------------------------------------------------------------
VitalPathDelay01 (
    OutSignal       => Q0int,
    OutSignalName   => "Q0int",
    OutTemp         => Q0_zd,
    Mode            => VitalTransport,
    GlitchData      => Q0_GlitchData,
    XOn             => XOn,
    MsgOn           => MsgOn,
    Paths           => (
        0 => (InputChangeTime  => CLKint'LAST_EVENT,
              PathDelay        => tpd_CLK_Q0,
              PathCondition    => TRUE)
    )
);
```

Figure 11.5   Models of component with negative constraints *(continued)*

```
VitalPathDelay01 (
    OutSignal       =>  Q1int,
    OutSignalName   =>  "Q1int",
    OutTemp         =>  Q1_zd,
    Mode            => VitalTransport,
    GlitchData      => Q1_GlitchData,
    XOn             => XOn,
    MsgOn           => MsgOn,
    Paths           => (
        0 => (InputChangeTime   => CLKint'LAST_EVENT,
              PathDelay         => tpd_CLK_Q0,
              PathCondition     => TRUE)
    )
);

VitalPathDelay01 (
    OutSignal       =>  Q2int,
    OutSignalName   =>  "Q2int",
    OutTemp         =>  Q2_zd,
    Mode            => VitalTransport,
    GlitchData      => Q2_GlitchData,
    XOn             => XOn,
    MsgOn           => MsgOn,
    Paths           => (
        0 => (InputChangeTime   => CLKint'LAST_EVENT,
              PathDelay         => tpd_CLK_Q0,
              PathCondition     => TRUE)
    )
);

VitalPathDelay01 (
    OutSignal       =>  Q3int,
    OutSignalName   =>  "Q3int",
    OutTemp         =>  Q3_zd,
    Mode            => VitalTransport,
    GlitchData      => Q3_GlitchData,
    XOn             => XOn,
    MsgOn           => MsgOn,
    Paths           => (
        0 => (InputChangeTime   => CLKint'LAST_EVENT,
              PathDelay         => tpd_CLK_Q0,
              PathCondition     => TRUE)
    )
);

VitalPathDelay01 (
    OutSignal       =>  SOUTint,
    OutSignalName   =>  "SOUTint",
    OutTemp         =>  Q0int_zd,
    Mode            => VitalTransport,
    GlitchData      => SOUT_GlitchData,
    XOn             => XOn,
    MsgOn           => MsgOn,
    Paths           => (
        0 => (InputChangeTime   => CLKint'LAST_EVENT,
              PathDelay         => tpd CLK_SOUT,
              PathCondition     => TRUE)
    )
);
```

Figure 11.5 Models of component with negative constraints *(continued)*

```
VitalPathDelay01 (
    OutSignal        =>  CL4int,
    OutSignalName    =>  "CL4int",
    OutTemp          =>  CL4_zd,
    Mode             => VitalTransport,
    GlitchData       => CL4_GlitchData,
    XOn              => XOn,
    MsgOn            => MsgOn,
    Paths            => (
       0 => (InputChangeTime    => CLKint'LAST_EVENT,
             PathDelay          => tpd_CLK_CL4,
             PathCondition      => TRUE),
       1 => (InputChangeTime    => RESET_ipd'LAST_EVENT,
             PathDelay          => tpd_RESET_CL4,
             PathCondition      => TRUE)
    )
);

VitalPathDelay01 (
    OutSignal        =>  CL8int,
    OutSignalName    =>  "CL8int",
    OutTemp          =>  CL8_zd,
    Mode             => VitalTransport,
    GlitchData       => CL8_GlitchData,
    XOn              => XOn,
    MsgOn            => MsgOn,
    Paths            => (
       0 => (InputChangeTime    => CLKint'LAST_EVENT,
             PathDelay          => tpd_CLK_CL8,
             PathCondition      => TRUE),
       1 => (InputChangeTime    => RESET_ipd'LAST_EVENT,
             PathDelay          => tpd_RESET_CL4,
             PathCondition      => TRUE)
    )
);

END PROCESS;

END vhdl_behavioral;
```

Figure 11.5 Models of component with negative constraints *(continued)*

the other input is driving in single-ended mode. This is done using a table from the FMF.ecl_utils package, ECL_diff_mode_tab.

```
Mode1 := ECL_diff_mode_tab (CLK_dly, CLKNeg_dly);
```

Then, a VITAL state table, ECL_clk_tab, also defined in the FMF.ecl_utils package, reads the mode and the two clock signals and outputs a signal delayed clock:

```
VitalStateTable (
    StateTable      => ECL_clk_tab,
    DataIn          => (CLK_dly, CLKNeg_dly, Mode1),
    Result          => CLKint_zd,
    PreviousDataIn  => PrevData
);
```

However, the result of a state table can only be a variable and a signal is needed. Because this is a VITAL_Level1 model, a direct signal assignment is not permitted. Therefore, a VitalPathDelay procedure is used to make the assignment from CLKint_zd to CLKint.

```
----------------------------------------------------------------------
-- (Dummy) Path Delay Section
----------------------------------------------------------------------
VitalPathDelay        (
      OutSignal       => CLKint,
      OutSignalName   => "CLKint",
      OutTemp         => CLKint_zd,
      GlitchData      => CLK_GlitchData,
      XOn             => XOn,
      MsgOn           => MsgOn,
      Paths           => (
           0 => (InputChangeTime => CLK_dly'LAST_EVENT,
                 PathDelay       => VitalZeroDelay,
                 PathCondition   => FALSE))
);
```

Because the intent is to make the assignment with zero delay, the FMF convention is to comment this as a "Dummy" path delay. From this point on, the only clock signal referenced is CLKint.

The Setup/Hold check,

```
VitalSetupHoldCheck (
      TestSignal      => SINint,
      TestSignalName  => "SINint",
      RefSignal       => CLKint,
      RefSignalName   => "CLK",
      RefDelay        => ticd_CLK,
      SetupHigh       => tsetup_SINA_CLK,
      SetupLow        => tsetup_SINA_CLK,
      HoldHigh        => thold_SINA_CLK,
      HoldLow         => thold_SINA_CLK,
      CheckEnabled    => TRUE,
      RefTransition   => '/',
      HeaderMsg       => InstancePath & "/eclps445",
      TimingData      => TD_SIN_CLK,
      XOn             => XOn,
      MsgOn           => MsgOn,
      Violation       => Tviol_SIN_CLK
);
```

includes the RefDelay parameter:

```
RefDelay        => ticd_CLK
```

If this parameter were to be omitted, the constraint check would still correctly detect violations. However, the error messages generated would state an incorrect time of occurrence of the violations.

The VITAL state tables that model the internal flip-flops in the component must read the delayed clock.

```
VitalStateTable (
    StateTable       => DFF_tab,
    DataIn           => (Pviol_CLK, CLKint, Q1int_zd),
    Result           => Q0int_zd,
    PreviousDataIn   => PrevDataQ0a
);
```

If the nondelayed clock was used, incorrect operation would result when the data input changed within the negative setup time of the clock.

Finally, the path delay procedure calls must also use the delayed clock.

```
VitalPathDelay01 (
    OutSignal       => Q0int,
    OutSignalName   => "Q0int",
    OutTemp         => Q0_zd,
    Mode            => VitalTransport,
    GlitchData      => Q0_GlitchData,
    XOn             => XOn,
    MsgOn           => MsgOn,
    Paths           => (
        0 => (InputChangeTime  => CLKint'LAST_EVENT,
              PathDelay        => tpd_CLK_Q0,
              PathCondition    => TRUE)
    )
);
```

The simulator expects the InputChangeTime to be based on the delayed clock and adjusts the value of tpd_CLK_Q0 to compensate.

11.3   How Simulators Handle Negative Constraints

Everything else we have discussed regarding VITAL has been based on the VITAL packages. The source code for these packages is included with most VHDL simulators and you can read it to understand how it works. This is not the case with negative constraints. For negative constraints to work special features must be built into the simulator.

The values of the ticd and tisd generics are not passed into the model during backannotation. In fact, they are not even in the SDF file. Negative constraint delays are computed by the VITAL compliant simulator during a special negative constraint calculation phase. This phase runs after the VITAL backannotation and

Table 11.1   Timing values before and after NCC

Timing Generic	Before NCC	After NCC
ticd_CLK	0 ps	100 ps
tsetup_SINA_CLK	−100 ps	0 ps
thold_SINA_CLK	450 ps	350 ps
tpd_CLK_Q0	2100 ps	2000 ps
tpd_CLK_SOUT	1150 ps	1050 ps

before the normal VHDL initialization. It is referred to as the Negative Constraint Calculation (NCC) phase.

For each level 0 instance found in the design netlist that defines a negative constraint timing generic, negative constraint calculations are performed. The values of some timing generics are computed and set and the values of others are adjusted. This is an iterative algorithm that uses the generic values set during previous steps.

Negative constraint calculation is performed in the following sequence:

1. Calculate internal clock delays

2. Calculate internal signal delays

3. Calculate biased propagation delays

4. Adjust propagation delays

5. Adjust timing constraint values for setup, hold, recovery, and removal

Although negative values may have been read in from the SDF file, at the end of the NCC phase all timing generics will have positive values.

In the case of the eclps445 model, the timing generics values change, as shown in Table 11.1.

11.4   Ramifications

Correct simulation of negative constraints depends on code built into the VITAL compliant simulator. If a model with negative constraints is run on a non-VITAL compliant simulator, both timing constraint checks and model functionality may be inaccurate. For example, if a model has negative setup constraints, the clock signal must have an internal delay. If the simulator does not supply that delay, a clock transition could register previous data instead of the correct data. In addition, if such a model is compiled with VITAL acceleration disabled, incorrect operation may result when simulated using a VITAL compliant simulator.

When developing VITAL models it is common practice to compile with VITAL acceleration disabled. Otherwise, the models are difficult to debug because the simulator cannot show line-by-line execution. Therefore, it is important to remember to recompile and reverify your models after completion. This is a good practice for all VITAL models, whether or not they have negative constraints.

11.5   Summary

Due to internal delays, real components may have negative timing constraints. They can be modeled accurately by accounting for the internal delays. VITAL compliant simulators have features that calculate values for the internal delays and apply them to special generics. The delay values are added to clocks and signals using a `SignalDelay` block and `VitalSignalDelay` procedure calls. Subsequent code should reference the delayed signals.

Models with negative constraints may not simulate correctly if the simulator does not run a `VITAL_Level1` Negative Constraint Calculation phase.

12

Timing Files and Backannotation

The technology-independent VITAL models upon which this book is based contain no timing values. To run an accurate simulation with timing we must have three things: a timing file, a way to extract from it the correct timing values and format them into SDF, and a way to backannotate those values into the simulation. In this chapter we look at how the timing values for a real component can be applied to a model in simulation.

12.1 Anatomy of a Timing File

Although there are many ways timing files can be written, we will look at the way the Free Model Foundry has chosen to write them. The format is simple and straightforward and based entirely on existing standards.

Figure 12.1 shows the FMF timing file for the STD01 model 2-input positive-NAND gate with open-collector output. The file utilizes XML, a small subset of Standard Generalized Markup Language (SGML) (ISO 8879). SDF code is embedded in this XML wrapper. The result is a file that can be easily parsed or written by either a computer or a human.

The timing file in Figure 12.1 provides timing values for four part numbers from Texas Instruments. There are values for two package types of SN7401 and two package types of SN74LS01. In the case of these parts, package type does not have an impact on the timing values. That is not always the case. Let us dissect this timing file section by section to understand what it means and why it is written this way.

12.1.1 Header

The first line,

```
<!DOCTYPE FTML SYSTEM "ftml.dtd">
```

is a standard SGML header naming the format of the file and its document type definition. From a practical perspective, this line is a formality of little consequence. Existing tools ignore it, but it could be read by some future software.

The next line

```
<FTML><HEAD><TITLE>FMF Timing for STD01 Parts</TITLE>
```

contains the opening tags and the title of the file. If it looks a bit like HTML, that is not a coincidence. HTML is another subset of SGML. The FTML tag opens the file. An FTML timing file has two primary sections, a HEAD and a BODY. The HEAD contains the title and the REVISION.HISTORY.

```
<REVISION.HISTORY>
version: | author:  | mod date: | changes made:
  V1.0     R. Munden   99 APR 04    Initial release
```

The revision history supplies the same type of version, author, date, and change list data found in the model header. Timing files may change over time, timing values for a new part number might be added, and sometimes a change to a model requires a corresponding modification to the model's timing file. The revision history tracks these changes.

```
<!DOCTYPE FTML SYSTEM "ftml.dtd">
<FTML><HEAD><TITLE>FMF Timing for STD01 Parts</TITLE>
<REVISION.HISTORY>
version: | author:  | mod date: | changes made:
  V1.0     R. Munden   99 APR 04    Initial release
</REVISION.HISTORY>
</HEAD>
<BODY>
<TIMESCALE>1ns</TIMESCALE>
<MODEL>STD01
<FMFTIME>
SN7401D<SOURCE>Texas Instruments SDLS026-Revised March 1988</SOURCE>
SN7401N<SOURCE>Texas Instruments SDLS026-Revised March 1988</SOURCE>
<COMMENT>The Values listed are for VCC=5V, CL=15pF, Ta=+25 Celsius</COMMENT>
<COMMENT>Min values are derived</COMMENT>
<TIMING>
(DELAY (ABSOLUTE
    (IOPATH A YNeg (18:35:55) (4:8:15))
  ))
</TIMING></FMFTIME>
<FMFTIME>
SN74LS01D<SOURCE>Texas Instruments SDLS026-Revised March 1988</SOURCE>
SN74LS01N<SOURCE>Texas Instruments SDLS026-Revised March 1988</SOURCE>
<COMMENT>The Values listed are for VCC=5V, CL=15pF, Ta=+25 Celsius</COMMENT>
<COMMENT>Min values are derived</COMMENT>
<TIMING>
(DELAY (ABSOLUTE
    (IOPATH A YNeg (8:17:32) (7:15:28))
  ))
</TIMING></FMFTIME>
</BODY></FTML>
```

Figure 12.1 Timing file for STD01 model

These tags

```
</REVISION.HISTORY>
</HEAD>
```

close those sections of the file.

12.1.2   Body

The next tag

```
<BODY>
```

opens the body section. The first line in the body section:

```
<TIMESCALE>1ns</TIMESCALE>
```

provides the timescale for the file. This corresponds to the timescale directive in an SDF file. Because software written to date assumes the timescale is 1 nanosecond without actually reading it, it is advisable to always use a timescale of 1 nanosecond. Timing values of less than 1 ns are written as decimals. Large values can be written as exponents. One second is 1E9.

The <MODEL> tag precedes the name of the model for which this file provides timing values.

12.1.3   FMFTIME

The next section, bracketed between the tags <FMFTIME> and </FMFTIME>, is the heart of the timing file. This section may appear one or more times in any file. In each appearance it will contain one or more lines listing a part number and the source of the data for that part number:

```
SN7401D<SOURCE>Texas Instruments SDLS026-Revised March 1988</SOURCE>
```

The part number is not necessarily the complete ordering number. It usually includes package type and timing but not temperature or packing (i.e., trays or tape and reel). Each of the consecutive part numbers must represent parts with identical timing. These parts may be different packages for similar parts from the same vendor or they may be parts from multiple vendors with identical timing specifications. The part numbers are used to select the timing values that are correct for each component instantiated in the netlist that is being backannotated.

Following the list of part numbers are optional comments regarding the test conditions under which the vendor supplied the timing values:

```
<COMMENT>The Values listed are for VCC=5V, CL=15pF, Ta=+25 Celsius</COMMENT>
<COMMENT>Min values are derived</COMMENT>
```

Comments are valuable for documentation but are not required to generate a usable SDF file. The comments often include disclaimers about data that had to be estimated because the vendor did not supply values.

Embedded within the <FMFTIME> tags is a single instance of the <TIMING> tag. Between the <TIMING> and </TIMING> tags resides the timing information for the listed part number(s) in SDF format.

```
<TIMING>
(DELAY (ABSOLUTE
    (IOPATH A YNeg (18:35:55) (4:8:15))
  ))
</TIMING>
```

The SDF data are copied without modification and inserted into the SDF file that is applied to the netlist.

The file ends with a series of closing tags: </FMFTIME></BODY></FTML>.

12.2   Separate Timing Specifications

If the vendor provides separate timing specifications for the part under different test conditions, the part's timing file may include several entries for the same part number. For example, some technologies in the 7400 series are specified at two or more operating voltages. Such parts would have multiple sections in their timing files, as shown in Figure 12.2. In this case, the primary operating voltage is appended to the part number to aid in selecting the correct set of values.

```
<FMFTIME>
74AHC244D_3V3<SOURCE>Philips Semiconductors Data Sheet 1999 Sep 28</SOURCE>
74AHC244PW_3V3<SOURCE>Philips Semiconductors Data Sheet 1999 Sep 28</SOURCE>
<COMMENT>The Values listed are for VCC=3.0V-3.6V, CL=50pF, Ta=-40 to +85 Celsius
</COMMENT>
<COMMENT>Typical values are at Ta=25C</COMMENT>
<TIMING>
(DELAY (ABSOLUTE
    (IOPATH A Y (1.0:7.0:13.5) (1.0:7.0:13.5))
    (IOPATH OENeg Y () () (1.0:10.0:16.0) (1.0:7.5:16.0) (1.0:10.0:16.0)
(1.0:7.5:16.0))
  ))
</TIMING></FMFTIME>
<FMFTIME>
74AHC244D_5V<SOURCE>Philips Semiconductors Data Sheet 1999 Sep 28</SOURCE>
74AHC244PW_5V<SOURCE>Philips Semiconductors Data Sheet 1999 Sep 28</SOURCE>
<COMMENT>The Values listed are for VCC=4.5V-5.5V, CL=50pF, Ta=-40 to +85 Celsius
</COMMENT>
<COMMENT>Typical values are at Ta=25C</COMMENT>
<TIMING>
(DELAY (ABSOLUTE
    (IOPATH A Y (1.0:5.0:8.5) (1.0:5.0:8.5))
    (IOPATH OENeg Y () () (1.0:7.0:10.5) (1.0:5.5:10.5) (1.0:7.0:10.5)
(1.0:5.5:10.5))
  ))
</TIMING></FMFTIME>
```

Figure 12.2 Timing file for part with voltage-dependent timing

12.3   Importing Timing Values

Some complex models use `tdevice` generics to import a timing value that is not directly related to a port. In the SDF file, the delays associated with these generics look as though they are being applied to a device (cell) within the model. They each have their own instance name and path. Because the path is not known at the time the timing file is created, a special variable %LABEL% is used. The `mk_sdf` script, discussed in the next section, substitutes the correct path for the variable when it generates the SDF file. Figure 12.3 shows a timing file for a FIFO model that uses `tdevice` generics.

12.4   Custom Timing Sections

It is also useful to create custom timing sections to provide a set of timing values that reflects the range of components that are allowed to be used in the manufacture of a product. Many companies require their engineers to select components available from two or more sources for each function in a design. The components

```
<MODEL>IDT72V241
<FMFTIME>
IDT72V241L10J<SOURCE>IDT data sheet February 1999</SOURCE>
IDT72V241L10PF<SOURCE>IDT data sheet February 1999</SOURCE>
<COMMENT>The Values listed are for VCC=3.0V to 3.6V, CL=30pF, Ta=0 to 70 Celsius
</COMMENT>
<COMMENT>Typical values are derived</COMMENT>
<TIMING>
   (DELAY (ABSOLUTE
      (IOPATH RCLK   Q0      (2:5:6.5) (2:5:6.5) (2:5:6.5) (0:5:6.5) (2:5:6.5)
(0:5:6.5))
      (IOPATH RSNeg  Q0      (2:7:10) (2:7:10) (2:7:10) (2:7:10) (2:7:10) (2:7:10))
      (IOPATH RSNeg  EFNeg  (2:7:10) (2:7:10))
      (IOPATH OENeg  Q0      () () (3:6:9) (0:6:9) (3:6:9) (0:6:9))
      (IOPATH WCLK   FFNeg  (2:4:6.5) (2:4:6.5))
      (IOPATH RCLK   EFNeg  (2:4:6.5) (2:4:6.5))
      (IOPATH RCLK   PAENeg (2:4:6.5) (2:4:6.5))
      (IOPATH WCLK   PAFNeg (2:4:6.5) (2:4:6.5))
   ))
   (TIMINGCHECK
      (PERIOD (posedge RCLK) (10))
      (PERIOD (posedge WCLK) (10))
      (WIDTH (posedge RCLK) (4.5))
      (WIDTH (negedge RCLK) (4.5))
      (SETUPHOLD DO  WCLK (3) (0.5))
      (SETUPHOLD REN1Neg RCLK (3) (0.5))
      (WIDTH (negedge RSNeg) (10))
      (SETUP REN1Neg RSNeg (8))
      (RECOVERY REN1Neg RSNeg (8))
   ))
   (CELL (CELLTYPE "VITALbuf")
      (INSTANCE %LABEL%/SKEW1) (DELAY  (ABSOLUTE ( DEVICE  (5) ) ) ) )
   (CELL (CELLTYPE "VITALbuf" )
      (INSTANCE %LABEL%/SKEW2) (DELAY  (ABSOLUTE ( DEVICE  (14) ) ) ) )
</TIMING></FMFTIME>
```

Figure 12.3   Timing file with `tdevice` cells

available from alternate sources frequently have slightly different timing specifications. It is common to place a custom entry in the timing file that represents the full range of timings that are specified across all vendors. Such an entry would include the fastest minimum propagation delays and the slowest maximum delays, along with the worst-case setup, hold, and pulsewidth requirements. If the design simulates without errors with those timing values, it is likely to work when built with any combination of component sources.

12.5   Generating Timing Files

A `tcl/tk` script named `vhd2ftm` (available for free download from the FMF Web site) is an example of a tool that can aid in the generation of timing files. The script reads a VHDL/VITAL model and produces a form on the screen that can be filled in. A screen shot of `vhd2ftm` using the `eclps445` model from Chapter 11 is shown in Figure 12.4. Once the form is filled, clicking the generate button writes two files. One has the timing file boilerplate and the other has the SDF data and the surrounding XML tags. It also clears the form, making it ready to enter another set of timing values for the same model.

12.6   Generating SDF Files

A single model may be interesting, but by itself is of limited utility. To be truly useful, models must be connected together in a netlist that represents a design. Netlists can be written by hand using a text editor. However, most people design boards by drawing schematics. They usually find it easier to netlist their schematic to VHDL than to write the netlist by hand. Although correct netlisting from a schematic requires integrating the models into the schematic system's library, most schematic tool vendors support this methodology.

When a component model is instantiated in a netlist, the value of its `Timing-Model` generic is set to match a part number in the model's timing file. It can be set by hand or, better yet, the value can be passed through from the schematic by the netlister.

Once a correct netlist has been created, the SDF file can be generated. Once again, this task could be done by hand, copying the appropriate sections from the model timing files into a file with the proper SDF header and references to each instance. Or you can run the perl script `mk_sdf`, which is also freely available on the FMF Web site. Some configuration is required to run `mk_sdf`, but documentation can be downloaded with the script. The configuration is primarily concerned with where in your file system you keep the timing files and how the tool can find them.

The `mk_sdf` script reads a board-level netlist and extracts the values of the `TimingModel` generics instance by instance. It then searches the timing files for part numbers that match those values. If it finds them, it copies the SDF code for those part numbers into an SDF file that correctly maps to each instance in the netlist. Instructions for using the `mk_sdf` script can be found in Chapter 15.

Figure 12.4   vhd2ftm screen shot

12.7   Backannotation and Hierarchy

The exact method of backannotation will depend on your simulator. However, some considerations will be common. The contents of an SDF file are hierarchical and based on a starting point. That starting point is the top level of the object for which the SDF file is written. For example, if you ran mk_sdf on a netlist, the SDF file starts at the point the netlist is instantiated in the testbench.

If there is a gate-level FPGA model in the netlist and it has its own SDF file, that file starts at the instantiation of the FPGA. Gate-level FPGA and ASIC netlists are generated by synthesis tools. An associated SDF file can be generated at the same time.

When you tell your simulator to read an SDF file, you must also specify where in the design the file is to be applied. In Figure 12.5 we see a typical design

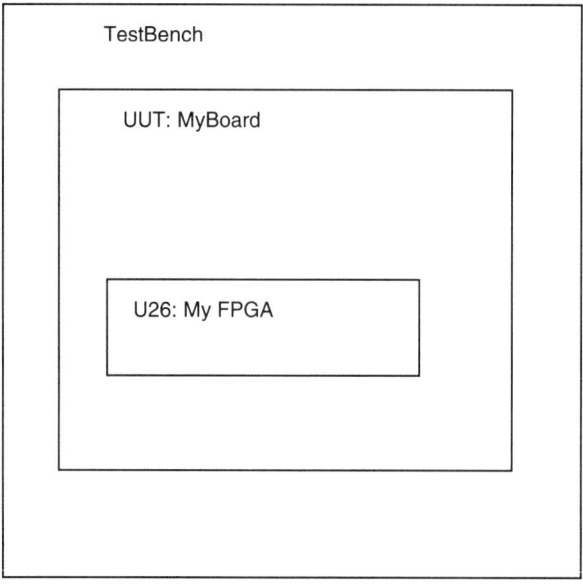

Figure 12.5   Design hierarchy

hierarchy. There is a testbench that instantiates a board design named MyBoard. The board is the Unit Under Test (in the testbench labeled UUT). The mk_sdf script has been run on the board netlist and has created an SDF file named myboard.sdf. In the board design is, among other parts, an FPGA name MyFPGA. It is instantiated with the label U26. In this case we are using a gate-level model of the FPGA. The vendor tool that output the gate-level model also output an SDF file named myfpga.sdf.

Let us assume the simulator being used is ModelSim. To start a simulation without timing (from the command line), we would issue the command

```
vsim testbench
```

The command vsim starts the simulator. The argument testbench specifies the design unit to simulate.

To simulate with SDF backannotation, a command line option is used. If we want typical timing for the components in board (other than the FPGA, which has its own SDF file), we use the command

```
vsim -sdftyp testbench/uut=myboard.sdf testbench
```

Here the option -sdftyp tells the simulator to perform a SDF backannotation using typical timing. The next argument tells it the backannotation is for instance UUT in the testbench and myboard.sdf is the name of the SDF file to be used.

To add backannotation of the FPGA with maximum timing values, the command is expanded to

```
vsim -sdftyp testbench/uut=myboard.sdf -sdfmax testbench/uut/u26=myfpga.sdf
   testbench
```

The added option -sdfmax tells the simulator to perform another backannotation using maximum timing values. This backannotation is to be applied to instance u26 in UUT (in the testbench). The SDF file to use is named myfpga.sdf.

More than one SDF file can be backannotated to a single object. If we had an SDF file with the interconnect delays for the board, it could also be applied to test-bench/UUT.

12.8   Summary

External timing files allow us to write technology-independent (timing) models. We use this modeling methodology because it reduces the number of models that must be written and maintained. There are many ways timing files could be written. This book shows the method developed by the Free Model Foundry. This method can be simply described as SDF embedded in an XML wrapper. Writing timing files can be made less tedious by using a tcl/tk program called vhd2ftm.

The external timing files are used to create an SDF file that can be read by a simulator. The preferred means of creating the SDF file is by running a perl script named mk_sdf.

If a simulation uses a backannotation file, the path to the instance being back-annotated must be supplied.

IV Advanced Modeling

Part IV takes the modeling basics learned in the previous chapters and applies them to modeling the more interesting and complex components that you will find in your board designs. In this final part we look at modeling several specific types of components, including memories and components with special features such as PLLs and bus-hold.

This part also applies what you learned about timing and generics and shows you how to use it in your FPGA RTL code. By doing this you can run board-level simulations with timing without resorting to the use of the gate-level representation of your FPGA design. Simulating can therefore be much faster than using the gate-level netlist.

Chapter 13 shows how to instantiate your FPGA RTL model in a VITAL wrapper that will add delays and timing constraints. This technique will work even if your RTL code is in Verilog.

Chapter 14 covers the issues specific to modeling memory components, including storage arrays. It looks at several techniques and shows the advantages and disadvantages of each. It also shows how to take advantage of certain features in the VITAL2000 memory package that make behavior modeling output retention easier.

Chapter 15 investigates the integration of component simulation models into the schematic capture environment. It discusses requirements for netlisting and for passing generic values from the schematic to the model. The examples here are vendor-specific but will help you understand whatever system you use.

In Chapter 16 the modeling of specific component features is discussed. The features include differential inputs, bus-hold, PLLs, and state machines. It also discusses how to use assertion statements in component models and how to modify model behavior with the `TimingModel` generic. It concludes with a discussion of modeling mixed-signal devices.

Chapter 17 covers writing testbenches for debugging and verifying your component models. It includes material on assertions and transactors.

13

Adding Timing to Your RTL Code

Now that you know how to model with timing and simulate all the off-the-shelf components on your board, wouldn't it be nice to add delays and timing constraints to the RTL model of your ASIC or FPGA rather than use the gate-level model?

These days, boards and systems are designed using schematics, and chips, ASICs, and FPGAs are designed using HDLs, usually VHDL or Verilog. From the schematic, a VHDL netlist can be generated. VHDL behavioral models of off-the-shelf components can be written based on the vendors' data sheets. If you started your chip design by writing a behavioral model, that could also be included and you could perform early simulation to test the correctness of your design specification. However, most engineers get their design specs from a text document. They start their chip designs by writing and simulating RTL code. The RTL code could be used in the board-level simulation, but without timing it might not behave in a manner indicative of the hardware you intend to build and would not accurately verify its interfaces to other components. This chapter demonstrates a technique for adding timing to your RTL code that is compatible with the other models you have created.

13.1 Using VITAL to Simulate Your RTL

Unless you have written a behavioral model of your FPGA, your only choices for board-level verification are gate level and RTL. Gate level is the most accurate, but you cannot get a gate-level model until you have completed the RTL model sufficiently to synthesize and run the vendor's place and route tool. Then, you will find that gate-level simulation is also the slowest and most memory-intensive way to verify an ASIC or FPGA.

RTL is much faster and uses much less memory than gate level. However, the RTL model has no timing information. It may not accurately describe board-level interfaces, such as differential inputs and outputs. Fortunately, you can embed the RTL in a VITAL wrapper and have the best of both worlds: RTL speed plus full timing. You can even start your simulations with the timing constraints you intend to use for synthesis to verify you are not overconstraining or underconstraining synthesis. After place and route, you can substitute timing values from the vendor's timing analyzer to reverify with realized timing.

There are other advantages to performing board-level simulation using the RTL model of your chip design:

- The interfaces between your design and the other components on the board become better understood.

- The chip design is verified earlier and without spending time on synthesis and place and route.

- The board design is verified earlier and can be released to layout sooner and with a higher degree of confidence.

- Onboard diagnostics can be developed and verified earlier in the design cycle.

Although writing a VITAL wrapper for a complex ASIC or FPGA design is not trivial, it can to some extent be automated. The bilingual capabilities of today's simulators allow this strategy to work whether the RTL code is written in VHDL or Verilog.

13.2   The Basic Wrapper

The RTL code used for synthesis describes the behavior of the component being designed but does not describe the timing. The timing is necessary to ensure the chip interfaces correctly with the rest of the system. We can add propagation delays and timing constraint checks to the RTL without changing it by instantiating it in a higher-level component. We call this higher-level component a wrapper because it does not change the functionality of the RTL model, it just wraps it in a set of path delay and timing check procedures. Figure 13.1 shows the overall structure of a VITAL wrapper.

As an example of how to create a timing wrapper, let us begin with a fictitious design named FPGA299. It is based on the 74AC299 8-bit universal shift register. This design is much smaller than a real-life FPGA but will illustrate most of the concepts. To better understand the design, a pin description list is shown in Table 13.1.

Table 13.1   FPGA299 pin description

Pin Names	Description
CLK	Clock Pulse Input (Active Edge Rising)
SR	Serial Data Input for Right Shift
SL	Serial Data Input for Left Shift
S0, S1	Mode Select Inputs
CLR_L	Asynchronous Master Reset Input (Active Low)
OE1_L, OE2_L	3-State Output Enable Inputs (Active Low)
IO	8-Bit Parallel Data Inputs or 3-State Parallel Outputs
Q0, Q7	Serial Outputs

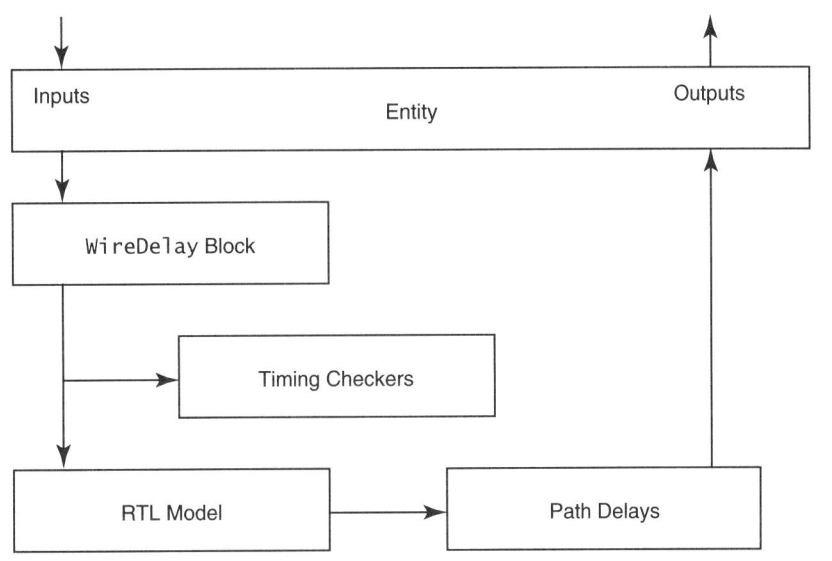

Figure 13.1   Structural view of a wrapper

We first look at a VHDL version of the design. The entity is

```
ENTITY fpga299 IS
    PORT (
            CLR_L        : IN     std_logic;
            OE1_L        : IN     std_logic;
            OE2_L        : IN     std_logic;
            S0           : IN     std_logic;
            S1           : IN     std_logic;
            CLK          : IN     std_logic;
            IO           : INOUT  std_logic_vector(7 downto 0);
            Q0           : OUT    std_logic;
            Q7           : OUT    std_logic;
            SL           : IN     std_logic;
            SR           : IN     std_logic
        );
    END fpga299;
```

In order to avoid naming conflicts, at the board level (in the schematic) we will call this component `chip299`.

We begin by writing an entity that will be compatible with the schematic symbol that represents our FPGA design at the board level. The wrapper will be a non-compliant VITAL model, so the file begins with LIBRARY and USE clauses that include the following VITAL packages:

```
-------------------------------------------------------------------------------
-- File Name: chip299.vhd
-------------------------------------------------------------------------------
-- Description: Timing wrapper for fpga299
-------------------------------------------------------------------------------
LIBRARY IEEE;    USE IEEE.std_logic_1164.ALL;
                 USE IEEE.VITAL_timing.ALL;
                 USE IEEE.VITAL_primitives.ALL;
LIBRARY FMF;     USE FMF.gen_utils.ALL;


-------------------------------------------------------------------------------
-- ENTITY DECLARATION
-------------------------------------------------------------------------------
ENTITY chip299 IS
  GENERIC (
    -- tipd delays: interconnect path delays
    tipd_CLRNeg             : VitalDelayType01 := VitalZeroDelay01;
    tipd_OE1Neg             : VitalDelayType01 := VitalZeroDelay01;
    tipd_OE2Neg             : VitalDelayType01 := VitalZeroDelay01;
    tipd_S0                 : VitalDelayType01 := VitalZeroDelay01;
    tipd_S1                 : VitalDelayType01 := VitalZeroDelay01;
    tipd_CLK                : VitalDelayType01 := VitalZeroDelay01;
    tipd_IO0                : VitalDelayType01 := VitalZeroDelay01;
    tipd_IO1                : VitalDelayType01 := VitalZeroDelay01;
    tipd_IO2                : VitalDelayType01 := VitalZeroDelay01;
    tipd_IO3                : VitalDelayType01 := VitalZeroDelay01;
    tipd_IO4                : VitalDelayType01 := VitalZeroDelay01;
    tipd_IO5                : VitalDelayType01 := VitalZeroDelay01;
    tipd_IO6                : VitalDelayType01 := VitalZeroDelay01;
    tipd_IO7                : VitalDelayType01 := VitalZeroDelay01;
    tipd_SL                 : VitalDelayType01 := VitalZeroDelay01;
    tipd_SR                 : VitalDelayType01 := VitalZeroDelay01;
```

The GENERIC list starts with the tipd generics for the interconnect delays. This allows us to backannotate the PCB wire delays. These delays are particularly important for designs that incorporate an internal phase locked loop with a feedback path that is on the board.

Next come the rest of the timing generics for pin-to-pin delays, setup, hold, pulsewidth, and any other timing constraints, along with the control parameters:

```
    -- tpd delays
    tpd_CLRNeg_IO0          : VitalDelayType01  := UnitDelay01;
    tpd_CLRNeg_Q0           : VitalDelayType01  := UnitDelay01;
    tpd_OE1Neg_IO0          : VitalDelayType01Z := UnitDelay01Z;
    tpd_S0_IO0              : VitalDelayType01Z := UnitDelay01Z;
    tpd_CLK_IO0             : VitalDelayType01  := UnitDelay01;
    tpd_CLK_Q0              : VitalDelayType01  := UnitDelay01;
```

```
-- tsetup values: setup times
tsetup_S0_CLK              : VitalDelayType := UnitDelay;
tsetup_SL_CLK              : VitalDelayType := UnitDelay;
tsetup_IO0_CLK             : VitalDelayType := UnitDelay;
-- thold values: hold times
thold_S0_CLK               : VitalDelayType := UnitDelay;
thold_SL_CLK               : VitalDelayType := UnitDelay;
thold_IO0_CLK              : VitalDelayType := UnitDelay;
-- trecovery values: release times
trecovery_CLRNeg_CLK       : VitalDelayType := UnitDelay;
-- tpw values: pulse widths
tpw_CLK_posedge            : VitalDelayType := UnitDelay;
tpw_CLK_negedge            : VitalDelayType := UnitDelay;
tpw_CLRNeg_negedge         : VitalDelayType := UnitDelay;
-- tperiod_min: minimum clock period = 1/max freq
tperiod_CLK_posedge        : VitalDelayType := UnitDelay;
-- generic control parameters
InstancePath      : STRING      := DefaultInstancePath;
TimingChecksOn    : BOOLEAN     := DefaultTimingChecks;
MsgOn             : BOOLEAN     := DefaultMsgOn;
XOn               : BOOLEAN     := DefaultXon;
-- For FMF SDF technology file usage
TimingModel       : STRING      := DefaultTimingModel
);
```

The generics and port list, indeed the entity, are written as if this was a model of an off-the-shelf component.

```
PORT (
    CLRNeg       : IN        std_ulogic := 'U';
    OE1Neg       : IN        std_ulogic := 'U';
    OE2Neg       : IN        std_ulogic := 'U';
    S0           : IN        std_ulogic := 'U';
    S1           : IN        std_ulogic := 'U';
    CLK          : IN        std_ulogic := 'U';
    IO0          : INOUT     std_ulogic := 'U';
    IO1          : INOUT     std_ulogic := 'U';
    IO2          : INOUT     std_ulogic := 'U';
    IO3          : INOUT     std_ulogic := 'U';
    IO4          : INOUT     std_ulogic := 'U';
    IO5          : INOUT     std_ulogic := 'U';
    IO6          : INOUT     std_ulogic := 'U';
    IO7          : INOUT     std_ulogic := 'U';
    Q0           : OUT       std_ulogic := 'U';
    Q7           : OUT       std_ulogic := 'U';
```

```
            SL              : IN        std_ulogic := 'U';
            SR              : IN        std_ulogic := 'U'
        );
        ATTRIBUTE VITAL_LEVEL0 of chip299 : ENTITY IS TRUE;
    END chip299;
```

The entity has a VITAL_LEVEL0 attribute set to TRUE.

Note that the FPGA has eight pins of mode INOUT. These present a special challenge. They are not a problem in a component model because the control logic that determines whether they are acting as drivers or receivers is in the model. But in a wrapper, that control logic is in the instantiated RTL and is not visible to the wrapper. A method for overcoming this obstacle of unknown directionality uses the std_logic_1164 resolution function and will be explained as we progress through the code.

The architecture of the wrapper is not VITAL compliant. Therefore, the VITAL attribute is omitted.

```
    -------------------------------------------------------------------------
    -- ARCHITECTURE DECLARATION
    -------------------------------------------------------------------------
    ARCHITECTURE vhdl_behavioral of chip299 IS

        CONSTANT partID         : STRING := "chip299";

    COMPONENT fpga299
        PORT (
            CLR_L           : IN     std_logic;
            OE1_L           : IN     std_logic;
            OE2_L           : IN     std_logic;
            S0              : IN     std_logic;
            S1              : IN     std_logic;
            CLK             : IN     std_logic;
            IO              : INOUT  std_logic_vector(7 downto 0);
            Q0              : OUT    std_logic;
            Q7              : OUT    std_logic;
            SL              : IN     std_logic;
            SR              : IN     std_logic
        );
    END COMPONENT;
```

The RTL model is declared as a component that will be instantiated further down in the wrapper. Then signals are declared:

```
    SIGNAL CLRNeg_ipd           : std_ulogic := 'U';
    SIGNAL OE1Neg_ipd           : std_ulogic := 'U';
    SIGNAL OE2Neg_ipd           : std_ulogic := 'U';
    SIGNAL S0_ipd               : std_ulogic := 'U';
    SIGNAL S1_ipd               : std_ulogic := 'U';
```

```
SIGNAL CLK_ipd            : std_ulogic := 'U';
SIGNAL IO0_ipd            : std_ulogic := 'U';
SIGNAL IO1_ipd            : std_ulogic := 'U';
SIGNAL IO2_ipd            : std_ulogic := 'U';
SIGNAL IO3_ipd            : std_ulogic := 'U';
SIGNAL IO4_ipd            : std_ulogic := 'U';
SIGNAL IO5_ipd            : std_ulogic := 'U';
SIGNAL IO6_ipd            : std_ulogic := 'U';
SIGNAL IO7_ipd            : std_ulogic := 'U';
SIGNAL SL_ipd             : std_ulogic := 'U';
SIGNAL SR_ipd             : std_ulogic := 'U';
SIGNAL IO_w               : std_logic_vector(7 downto 0) := (others => 'Z');
```

A configuration specification is required and supplied:

```
FOR ALL : fpga299 USE ENTITY work.fpga299(rtl);
```

The standard wire delay block is employed to apply interconnect delays to the input pins.

```
BEGIN
  ----------------------------------------------------------------------------
  -- Wire Delays
  ----------------------------------------------------------------------------
  WireDelay : BLOCK
  BEGIN
       w_1 : VitalWireDelay (CLRNeg_ipd, CLRNeg, tipd_CLRNeg);
       w_2 : VitalWireDelay (OE1Neg_ipd, OE1Neg, tipd_OE1Neg);
       w_3 : VitalWireDelay (OE2Neg_ipd, OE2Neg, tipd_OE2Neg);
       w_4 : VitalWireDelay (S0_ipd, S0, tipd_S0);
       w_5 : VitalWireDelay (S1_ipd, S1, tipd_S1);
       w_6 : VitalWireDelay (CLK_ipd, CLK, tipd_CLK);
       w_8 : VitalWireDelay (IO0_ipd, IO0, tipd_IO0);
       w_9 : VitalWireDelay (IO1_ipd, IO1, tipd_IO1);
       w_10 : VitalWireDelay (IO2_ipd, IO2, tipd_IO2);
       w_11 : VitalWireDelay (IO3_ipd, IO3, tipd_IO3);
       w_12 : VitalWireDelay (IO4_ipd, IO4, tipd_IO4);
       w_13 : VitalWireDelay (IO5_ipd, IO5, tipd_IO5);
       w_14 : VitalWireDelay (IO6_ipd, IO6, tipd_IO6);
       w_15 : VitalWireDelay (IO7_ipd, IO7, tipd_IO7);
       w_18 : VitalWireDelay (SL_ipd, SL, tipd_SL);
       w_19 : VitalWireDelay (SR_ipd, SR, tipd_SR);
  END BLOCK;
```

Getting back to the unknown directionality problem of the INOUT ports, we need a way to tell whether the port is being driven from the RTL model or an external device. We know the RTL model will only output strong signals, '0', '1', and 'Z'. So the next group of assignments takes the external inputs and converts them to weak signals, 'L', 'H', and 'Z'.

```
IO_w(0)  <= To_UXLHZ(IO0_ipd);
IO_w(1)  <= To_UXLHZ(IO1_ipd);
IO_w(2)  <= To_UXLHZ(IO2_ipd);
IO_w(3)  <= To_UXLHZ(IO3_ipd);
IO_w(4)  <= To_UXLHZ(IO4_ipd);
IO_w(5)  <= To_UXLHZ(IO5_ipd);
IO_w(6)  <= To_UXLHZ(IO6_ipd);
IO_w(7)  <= To_UXLHZ(IO7_ipd);
```

The assignments use a function found in the FMF.gen_utils package named To_UXLHZ. These function calls result in the assignment of weak signal values to each bit of the signal IO_w.

The RTL model has a vectored port. Most FPGA and ASIC models will have several vectored ports. However, at the board level the wrapper has scalar ports. A block statement is used to make the scalar-to-vector conversion that enables the advantageous use of vectored signals:

```
------------------------------------------------------------------------------
-- Main Behavior Block
------------------------------------------------------------------------------
Behavior : BLOCK

PORT(
        CLRIn       : IN      std_logic;
        OE1In       : IN      std_logic;
        OE2In       : IN      std_logic;
        S0In        : IN      std_logic;
        S1In        : IN      std_logic;
        CLkIn       : IN      std_logic;
        IOIn        : IN      std_logic_vector(7 downto 0);
        IOOut       : OUT     std_logic_vector(7 downto 0);
        Q0Out       : OUT     std_logic;
        Q7Out       : OUT     std_logic;
        SLIn        : IN      std_logic;
        SRIn        : IN      std_logic
);
PORT MAP (
        CLRIn  => CLRNeg_ipd,
        OE1In  => OE1Neg_ipd,
        OE2In  => OE2Neg_ipd,
        S0In   => S0_ipd,
        S1In   => S1_ipd,
        CLKIn  => CLK_ipd,
```

```
            IOIn  => IO_w,
            IOOut(0)  => IO0,
            IOOut(1)  => IO1,
            IOOut(2)  => IO2,
            IOOut(3)  => IO3,
            IOOut(4)  => IO4,
            IOOut(5)  => IO5,
            IOOut(6)  => IO6,
            IOOut(7)  => IO7,
            Q0Out  => Q0,
            Q7Out  => Q7,
            SLIn  => SL_ipd,
            SRIn  => SR_ipd
        );
```

Within the block, all INOUT ports are split into separate IN and OUT ports. This provides some simplification.

The zero delay signals are declared:

```
    SIGNAL IO_zd     : std_logic_vector(7 downto 0);
    SIGNAL Q0_zd     : std_ulogic;
    SIGNAL Q7_zd     : std_ulogic;
```

Because they are used both inside and outside the timing process, they are declared as signals in a wrapper instead of as variables, as would usually be the case in a component model.

Once all the declarations are completed the RTL model is instantiated:

```
    BEGIN
        -- connect VHDL RTL model
        fpga299_1 : fpga299
        PORT MAP(
            CLR_L     => CLRIn,
            OE1_L     => OE1In,
            OE2_L     => OE2In,
            S0        => S0In,
            S1        => S1In,
            CLK       => CLKIn,
            IO        => IO_w,
            Q0        => Q0_zd,
            Q7        => Q7_zd,
            SL        => SLIn,
            SR        => SRIn
        );
```

Input ports are mapped to the block IN ports and outputs are mapped to the block OUT ports. The IO port, of mode INOUT in the RTL model, is mapped to the IO_w signal created earlier. Values being read from the outside on this signal will always be weak, and values being driven from the RTL will always be strong. Therefore, any value originating from the RTL model will override the value coming from outside, unless the RTL model is driving 'Z'. This is an important characteristic and will be exploited further down the code.

Now comes the main process. In a model this would be the behavioral process. In a wrapper, all the behavior is described in the RTL model, so there is little left for this process other than timing constraints and path delays.

As always, the process begins with a sensitivity list and variable declarations:

```
------------------------------------------------------------------------------
-- Main Behavior Process
------------------------------------------------------------------------------
TIMING : PROCESS (CLKIn, CLRIn, OE1In, OE2In, S0In, S1In, IO_w, SLIn, SRIn)

    -- Timing Check Variables
    VARIABLE Tviol_SL_CLK     : X01 := '0';
    VARIABLE TD_SL_CLK        : VitalTimingDataType;

    VARIABLE Tviol_SR_CLK     : X01 := '0';
    VARIABLE TD_SR_CLK        : VitalTimingDataType;

    VARIABLE Tviol_S0_CLK     : X01 := '0';
    VARIABLE TD_S0_CLK        : VitalTimingDataType;

    VARIABLE Tviol_S1_CLK     : X01 := '0';
    VARIABLE TD_S1_CLK        : VitalTimingDataType;

    VARIABLE Tviol_IO_CLK     : X01 := '0';
    VARIABLE TD_IO_CLK        : VitalTimingDataType;

    VARIABLE Rviol_CLRNeg_CLK : X01 := '0';
    VARIABLE TD_CLRNeg_CLK    : VitalTimingDataType;

    VARIABLE PD_CLK           : VitalPeriodDataType := VitalPeriodDataInit;
    VARIABLE Pviol_CLK        : X01 := '0';
    VARIABLE PD_CLRNeg        : VitalPeriodDataType := VitalPeriodDataInit;
    VARIABLE Pviol_CLRNeg     : X01 := '0';

    VARIABLE Violation        : X01 := '0';
    -- Output Glitch Detection Variables
    VARIABLE Q0_GlitchData    : VitalGlitchDataType;
    VARIABLE Q7_GlitchData    : VitalGlitchDataType;
```

They are followed by the timing check section with its constraint procedure calls:

```
BEGIN
    ------------------------------------------------------------------------------
    -- Timing Check Section
    ------------------------------------------------------------------------------
    IF (TimingChecksOn) THEN
```

```
VitalSetupHoldCheck (
      TestSignal              => S0In,
      TestSignalName          => "S0",
      RefSignal               => CLKIn,
      RefSignalName           => "CLK",
      SetupHigh               => tsetup_S0_CLK,
      SetupLow                => tsetup_S0_CLK,
      HoldHigh                => thold_S0_CLK,
      HoldLow                 => thold_S0_CLK,
      CheckEnabled            => true,
      RefTransition           => '/',
      HeaderMsg               => InstancePath & partID,
      TimingData              => TD_S0_CLK,
      XOn                     => XOn,
      MsgOn                   => MsgOn,
      Violation               => Tviol_S0_CLK
);

VitalSetupHoldCheck (
      TestSignal              => S1In,
      TestSignalName          => "S1",
      RefSignal               => CLKIn,
      RefSignalName           => "CLK",
      SetupHigh               => tsetup_S0_CLK,
      SetupLow                => tsetup_S0_CLK,
      HoldHigh                => thold_S0_CLK,
      HoldLow                 => thold_S0_CLK,
      CheckEnabled            => true,
      RefTransition           => '/',
      HeaderMsg               => InstancePath & partID,
      TimingData              => TD_S1_CLK,
      XOn                     => XOn,
      MsgOn                   => MsgOn,
      Violation               => Tviol_S1_CLK
);

VitalSetupHoldCheck (
      TestSignal              => SLIn,
      TestSignalName          => "SL",
      RefSignal               => CLKIn,
      RefSignalName           => "CLK",
      SetupHigh               => tsetup_SL_CLK,
      SetupLow                => tsetup_SL_CLK,
      HoldHigh                => thold_SL_CLK,
      HoldLow                 => thold_SL_CLK,
```

```
        CheckEnabled            => true,
        RefTransition           => '/',
        HeaderMsg               => InstancePath & partID,
        TimingData              => TD_SL_CLK,
        XOn                     => XOn,
        MsgOn                   => MsgOn,
        Violation               => Tviol_SL_CLK
    );

VitalSetupHoldCheck (
        TestSignal              => SRIn,
        TestSignalName          => "SR",
        RefSignal               => CLKIn,
        RefSignalName           => "CLK",
        SetupHigh               => tsetup_SL_CLK,
        SetupLow                => tsetup_SL_CLK,
        HoldHigh                => thold_SL_CLK,
        HoldLow                 => thold_SL_CLK,
        CheckEnabled            => true,
        RefTransition           => '/',
        HeaderMsg               => InstancePath & partID,
        TimingData              => TD_SR_CLK,
        XOn                     => XOn,
        MsgOn                   => MsgOn,
        Violation               => Tviol_SR_CLK
    );

VitalSetupHoldCheck (
        TestSignal              => IO_w,
        TestSignalName          => "IO",
        RefSignal               => CLKIn,
        RefSignalName           => "CLK",
        SetupHigh               => tsetup_IO0_CLK,
        SetupLow                => tsetup_IO0_CLK,
        HoldHigh                => thold_IO0_CLK,
        HoldLow                 => thold_IO0_CLK,
        CheckEnabled            => true,
        RefTransition           => '/',
        HeaderMsg               => InstancePath & partID,
        TimingData              => TD_IO_CLK,
        XOn                     => XOn,
        MsgOn                   => MsgOn,
        Violation               => Tviol_IO_CLK
    );
```

```
VitalRecoveryRemovalCheck (
        TestSignal          => CLRIn,
        TestSignalName      => "CLRNeg",
        RefSignal           => CLKIn,
        RefSignalName       => "CLK",
        Recovery            => trecovery_CLRNeg_CLK,
        ActiveLow           => TRUE,
        CheckEnabled        => TRUE,
        RefTransition       => '/',
        HeaderMsg           => InstancePath & partID,
        TimingData          => TD_CLRNeg_CLK,
        XOn                 => XOn,
        MsgOn               => MsgOn,
        Violation           => Rviol_CLRNeg_CLK
);
VitalPeriodPulseCheck (
        TestSignal          => CLKIn,
        TestSignalName      => "CLK_ipd",
        Period              => tperiod_CLK_posedge,
        PulseWidthHigh      => tpw_CLK_posedge,
        PulseWidthLow       => tpw_CLK_negedge,
        CheckEnabled        => TRUE,
        HeaderMsg           => InstancePath & partID,
        PeriodData          => PD_CLK,
        XOn                 => XOn,
        MsgOn               => MsgOn,
        Violation           => Pviol_CLK
);
VitalPeriodPulseCheck (
        TestSignal          => CLRIn,
        TestSignalName      => "CLRNeg",
        PulseWidthLow       => tpw_CLRNeg_negedge,
        CheckEnabled        => TRUE,
        HeaderMsg           => InstancePath & partID,
        PeriodData          => PD_CLRNeg,
        XOn                 => XOn,
        MsgOn               => MsgOn,
        Violation           => Pviol_CLRNeg
);
    END IF;
```

Because **IO_w** was used in its vector form in its setup/hold check, only one procedure call was required, rather than eight.

The final element in supporting the IO bus is the loop:

```
FOR i IN IO_w'range LOOP
    IF IO_w(i) = '1' OR IO_w(i) = '0' THEN
        IO_zd(i) <= IO_w(i);
    ELSE
        IO_zd(i) <= 'Z';
    END IF;
END LOOP;
```

In the loop the strength of each bit in IO_w is tested. If the bit has a strong value it must be coming from the RTL model, so it is assigned to IO_zd. If it has a weak value the RTL model is driving 'Z', so IO_zd gets 'Z'.

What remains of the wrapper is the path delays. The two scalar outputs have their path delay procedure calls within the timing process:

```
-------------------------------------------------------------------------------
-- Path Delay Section
-------------------------------------------------------------------------------
VitalPathDelay01 (
    OutSignal          => Q0Out,
    OutSignalName      => "Q0",
    OutTemp            => Q0_zd,
    Paths =>             (
        0 => (InputChangeTime   => CLRIn'LAST_EVENT,
              PathDelay         => tpd_CLRNeg_Q0,
              PathCondition     => true),
        1 => (InputChangeTime   => CLKIn'LAST_EVENT,
              PathDelay         => tpd_CLK_Q0,
              PathCondition     => true ) ),
    GlitchData         => Q0_GlitchData );

VitalPathDelay01 (
    OutSignal          => Q7Out,
    OutSignalName      => "Q7",
    OutTemp            => Q7_zd,
    Paths              => (
        0 => (InputChangeTime   => CLRIn'LAST_EVENT,
              PathDelay         => tpd_CLRNeg_Q0,
              PathCondition     => true),
        1 => (InputChangeTime   => CLKIn'LAST_EVENT,
              PathDelay         => tpd_CLK_Q0,
              PathCondition     => true ) ),
    GlitchData         => Q7_GlitchData );

END PROCESS;
```

In order to reduce the code size, the IO bus has its path delay in a generate statement outside the process:

```
IO_OUT : FOR i IN IO_zd'range GENERATE
    PROCESS(IO_zd(i))
        VARIABLE IO_GlitchData  : VitalGlitchDataType;


    BEGIN
    VitalPathDelay01Z (
        OutSignal       => IOOut(i),
        OutSignalName   => "IO",
        OutTemp         => IO_zd(i),
        Paths           => (
            0 => (InputChangeTime => CLRIn'LAST_EVENT,
                    PathDelay       =>
                                VitalExtendToFillDelay(tpd_CLRNeg_IOO),
                    PathCondition   => true),
            1 => (InputChangeTime => CLKIn'LAST_EVENT,
                    PathDelay       =>
                                VitalExtendToFillDelay(tpd_CLK_IOO),
                    PathCondition   => true ),
            2 => (InputChangeTime => OE1In'LAST_EVENT,
                    PathDelay       => tpd_OE1Neg_IOO,
                    PathCondition   => true ),
            3 => (InputChangeTime => OE2In'LAST_EVENT,
                    PathDelay       => tpd_OE1Neg_IOO,
                    PathCondition   => true ),
            4 => (InputChangeTime => S0In'LAST_EVENT,
                    PathDelay       => tpd_S0_IOO,
                    PathCondition   => true ),
            5 => (InputChangeTime => S1In'LAST_EVENT,
                    PathDelay       => tpd_S0_IOO,
                    PathCondition   => true ) ),
        GlitchData      => IO_GlitchData );
    END PROCESS;
END GENERATE IO_OUT;


    END BLOCK;


END vhdl_behavioral;
```

This is a relatively simple wrapper for a very small FPGA. Still, it came to 480 lines. Do not despair, as most of it can be generated by a perl script or can be cut and pasted from other models.

13.3   A Wrapper for Verilog RTL

It is possible your RTL model is written in Verilog. Verilog works fine for synthesis, but you may find VHDL is better supported for board-level verification. If so, don't worry. The exact same wrapper will work for Verilog RTL. As long as you have a bilingual simulator and you make the port names in the component declaration match those in your Verilog module, there is no difference:

```
COMPONENT fpga299
    PORT (
            CLR_L           : IN      std_logic;
            OE1_L           : IN      std_logic;
            OE2_L           : IN      std_logic;
            S0              : IN      std_logic;
            S1              : IN      std_logic;
            CLK             : IN      std_logic;
            IO              : INOUT   std_logic_vector(7 downto 0);
            Q0              : OUT     std_logic;
            Q7              : OUT     std_logic;
            SL              : IN      std_logic;
            SR              : IN      std_logic
        );

    END COMPONENT;
```

Just make sure to compile the Verilog into the same work library as the wrapper.

Some simulation packages have utilities to make the job easier. ModelSim, for example, has a utility called **vgencomp** that will read your compiled Verilog model and generate a matching VHDL component declaration.

13.4   Modeling Delays in Designs with Internal Clocks

In the example wrapper, all path delays were timed to a transition on an input port. This will not work for many modern ASIC/FPGA designs. Many designs today take advantage of internal DLLs, PLLs, or clock multipliers. If an output is clocked by an internally generated signal, another strategy is needed for correctly delaying the output signals.

In most cases the FPGA timing tool will report the delay of a signal relative to some input. However, that input may not be the signal that actually clocks the output. The following code is from the tco section of a timing results file (.tao) file from an Altera design:

```
Path Number       : 175
Slack             : 0.008 ns
Required tco      : 4.000 ns
Actual tco        : 3.992 ns
```

```
Source Name        : mDFB:DFB|mDRAM80:mDFBMem|mDRAM80IO:mDRAM80IO|cke
Destination Name   : dfb_cke
Source Clock Name  : acqdet_clkin
```

Although the source clock is said to be acqdet_clkin, in this design it is really a PLL-generated clock based on acqdet_clkin but running at twice the frequency.

Because an RTL model has no internal delays, that delay value can be applied to the signal referencing only itself. The following code shows a path delay statement that references a single signal:

```
VitalPathDelay01 (
      OutSignal         => DFBCKEO,
      OutSignalName     => "DFBCKEO",
      OutTemp           => DFBCKEO_zd,
      GlitchData        => DFBCKEO_GlitchData,
      XOn               => XOn,
      MsgOn             => MsgOn,
      Paths             => (
            0 => (InputChangeTime   => DFBCKEO_zd'LAST_EVENT,
                  PathDelay         => tpd_ACQDETCLKIN_DFBCKEO,
                  PathCondition     => TRUE)
      )
   );
```

In this procedure call, DFBCKEO_zd is the input signal and the event on which the delay will be based. If the InputChangeTime was specified as ACQDETCLKIN¢LAST_ EVENT, incorrect outputs would result.

The generic tpd_ACQDETCLKIN_DFBCKEO is backannotated through SDF with a value of 3.992 ns. It is not possible to use internal states to control path selection in a wrapper.

13.5 Caveats

The implementation of timing constraints in a wrapper is subject to some restrictions. First, internal state cannot be used to enable or disable timing checks. They also may not be used in the selection of timing paths. Likewise, internal signals are not accessible for use in timing checks or path selection.

In many models it is possible to use the violation flags to control some aspect of a model's behavior, such as corrupting memory if there is a violation during a write cycle. Similar opportunities may not exist in a wrapper.

In ASIC/FPGA design, it is common that differential inputs and/or outputs are handled by I/O cell selection. The differential signals do not appear as ports in the RTL model. However, the differential I/O will be included in the schematic. In such cases the wrapper must serve as a mediator between the schematic and the RTL model.

For differential outputs the solution is as simple as generating the compliment through inversion:

```
DFBCK1N_zd  <= not(DFBCK1int);
DFBCK1_zd  <= DFBCK1int;
DFBCK0N_zd  <= not(DFBCK0int);
DFBCK0_zd  <= DFBCK0int;
```

The solutions range from just passing through one of the two signals to more complex schemes that check that both signals are active and compliment each other. The best solution depends on the design and verification requirements.

13.6   Summary

The complexity of the interfaces of today's ASICs and FPGAs make board-level simulation desirable if not a necessity. Gate-level simulation is too slow and occurs too late in the design process to fill this need. In this chapter, a method has been presented that allows timing constraints and propagation delays to be wrapped around the RTL code of an ASIC/FPGA so they may be used in board-level verification before chip-level place and route has been performed.

The method closely resembles the modeling of an off-the-shelf component, except the RTL model is instantiated instead of coding a behavioral model. Special care must be taken in dealing with ports of mode INOUT and with outputs clocked by internally generated signals.

14 Modeling Memories

Memories are among the most frequently modeled components. How they are modeled can determine not just the performance, but the very practicality of board-level simulation.

Many boards have memory on them, and FPGA designs frequently interface to one or more types of memory. These are not the old asynchronous static RAMs of more innocent times. These memories are pipelined Zero Bus Turnaround (ZBT) synchronous static RAMs (SSRAM), multibanked Synchronous Dynamic RAMs (SDRAM), or Double Data Rate (DDR) DRAMs. The list goes on and complexities go up. Verify the interfaces or face the consequences.

Just as the memories have become complex, so have the models. There are several issues specific to memory models. How they are dealt with will determine the accuracy, performance, and resource requirements of the models.

14.1   Memory Arrays

There are a number of ways memory arrays can be modeled. The most obvious and commonly used is an array of bits. This is the method that most closely resembles the way a memory component is constructed. Because the model's ports are of type `std_ulogic`, we can create an array of type `std_logic_vector` for our memory:

```
TYPE MemStore IS ARRAY (0 to 255) OF STD_LOGIC_VECTOR(7 DOWNTO 0);
```

This has the advantage of allowing reads and writes to the array without any type conversions.

However, using an `std_logic_vector` array is expensive in terms of simulation memory. `Std_logic` is a 9-value type, which is more values than we need or can use. A typical VHDL simulator will use 1 B of simulation memory for each `std_logic` bit. A 1 megabit memory array will consume about 1 MB of computer memory. At that rate, the amount of memory in a design can be too large to simulate.

In real hardware, memory can contain only 1s and 0s. That might suggest the use of type `bit_vector`, but the point of simulation is to debug and verify a design in an environment that makes it easier than debugging real hardware. It is useful if a read from a memory location that has never been written to gives a unique result. Although real hardware may contain random values, 'U's are more informative for simulation because they make it easy to see that an uninitialized location has been accessed. Likewise, if a timing violation occurs during a memory write, the simulation model usually emits a warning message. Ideally that location should also contain an invalid word that is recognizable as such. On reading that corrupt location, the user should see 'X's.

So it seems type UX01 would provide all the values required. We could declare our memory array:

```
TYPE MemStore IS ARRAY (0 to 255) OF UX01_VECTOR(7 DOWNTO 0);
```

Unfortunately, because UX01 is a subtype of `std_logic`, most simulators use the same amount of space to store type UX01 as they do to store `std_logic`.

14.1.1   The Shelor Method

One option for modeling memories is the Shelor Method. In 1996, Charles Shelor wrote an article in the *VHDL Times* as part of his VHDL Designer column [7], in which he discussed the problem of modeling large memories. He described several possible storage mechanisms. The method presented here is the one he favored.

Shelor noted that by converting the vector to a number of type natural we can store values in much less space. Of course, there are limitations. The largest integer guaranteed by the VHDL standard is $2^{31} - 1$, meaning a 30-bit word is the most you can safely model. It turns out this is not a problem. Few memory components use word sizes larger than 18 bits and most are either 8 or 9 bits wide.

So a range of 0 to 255 is sufficient for an 8-bit word, but that assumes every bit is either a 1 or a 0. It would be good to also allow words to be uninitialized or corrupted. To do so, just extend the range down to –2.

A generic memory array declaration is

```
-- Memory array declaration
TYPE MemStore IS ARRAY (0 to TotalLOC) OF INTEGER
                RANGE -2 TO MaxData;
```

where –2 is an uninitialized word, –1 is a corrupt word, and 0 to MaxData are valid data. This method does not lend itself to manipulation of individual bits, but that is rarely called for in a component model.

Simulators tested store integers in 4 B, so each word of memory, up to 18 bits, will occupy 4 B of simulator memory. This is a considerable improvement over using arrays of `std_logic_vector`.

```
                  -- VITAL Memory Declaration
                  VARIABLE Memdat : VitalMemoryDataType :=
                      VitalDeclareMemory (
                          NoOfWords              => TotalLOC,
                          NoOfBitsPerWord        => DataWidth,
                          NoOfBitsPerSubWord     => DataWidth,
                          MemoryLoadFile         => MemLoadFileName,
                          BinaryLoadFile         => FALSE
                      );
```

Figure 14.1 VITAL2000 memory array declaration

14.1.2 The VITAL_Memory Package

Another option is to use the VITAL_Memory package released with VITAL2000. This package has an extensive array of features specific to memory modeling, including a method of declaring a memory array that results in a specific form of storage. In Figure 14.1, a memory array using the VITAL2000 memory package is declared.

In Figure 14.1 a procedure call is used to create the memory array.

The storage efficiency is very good, a 1 B word occupies only 2 B of memory, but this holds true only for 8-bit words. A 9-bit word occupies 4 B of memory, which is the same as the Shelor method.

14.2 Modeling Memory Functionality

There are two distinct ways of modeling memory functionality. One is to use standard VHDL behavioral modeling methods. The other is to use the VITAL2000 memory package. Some features of the memory package can be used in behavioral models.

Let us look at how to model a generic SRAM using each method. The component modeled is a 4 MB SRAM with an 8-bit word width.

14.2.1 Using the Behavioral (Shelor) Method

The model entity has the same general features as previous models. It begins with copyright, history, description, and library declarations:

```
----------------------------------------------------------------------
-- File Name: sram4m8.vhd
----------------------------------------------------------------------
-- Copyright (C) 2001 Free Model Foundry; http://vhdl.org/fmf/
--
-- This program is free software; you can redistribute it and/or modify
-- it under the terms of the GNU General Public License version 2 as
```

```
            -- published by the Free Software Foundation.
            --
            -- MODIFICATION HISTORY:
            --
            -- version: |   author:  |   mod date:  |  changes made:
            --   V1.0       R. Munden     01 MAY 27      Initial release
            --
            -------------------------------------------------------------------------------
            -- PART DESCRIPTION:
            --
            -- Library:     MEM
            -- Technology:  not ECL
            -- Part:        SRAM4M8
            --
            -- Description: 4M X 8 SRAM
            -------------------------------------------------------------------------------

            LIBRARY IEEE;   USE IEEE.std_logic_1164.ALL;
                            USE IEEE.VITAL_timing.ALL;
                            USE IEEE.VITAL_primitives.ALL;
            LIBRARY FMF;    USE FMF.gen_utils.ALL;
                            USE FMF.conversions.ALL;
```

The library declarations include the FMF conversions library. It is needed for converting between `std_logic_vector` and integer types.

The generic declarations contain the usual interconnect, path delay, setup and hold, and pulse width generics, as well as the usual control parameters:

```
            -------------------------------------------------------------------------------
            -- ENTITY DECLARATION
            -------------------------------------------------------------------------------
            ENTITY sram4m8 IS
              GENERIC (
                -- tipd delays: interconnect path delays
                tipd_OENeg           : VitalDelayType01 := VitalZeroDelay01;
                tipd_WENeg           : VitalDelayType01 := VitalZeroDelay01;
                tipd_CENeg           : VitalDelayType01 := VitalZeroDelay01;
                tipd_CE              : VitalDelayType01 := VitalZeroDelay01;
                tipd_D0              : VitalDelayType01 := VitalZeroDelay01;
                tipd_D1              : VitalDelayType01 := VitalZeroDelay01;
                tipd_D2              : VitalDelayType01 := VitalZeroDelay01;
                tipd_D3              : VitalDelayType01 := VitalZeroDelay01;
                tipd_D4              : VitalDelayType01 := VitalZeroDelay01;
                tipd_D5              : VitalDelayType01 := VitalZeroDelay01;
                tipd_D6              : VitalDelayType01 := VitalZeroDelay01;
                tipd_D7              : VitalDelayType01 := VitalZeroDelay01;
```

```
        tipd_A0                 : VitalDelayType01 := VitalZeroDelay01;
        tipd_A1                 : VitalDelayType01 := VitalZeroDelay01;
        tipd_A2                 : VitalDelayType01 := VitalZeroDelay01;
        tipd_A3                 : VitalDelayType01 := VitalZeroDelay01;
        tipd_A4                 : VitalDelayType01 := VitalZeroDelay01;
        tipd_A5                 : VitalDelayType01 := VitalZeroDelay01;
        tipd_A6                 : VitalDelayType01 := VitalZeroDelay01;
        tipd_A7                 : VitalDelayType01 := VitalZeroDelay01;
        tipd_A8                 : VitalDelayType01 := VitalZeroDelay01;
        tipd_A9                 : VitalDelayType01 := VitalZeroDelay01;
        tipd_A10                : VitalDelayType01 := VitalZeroDelay01;
        tipd_A11                : VitalDelayType01 := VitalZeroDelay01;
        tipd_A12                : VitalDelayType01 := VitalZeroDelay01;
        tipd_A13                : VitalDelayType01 := VitalZeroDelay01;
        tipd_A14                : VitalDelayType01 := VitalZeroDelay01;
        tipd_A15                : VitalDelayType01 := VitalZeroDelay01;
        tipd_A16                : VitalDelayType01 := VitalZeroDelay01;
        tipd_A17                : VitalDelayType01 := VitalZeroDelay01;
        tipd_A18                : VitalDelayType01 := VitalZeroDelay01;
        tipd_A19                : VitalDelayType01 := VitalZeroDelay01;
        tipd_A20                : VitalDelayType01 := VitalZeroDelay01;
        tipd_A21                : VitalDelayType01 := VitalZeroDelay01;
        -- tpd delays
        tpd_OENeg_D0                    : VitalDelayType01Z := UnitDelay01Z;
        tpd_CENeg_D0                    : VitalDelayType01Z := UnitDelay01Z;
        tpd_A0_D0                       : VitalDelayType01  := UnitDelay01;
        -- tpw values: pulse widths
        tpw_WENeg_negedge               : VitalDelayType    := UnitDelay;
        tpw_WENeg_posedge               : VitalDelayType    := UnitDelay;
        -- tsetup values: setup times
        tsetup_D0_WENeg                 : VitalDelayType    := UnitDelay;
        tsetup_D0_CENeg                 : VitalDelayType    := UnitDelay;
        -- thold values: hold times
        thold_D0_WENeg                  : VitalDelayType    := UnitDelay;
        thold_D0_CENeg                  : VitalDelayType    := UnitDelay;
        -- generic control parameters
        InstancePath    : STRING      := DefaultInstancePath;
        TimingChecksOn  : BOOLEAN     := DefaultTimingChecks;
        MsgOn           : BOOLEAN     := DefaultMsgOn;
        XOn             : BOOLEAN     := DefaultXOn;
        SeverityMode    : SEVERITY_LEVEL := WARNING;
        -- For FMF SDF technology file usage
        TimingModel     : STRING      := DefaultTimingModel
    );
```

The port declarations are equally straightforward:

```
PORT (
    A0          : IN     std_ulogic := 'X';
    A1          : IN     std_ulogic := 'X';
    A2          : IN     std_ulogic := 'X';
    A3          : IN     std_ulogic := 'X';
    A4          : IN     std_ulogic := 'X';
    A5          : IN     std_ulogic := 'X';
    A6          : IN     std_ulogic := 'X';
    A7          : IN     std_ulogic := 'X';
    A8          : IN     std_ulogic := 'X';
    A9          : IN     std_ulogic := 'X';
    A10         : IN     std_ulogic := 'X';
    A11         : IN     std_ulogic := 'X';
    A12         : IN     std_ulogic := 'X';
    A13         : IN     std_ulogic := 'X';
    A14         : IN     std_ulogic := 'X';
    A15         : IN     std_ulogic := 'X';
    A16         : IN     std_ulogic := 'X';
    A17         : IN     std_ulogic := 'X';
    A18         : IN     std_ulogic := 'X';
    A19         : IN     std_ulogic := 'X';
    A20         : IN     std_ulogic := 'X';
    A21         : IN     std_ulogic := 'X';

    D0          : INOUT  std_ulogic := 'X';
    D1          : INOUT  std_ulogic := 'X';
    D2          : INOUT  std_ulogic := 'X';
    D3          : INOUT  std_ulogic := 'X';
    D4          : INOUT  std_ulogic := 'X';
    D5          : INOUT  std_ulogic := 'X';
    D6          : INOUT  std_ulogic := 'X';
    D7          : INOUT  std_ulogic := 'X';

    OENeg       : IN     std_ulogic := 'X';
    WENeg       : IN     std_ulogic := 'X';
    CENeg       : IN     std_ulogic := 'X';
    CE          : IN     std_ulogic := 'X'
);
    ATTRIBUTE VITAL_LEVEL0 of sram4m8 : ENTITY IS TRUE;
END sram4m8;
```

Up to this point, the behavioral model and the VITAL2000 model are identical. Next comes the `VITAL_LEVEL0` architecture, beginning with the constant and signal declarations:

```
-----------------------------------------------------------------------------
-- ARCHITECTURE DECLARATION
-----------------------------------------------------------------------------
ARCHITECTURE vhdl_behavioral of sram4m8 IS
    ATTRIBUTE VITAL_LEVEL0 of vhdl_behavioral : ARCHITECTURE IS TRUE;

    -----------------------------------------------------------------------
    -- Note that this model uses the Shelor method of modeling large memory
    -- arrays. Data is stored as type INTEGER with the value -2 representing
    -- an uninitialized location and the value -1 representing a corrupted
    -- location.
    -----------------------------------------------------------------------
    CONSTANT partID         : STRING := "SRAM 4M X 8";
    CONSTANT MaxData        : NATURAL := 255;
    CONSTANT TotalLOC       : NATURAL := 4194303;
    CONSTANT HiAbit         : NATURAL := 21;
    CONSTANT HiDbit         : NATURAL := 7;
    CONSTANT DataWidth      : NATURAL := 8;

    SIGNAL D0_ipd           : std_ulogic := 'U';
    SIGNAL D1_ipd           : std_ulogic := 'U';
    SIGNAL D2_ipd           : std_ulogic := 'U';
    SIGNAL D3_ipd           : std_ulogic := 'U';
    SIGNAL D4_ipd           : std_ulogic := 'U';
    SIGNAL D5_ipd           : std_ulogic := 'U';
    SIGNAL D6_ipd           : std_ulogic := 'U';
    SIGNAL D7_ipd           : std_ulogic := 'U';
    SIGNAL D8_ipd           : std_ulogic := 'U';

    SIGNAL A0_ipd           : std_ulogic := 'U';
    SIGNAL A1_ipd           : std_ulogic := 'U';
    SIGNAL A2_ipd           : std_ulogic := 'U';
    SIGNAL A3_ipd           : std_ulogic := 'U';
    SIGNAL A4_ipd           : std_ulogic := 'U';
    SIGNAL A5_ipd           : std_ulogic := 'U';
    SIGNAL A6_ipd           : std_ulogic := 'U';
    SIGNAL A7_ipd           : std_ulogic := 'U';
    SIGNAL A8_ipd           : std_ulogic := 'U';
    SIGNAL A9_ipd           : std_ulogic := 'U';
    SIGNAL A10_ipd          : std_ulogic := 'U';
    SIGNAL A11_ipd          : std_ulogic := 'U';
    SIGNAL A12_ipd          : std_ulogic := 'U';
    SIGNAL A13_ipd          : std_ulogic := 'U';
    SIGNAL A14_ipd          : std_ulogic := 'U';
    SIGNAL A15_ipd          : std_ulogic := 'U';
    SIGNAL A16_ipd          : std_ulogic := 'U';
```

```
SIGNAL A17_ipd              : std_ulogic := 'U';
SIGNAL A18_ipd              : std_ulogic := 'U';
SIGNAL A19_ipd              : std_ulogic := 'U';
SIGNAL A20_ipd              : std_ulogic := 'U';
SIGNAL A21_ipd              : std_ulogic := 'U';

SIGNAL OENeg_ipd            : std_ulogic := 'U';
SIGNAL WENeg_ipd            : std_ulogic := 'U';
SIGNAL CENeg_ipd            : std_ulogic := 'U';
SIGNAL CE_ipd               : std_ulogic := 'U';
```

Manufacturers design memories in families, with the members differing only in depth and width. The constants declared here are used to enable the creation of models of memories within a family with a minimum amount of editing.

After the declarations, the architecture begins with the wire delay block:

```
BEGIN
   ----------------------------------------------------------------------------
   -- Wire Delays
   ----------------------------------------------------------------------------
   WireDelay : BLOCK
   BEGIN
      w_1: VitalWireDelay (OENeg_ipd, OENeg, tipd_OENeg);
      w_2: VitalWireDelay (WENeg_ipd, WENeg, tipd_WENeg);
      w_3: VitalWireDelay (CENeg_ipd, CENeg, tipd_CENeg);
      w_4: VitalWireDelay (CE_ipd, CE, tipd_CE);
      w_5: VitalWireDelay (D0_ipd, D0, tipd_D0);
      w_6: VitalWireDelay (D1_ipd, D1, tipd_D1);
      w_7: VitalWireDelay (D2_ipd, D2, tipd_D2);
      w_8: VitalWireDelay (D3_ipd, D3, tipd_D3);
      w_9: VitalWireDelay (D4_ipd, D4, tipd_D4);
      w_10: VitalWireDelay (D5_ipd, D5, tipd_D5);
      w_11: VitalWireDelay (D6_ipd, D6, tipd_D6);
      w_12: VitalWireDelay (D7_ipd, D7, tipd_D7);
      w_13: VitalWireDelay (A0_ipd, A0, tipd_A0);
      w_14: VitalWireDelay (A1_ipd, A1, tipd_A1);
      w_15: VitalWireDelay (A2_ipd, A2, tipd_A2);
      w_16: VitalWireDelay (A3_ipd, A3, tipd_A3);
      w_17: VitalWireDelay (A4_ipd, A4, tipd_A4);
      w_18: VitalWireDelay (A5_ipd, A5, tipd_A5);
      w_19: VitalWireDelay (A6_ipd, A6, tipd_A6);
      w_20: VitalWireDelay (A7_ipd, A7, tipd_A7);
      w_21: VitalWireDelay (A8_ipd, A8, tipd_A8);
      w_22: VitalWireDelay (A9_ipd, A9, tipd_A9);
      w_23: VitalWireDelay (A10_ipd, A10, tipd_A10);
      w_24: VitalWireDelay (A11_ipd, A11, tipd_A11);
```

```
      w_25: VitalWireDelay (A12_ipd, A12, tipd_A12);
      w_26: VitalWireDelay (A13_ipd, A13, tipd_A13);
      w_27: VitalWireDelay (A14_ipd, A14, tipd_A14);
      w_28: VitalWireDelay (A15_ipd, A15, tipd_A15);
      w_29: VitalWireDelay (A16_ipd, A16, tipd_A16);
      w_30: VitalWireDelay (A17_ipd, A17, tipd_A17);
      w_31: VitalWireDelay (A18_ipd, A18, tipd_A18);
      w_32: VitalWireDelay (A19_ipd, A19, tipd_A19);
      w_33: VitalWireDelay (A20_ipd, A20, tipd_A20);
      w_34: VitalWireDelay (A21_ipd, A21, tipd_A21);

   END BLOCK;
```

The model's ports have been declared as scalars to facilitate the model's use in a schematic capture environment and to provide a means of backannotating individual interconnect delays. However, in modeling the function of the memory it is more convenient to use vectors. The mapping from scalar to vector and back is best done within a block.

The block opens with a set of port declarations:

```
-------------------------------------------------------------------------------
-- Main Behavior Block
-------------------------------------------------------------------------------
Behavior: BLOCK

    PORT (
            AddressIn       : IN    std_logic_vector(HiAbit downto 0);
            DataIn          : IN    std_logic_vector(HiDbit downto 0);
            DataOut         : OUT   std_logic_vector(HiDbit downto 0);
            OENegIn         : IN    std_ulogic := 'U';
            WENegIn         : IN    std_ulogic := 'U';
            CENegIn         : IN    std_ulogic := 'U';
            CEIn            : IN    std_ulogic := 'U'
    );
```

Note that the data ports that were of type INOUT in the entity are split into separate ports of type IN and type OUT in the block. Also note that the address and data buses are sized using the constants declared at the top of the architecture.

The next step is the port map:

```
PORT MAP (
  DataOut(0)  => D0,
  DataOut(1)  => D1,
  DataOut(2)  => D2,
  DataOut(3)  => D3,
  DataOut(4)  => D4,
  DataOut(5)  => D5,
  DataOut(6)  => D6,
```

```
    DataOut(7) => D7,
    DataIn(0) => D0_ipd,
    DataIn(1) => D1_ipd,
    DataIn(2) => D2_ipd,
    DataIn(3) => D3_ipd,
    DataIn(4) => D4_ipd,
    DataIn(5) => D5_ipd,
    DataIn(6) => D6_ipd,
    DataIn(7) => D7_ipd,
    AddressIn(0) => A0_ipd,
    AddressIn(1) => A1_ipd,
    AddressIn(2) => A2_ipd,
    AddressIn(3) => A3_ipd,
    AddressIn(4) => A4_ipd,
    AddressIn(5) => A5_ipd,
    AddressIn(6) => A6_ipd,
    AddressIn(7) => A7_ipd,
    AddressIn(8) => A8_ipd,
    AddressIn(9) => A9_ipd,
    AddressIn(10) => A10_ipd,
    AddressIn(11) => A11_ipd,
    AddressIn(12) => A12_ipd,
    AddressIn(13) => A13_ipd,
    AddressIn(14) => A14_ipd,
    AddressIn(15) => A15_ipd,
    AddressIn(16) => A16_ipd,
    AddressIn(17) => A17_ipd,
    AddressIn(18) => A18_ipd,
    AddressIn(19) => A19_ipd,
    AddressIn(20) => A20_ipd,
    AddressIn(21) => A21_ipd,
    OENegIn => OENeg_ipd,
    WENegIn => WENeg_ipd,
    CENegIn => CENeg_ipd,
    CEIn     => CE_ipd
 );
```

In the port map, inputs to the block are associated with the delayed versions of the model's input ports. The block outputs are associated directly with the model's output ports. There is only one signal declaration in this block:

```
    SIGNAL D_zd    : std_logic_vector(HiDbit DOWNTO 0);
```

It is the zero delay data output.

The behavioral section of the model consists of a process. The process's sensitivity list includes all the input signals declared in the block:

```
BEGIN
-----------------------------------------------------------------------
-- Behavior Process
-----------------------------------------------------------------------
Behavior : PROCESS (OENegIn, WENegIn, CENegIn, CEIn, AddressIn, DataIn)

   -- Timing Check Variables
   VARIABLE Tviol_DO_WENeg : X01 := '0';
   VARIABLE TD_DO_WENeg    : VitalTimingDataType;

   VARIABLE Tviol_DO_CENeg : X01 := '0';
   VARIABLE TD_DO_CENeg    : VitalTimingDataType;

   VARIABLE Pviol_WENeg    : X01 := '0';
   VARIABLE PD_WENeg       : VitalPeriodDataType := VitalPeriodDataInit;
```

It is followed by the declarations for the timing check variables.

The memory array and functionality results variables are declared as follows:

```
-- Memory array declaration
TYPE MemStore IS ARRAY (0 to TotalLOC) OF INTEGER
                   RANGE -2 TO MaxData;

-- Functionality Results Variables
VARIABLE Violation : X01 := '0';
VARIABLE DataDrive : std_logic_vector(HiDbit DOWNTO 0)
                   := (OTHERS => 'X');
VARIABLE DataTemp  : INTEGER RANGE -2 TO MaxData := -2;
VARIABLE Location  : NATURAL RANGE 0 TO TotalLOC := 0;
VARIABLE MemData   : MemStore;
```

Again, these take advantage of the constants declared at the top of the architecture.

To avoid the need to test the control inputs for weak signals ('L' and 'H'), "no weak value" variables are declared for them:

```
-- No Weak Values Variables
VARIABLE OENeg_nwv : UX01 := 'U';
VARIABLE WENeg_nwv : UX01 := 'U';
VARIABLE CENeg_nwv : UX01 := 'U';
VARIABLE CE_nwv    : UX01 := 'U';
```

and they are converted at the beginning of the process body:

```
BEGIN
   OENeg_nwv   := To_UX01 (s => OENegIn);
   WENeg_nwv   := To_UX01 (s => WENegIn);
   CENeg_nwv   := To_UX01 (s => CENegIn);
   CE_nwv      := To_UX01 (s => CEIn);
```

The timing check section comes near the top of the process body, as usual:

```
------------------------------------------------------------------------------
-- Timing Check Section
------------------------------------------------------------------------------
IF (TimingChecksOn) THEN

    VitalSetupHoldCheck (
         TestSignal      => DataIn,
         TestSignalName  => "Data",
         RefSignal       => WENeg,
         RefSignalName   => "WENeg",
         SetupHigh       => tsetup_D0_WENeg,
         SetupLow        => tsetup_D0_WENeg,
         HoldHigh        => thold_D0_WENeg,
         HoldLow         => thold_D0_WENeg,
         CheckEnabled    => (CENeg ='0' and CE ='1'and OENeg ='1'),
         RefTransition   => '/',
         HeaderMsg       => InstancePath & PartID,
         TimingData      => TD_D0_WENeg,
         XOn             => XOn,
         MsgOn           => MsgOn,
         Violation       => Tviol_D0_WENeg );

    VitalSetupHoldCheck (
         TestSignal      => DataIn,
         TestSignalName  => "Data",
         RefSignal       => CENeg,
         RefSignalName   => "CENeg",
         SetupHigh       => tsetup_D0_CENeg,
         SetupLow        => tsetup_D0_CENeg,
         HoldHigh        => thold_D0_CENeg,
         HoldLow         => thold_D0_CENeg,
         CheckEnabled    => (WENeg ='0' and OENeg ='1'),
         RefTransition   => '/',
         HeaderMsg       => InstancePath & PartID,
         TimingData      => TD_D0_CENeg,
         XOn             => XOn,
         MsgOn           => MsgOn,
         Violation       => Tviol_D0_CENeg );

    VitalPeriodPulseCheck (
         TestSignal      => WENegIn,
         TestSignalName  => "WENeg",
         PulseWidthLow   => tpw_WENeg_negedge,
         PeriodData      => PD_WENeg,
         XOn             => XOn,
         MsgOn           => MsgOn,
         Violation       => Pviol_WENeg,
         HeaderMsg       => InstancePath & PartID,
         CheckEnabled    => TRUE );
```

```
        Violation := Pviol_WENeg OR Tviol_DO_WENeg OR Tviol_DO_CENeg;

    ASSERT Violation = '0'
            REPORT InstancePath & partID & ": simulation may be" &
                    " inaccurate due to timing violations"
            SEVERITY SeverityMode;

    END IF; -- Timing Check Section
```

This section is much shorter than it would have been if scalar signals had been used for the data bus instead of vectored signals.

An assertion statement is used to warn the user whenever a timing violation occurs. Corrupt data will also be written to memory when this happens, but the user could find it difficult to determine the source of the corruption without the assertion statement. The model's logic is in the functional section. The section begins by setting `DataDrive`, the output variable, to high impedance. Then, if the component is selected for either a read or a write operation the value of the address bus is translated to a natural and assigned to the variable `Location`:

```
    --------------------------------------------------------------------------
    -- Functional Section
    --------------------------------------------------------------------------
    DataDrive := (OTHERS => 'Z');

    IF (CE_nwv = '1' AND CENeg_nwv = '0') THEN
        IF (OENeg_nwv = '0' OR WENeg_nwv = '0') THEN

            Location := To_Nat(AddressIn);

            IF (OENeg_nwv = '0' AND WENeg_nwv = '1') THEN
                DataTemp := MemData(Location);
                IF DataTemp >= 0 THEN
                    DataDrive := To_slv(DataTemp, DataWidth);
                ELSIF DataTemp = -2 THEN
                    DataDrive := (OTHERS => 'U');
                ELSE
                    DataDrive := (OTHERS => 'X');
                END IF;
            ELSIF (WENeg_nwv = '0') THEN
                IF Violation = '0' THEN
                    DataTemp := To_Nat(DataIn);
                ELSE
                    DataTemp := -1;
                END IF;
                MemData(Location) := DataTemp;
            END IF;
        END IF;
    END IF;
```

If the operation is a read, the `Location` variable is used as an index to the memory array and the contents of that location assigned to `DataTemp`. `DataTemp` is tested to see if it contains a nonnegative number. If so, it is valid and assigned to `DataDrive`. If not, a –2 indicates the location is uninitialized and 'U's are assigned to `DataDrive`. Anything else (–1) causes 'X's to be assigned.

If the operation is a write and there is no timing violation, the value of the data bus is converted to a natural and assigned to `DataTemp`. If there is a timing violation, `DataTemp` is assigned a –1.

Finally, the element of the memory array indexed by `Location` is assigned the value of `DataTemp`.

At the end of the process, the zero delay signal gets the value of `DataDrive`:

```
  -------------------------------------------------------------------------
  -- Output Section
  -------------------------------------------------------------------------
  D_zd  <= DataDrive;

END PROCESS;
```

The model concludes with the output path delay. Because the output is a bus (within the block), a generate statement is used to shorten the model:

```
     -------------------------------------------------------------------------
     -- Path Delay Processes generated as a function of data width
     -------------------------------------------------------------------------
     DataOut_Width : FOR i IN HiDbit DOWNTO 0 GENERATE
       DataOut_Delay : PROCESS (D_zd(i))
         VARIABLE D_GlitchData:VitalGlitchDataArrayType(HiDbit Downto 0);
       BEGIN
         VitalPathDelay01Z (
           OutSignal        => DataOut(i),
           OutSignalName    => "Data",
           OutTemp          => D_zd(i),
           Mode             => OnEvent,
           GlitchData       => D_GlitchData(i),
           Paths            => (
             0  => (InputChangeTime  => OENeg_ipd'LAST_EVENT,
                    PathDelay        => tpd_OENeg_D0,
                    PathCondition    => TRUE),
               1  => (InputChangeTime  => CENeg_ipd'LAST_EVENT,
                      PathDelay        => tpd_CENeg_D0,
                      PathCondition    => TRUE),
                 2  => (InputChangeTime  => AddressIn'LAST_EVENT,
                        PathDelay => VitalExtendToFillDelay(tpd_A0_D0),
                        PathCondition    => TRUE)
             )
           );
```

```
        END PROCESS;
      END GENERATE;

    END BLOCK;
  END vhdl_behavioral;
```

Once again, one of the constants from the beginning of the architecture is used to control the size of the generate. Using a generate statement requires a separate process for the path delays, because although a process may reside within a generate, a generate statement may not be placed within a process.

14.2.2   Using the VITAL2000 Method

The VITAL2000 style memory model uses the same entity as the behavioral model, so the entity will not be repeated here. The first difference is the VITAL attribute in the architecture:

```
---------------------------------------------------------------------------
-- ARCHITECTURE DECLARATION
---------------------------------------------------------------------------
ARCHITECTURE vhdl_behavioral of sram4m8v2 IS
   ATTRIBUTE VITAL_LEVEL1_MEMORY of vhdl_behavioral : ARCHITECTURE IS TRUE;

   ---------------------------------------------------------------------------
   -- Note that this model uses the VITAL2000 method of modeling large memory
   -- arrays.
   ---------------------------------------------------------------------------
```

Here the attribute is **VITAL_LEVEL1_MEMORY**. It is required to get the full compiler benefits of a VITAL memory model.

From here the architecture is identical to the behavioral model for a while. The same constants and signals are declared. The wire delay block is the same. The behavior block is declared with the same ports and same port map. But the behavior process is completely different. It begins with the process declaration and sensitivity list:

```
BEGIN
   ---------------------------------------------------------------------------
   -- Behavior Process
   ---------------------------------------------------------------------------
   MemoryBehavior : PROCESS (OENegIn, WENegIn, CENegIn, CEIn, AddressIn,
                             DataIn)
```

Then comes the declaration of a constant of type `VitalMemoryTableType`:

```
CONSTANT Table_generic_sram :       VitalMemoryTableType    :=    (

-- -------------------------------------------------------------------------
--    CE, CEN, OEN, WEN, Addr, DI, act, DO
-- -------------------------------------------------------------------------
-- Address initiated read
   ( '1', '0', '0', '1', 'G', '-', 's', 'm' ),
   ( '1', '0', '0', '1', 'U', '-', 's', 'l' ),
```

```
-- Output Enable initiated read
   ( '1', '0', 'N', '1', 'g', '-', 's', 'm' ),
   ( '1', '0', 'N', '1', 'u', '-', 's', 'l' ),
   ( '1', '0', '0', '1', 'g', '-', 's', 'm' ),

-- CE initiated read
   ( 'P', '0', '0', '1', 'g', '-', 's', 'm' ),
   ( 'P', '0', '0', '1', 'u', '-', 's', 'l' ),

-- CEN initiated read
   ( '1', 'N', '0', '1', 'g', '-', 's', 'm' ),
   ( '1', 'N', '0', '1', 'u', '-', 's', 'l' ),

-- Write Enable Implicit Read
   ( '1', '0', '0', 'P', '-', '-', 's', 'M' ),

-- Write Enable initiated Write
   ( '1', '0', '1', 'N', 'g', '-', 'w', 'S' ),
   ( '1', '0', '1', 'N', 'u', '-', 'c', 'S' ),

-- CE initiated Write
   ( 'P', '0', '1', '0', 'g', '-', 'w', 'S' ),
   ( 'P', '0', '1', '0', 'u', '-', 'c', 'S' ),

-- CEN initiated Write
   ( '1', 'N', '1', '0', 'g', '-', 'w', 'Z' ),
   ( '1', 'N', '1', '0', 'u', '-', 'c', 'Z' ),

-- Address change during write
   ( '1', '0', '1', '0', '*', '-', 'c', 'Z' ),
   ( '1', '0', '1', 'X', '*', '-', 'c', 'Z' ),

-- data initiated Write
   ( '1', '0', '1', '0', 'g', '*', 'w', 'Z' ),
   ( '1', '0', '1', '0', 'u', '-', 'c', 'Z' ),
   ( '1', '0', '-', 'X', 'g', '*', 'e', 'e' ),
   ( '1', '0', '-', 'X', 'u', '*', 'c', 'S' ),

-- if WEN is X
   ( '1', '0', '1', 'r', 'g', '*', 'e', 'e' ),
   ( '1', '0', '1', 'r', 'u', '*', 'c', 'l' ),
   ( '1', '0', '-', 'r', 'g', '*', 'e', 'S' ),
   ( '1', '0', '-', 'r', 'u', '*', 'c', 'S' ),
   ( '1', '0', '1', 'f', 'g', '*', 'e', 'e' ),
   ( '1', '0', '1', 'f', 'u', '*', 'c', 'l' ),
   ( '1', '0', '-', 'f', 'g', '*', 'e', 'S' ),
   ( '1', '0', '-', 'f', 'u', '*', 'c', 'S' ),
```

```
-- OEN is unasserted
( '-', '-', '1', '-', '-', '-', 's', 'Z' ),
( '1', '0', 'P', '-', '-', '-', 's', 'Z' ),
( '1', '0', 'r', '-', '-', '-', 's', '1' ),
( '1', '0', 'f', '-', '-', '-', 's', '1' ),
( '1', '0', '1', '-', '-', '-', 's', 'Z' )
);
```

This table entirely defines the function of the memory model. The mechanics of a VITAL memory table are described in Chapter 7, but let us look at the table and see how it compares to the code in the behavioral model.

Columns CE, CEN, OEN, and WEN are the direct inputs. They are the control signals. Columns Addr and DI are the interpreted inputs. They represent the address and data buses, respectively. The act column specifies the memory action and the DO column specifies the output action. The table is searched from top to bottom until a match is found.

In the first section, an address initiated read is described:

```
-- -------------------------------------------------------------------
--   CE, CEN, OEN, WEN, Addr, DI, act, DO
-- -------------------------------------------------------------------
-- Address initiated read
( '1', '0', '0', '1', 'G', '-', 's', 'm' ),
( '1', '0', '0', '1', 'U', '-', 's', '1' ),
```

For either line to be selected, CE and WEN, the chip enable and write enable, must be high. Write enable is an active low signal. In addition, CEN and OEN, also active low signals, must be low. If the address bus transitions to any good value (no 'X's), the memory location indexed by the address bus retains its previous value and the output bus gets the value of the memory location. Otherwise, if the address bus transitions to any unknown value (any bit is 'X'), the output gets a corrupt ('X') value.

If there is no match in the first section, the next section, describing an output enable initiated read, is searched:

```
-- -------------------------------------------------------------------
--   CE, CEN, OEN, WEN, Addr, DI, act, DO
-- -------------------------------------------------------------------
-- Output Enable initiated read
( '1', '0', 'N', '1', 'g', '-', 's', 'm' ),
( '1', '0', 'N', '1', 'u', '-', 's', '1' ),
( '1', '0', '0', '1', 'g', '-', 's', 'm' ),
```

The two chip enables, CE and CEN, must be active and WEN inactive. If there is a falling edge or a low on OEN (1st and 3d lines) and the address bus has a good and steady (no transition) value, the value of the memory location is placed on the output. Otherwise, if OEN is falling (but steady) and the address is unknown (2nd line), the output is corrupted.

If there is no match in the previous sections, the next section, describing a chip enable initiated read is searched:

```
--  ---------------------------------------------------------------------------
--    CE, CEN, OEN, WEN, Addr, DI, act, DO
--  ---------------------------------------------------------------------------
--  CE initiated read
    ( 'P', '0', '0', '1', 'g', '-', 's', 'm' ),
    ( 'P', '0', '0', '1', 'u', '-', 's', 'l' ),
```

CEN and OEN must be active and WEN inactive. If there is a rising edge on CE and the address bus has a good and steady (no transition) value, the value of the memory location is placed on the output. Otherwise, if the address is unknown (but steady), the output is corrupted.

If there is no match in the previous sections, the next section, describing a chip enable initiated read is searched:

```
--  ---------------------------------------------------------------------------
--    CE, CEN, OEN, WEN, Addr, DI, act, DO
--  ---------------------------------------------------------------------------
--  CEN initiated read
    ( '1', 'N', '0', '1', 'g', '-', 's', 'm' ),
    ( '1', 'N', '0', '1', 'u', '-', 's', 'l' ),
```

CE and OEN must be active and WEN inactive. If there is a falling edge on CEN and the address bus has a good and steady (no transition) value, the value of the memory location is placed on the output. Otherwise, if the address is unknown (but steady), the output is corrupted.

The sections of the table just described correspond to the following lines of the behavioral model:

```
IF (CE_nwv = '1' AND CENeg_nwv = '0') THEN
    IF (OENeg_nwv = '0' OR WENeg_nwv = '0') THEN

        Location := To_Nat(AddressIn);

        IF (OENeg_nwv = '0' AND WENeg_nwv = '1') THEN
            DataTemp := MemData(Location);
            IF DataTemp >= 0 THEN
                DataDrive := To_slv(DataTemp, DataWidth);
            ELSIF DataTemp = -2 THEN
                DataDrive := (OTHERS => 'U');
            ELSE
                DataDrive := (OTHERS => 'X');
            END IF;
```

There are some differences. Although the VITAL model appears to be more complex, it does not require type conversions or special handling of uninitialized or corrupt locations. If a timing violation occurs during a read, the VITAL model will output 'X's. The behavioral model will send a warning message to the user but place valid data on the output bus. The VITAL modeling method provides more precise control of model behavior. How often that level of precision is required remains to be seen.

Continuing with the rest of the VITAL model process declarations, we have the following:

```
CONSTANT OENeg_D_Delay : VitalDelayArrayType01Z (HiDbit downto 0) :=
        (OTHERS => tpd_OENeg_DO);

CONSTANT CENeg_D_Delay : VitalDelayArrayType01Z (HiDbit downto 0) :=
        (OTHERS => tpd_CENeg_DO);

CONSTANT Addr_D_Delay : VitalDelayArrayType01 (175 downto 0) :=
        (OTHERS => tpd_AO_DO);
```

These constants are arrays of delays. A different delay could be assigned to every path from each input to each output. For the address to data out path, there are 22 inputs and 8 outputs. This method allows the assignment of 176 different delay values for address to data out. Of course, the same thing is possible in a behavior model; it would just take much more code. Although such detailed timing may be useful for memory embedded in an ASIC, the intended target of the VITAL_Memory package, it is rarely, if ever, required in component modeling.

The declaration for the timing check variables is the same as in the behavioral model:

```
-- Timing Check Variables
VARIABLE Tviol_DO_WENeg : X01 := '0';
VARIABLE TD_DO_WENeg    : VitalTimingDataType;

VARIABLE Tviol_DO_CENeg : X01 := '0';
VARIABLE TD_DO_CENeg    : VitalTimingDataType;

VARIABLE Pviol_WENeg    : X01 := '0';
VARIABLE PD_WENeg       : VitalPeriodDataType := VitalPeriodDataInit;
```

Although the VITAL_Memory package has its own set of timing check procedures, they are not used in this model. The more generic procedures are adequate in this case and easier to work with.

The memory declaration follows:

```
        -- VITAL Memory Declaration
        VARIABLE Memdat : VitalMemoryDataType :=
          VitalDeclareMemory (
              NoOfWords           => TotalLOC,
              NoOfBitsPerWord     => DataWidth,
              NoOfBitsPerSubWord  => DataWidth,
--            MemoryLoadFile      => MemLoadFileName,
              BinaryLoadFile      => FALSE
          );
```

The VITAL_Memory package uses a procedure call for the memory array declaration. Included in the procedure is the ability to preload part or all of the memory array from a binary or ASCII file. Preloading memories is discussed later in this chapter.

The functionality results variables

```
-- Functionality Results Variables
VARIABLE Violation  : X01 := '0';

VARIABLE D_zd      : std_logic_vector(HiDbit DOWNTO 0);

VARIABLE Prevcntls     : std_logic_vector(0 to 3);
VARIABLE PrevData      : std_logic_vector(HiDbit downto 0);
VARIABLE Prevaddr      : std_logic_vector(HiAbit downto 0);
VARIABLE PFlag         : VitalPortFlagVectorType(0 downto 0);
VARIABLE Addrvalue     : VitalAddressValueType;
VARIABLE OENegChange   : TIME := 0 ns;
VARIABLE CENegChange   : TIME := 0 ns;
VARIABLE AddrChangeArray : VitalTimeArrayT(HiAbit downto 0);

VARIABLE D_GlitchData  : VitalGlitchDataArrayType(HiDbit Downto 0);
VARIABLE DSchedData    : VitalMemoryScheduleDataVectorType
                         (HiDbit Downto 0);
```

include several that are specific to VITAL_Memory models, as discussed in Chapter 7.

The process begins with a timing check section similar to the one in the behavior model:

```
BEGIN
   ---------------------------------------------------------------------
   -- Timing Check Section
   ---------------------------------------------------------------------
   IF (TimingChecksOn) THEN

     VitalSetupHoldCheck (
        TestSignal        => DataIn,
        TestSignalName    => "Data",
        RefSignal         => WENeg,
        RefSignalName     => "WENeg",
        SetupHigh         => tsetup_DO_WENeg,
        SetupLow          => tsetup_DO_WENeg,
        HoldHigh          => thold_DO_WENeg,
        HoldLow           => thold_DO_WENeg,
        CheckEnabled      => (CENeg = '0' and CE = '1'and OENeg = '1'),
        RefTransition     => '/',
        HeaderMsg         => InstancePath & PartID,
        TimingData        => TD_DO_WENeg,
        XOn               => XOn,
```

```
    MsgOn             => MsgOn,
    Violation         => Tviol_DO_WENeg );

VitalSetupHoldCheck (
    TestSignal        => DataIn,
    TestSignalName    => "Data",
    RefSignal         => CENeg,
    RefSignalName     => "CENeg",
    SetupHigh         => tsetup_DO_CENeg,
    SetupLow          => tsetup_DO_CENeg,
    HoldHigh          => thold_DO_CENeg,
    HoldLow           => thold_DO_CENeg,
    CheckEnabled      => (WENeg = '0' and OENeg = '1'),
    RefTransition     => '/',
    HeaderMsg         => InstancePath & PartID,
    TimingData        => TD_DO_CENeg,
    XOn               => XOn,
    MsgOn             => MsgOn,
    Violation         => Tviol_DO_CENeg );

VitalPeriodPulseCheck (
    TestSignal        => WENegIn,
    TestSignalName    => "WENeg",
    PulseWidthLow     => tpw_WENeg_negedge,
    PeriodData        => PD_WENeg,
    XOn               => XOn,
    MsgOn             => MsgOn,
    Violation         => Pviol_WENeg,
    HeaderMsg         => InstancePath & PartID,
    CheckEnabled      => TRUE );

Violation := Pviol_WENeg OR Tviol_DO_WENeg OR Tviol_DO_CENeg;

ASSERT Violation = '0'
    REPORT InstancePath & partID & ": simulation may be" &
            " inaccurate due to timing violations"
    SEVERITY SeverityMode;

END IF; -- Timing Check Section
```

The functionality section contains only a single call to the VitalMemoryTable procedure:

```
------------------------------------------------------------------------
-- Functional Section
------------------------------------------------------------------------

VitalMemoryTable (
    DataOutBus        => D_zd,
```

```
            MemoryData        => Memdat,
            PrevControls      => Prevcntls,
            PrevDataInBus     => Prevdata,
            PrevAddressBus    => Prevaddr,
            PortFlag          => PFlag,
            Controls          => (CEIn, CENegIn, OENegIn, WENegIn),
            DataInBus         => DataIn,
            AddressBus        => AddressIn,
            AddressValue      => Addrvalue,
            MemoryTable       => Table_generic_sram
        );
```

The model concludes with the output section. There are procedure calls to three different VITAL_Memory procedures in this section, which are the subject of the next section of this chapter:

```
        ------------------------------------------------------------------
        -- Output Section
        ------------------------------------------------------------------

        VitalMemoryInitPathDelay (
            ScheduleDataArray      =>        DSchedData,
            OutputDataArray        =>        D_zd
        );

        VitalMemoryAddPathDelay (                          -- #11
            ScheduleDataArray      =>        DSchedData,
            InputSignal            =>        AddressIn,
            OutputSignalName       =>        "D",
            InputChangeTimeArray   =>        AddrChangeArray,
            PathDelayArray         =>        Addr_D_Delay,
            ArcType                =>        CrossArc,
            PathCondition          =>        true
        );

        VitalMemoryAddPathDelay (                          -- #14
            ScheduleDataArray      =>        DSchedData,
            InputSignal            =>        OENegIn,
            OutputSignalName       =>        "D",
            InputChangeTime        =>        OENegChange,
            PathDelayArray         =>        OENeg_D_Delay,
            ArcType                =>        CrossArc,
            PathCondition          =>        true,
            OutputRetainFlag       =>        false
        );

        VitalMemoryAddPathDelay (                          -- #14
            ScheduleDataArray      =>        DSchedData,
```

```
                    InputSignal              =>        CENegIn,
                    OutputSignalName         =>        "D",
                    InputChangeTime          =>        CENegChange,
                    PathDelayArray           =>        CENeg_D_Delay,
                    ArcType                  =>        CrossArc,
                    PathCondition            =>        true,
                    OutputRetainFlag         =>        false
                );

            VitalMemorySchedulePathDelay (
                    OutSignal                =>        DataOut,
                    OutputSignalName         =>        "D",
                    ScheduleDataArray        =>        DSchedData
                );

          END PROCESS;

        END BLOCK;
      END vhdl_behavioral;
```

The comment #14 is to remind the author and anyone who has to maintain the model which overloading of the VitalMemoryAddPathDelay procedure is being called.

14.3   VITAL_Memory Path Delays

As mentioned in Chapter 3, the VITAL_Memory package has its own set of path delay procedures. There are three procedures that replace the VitalPathDelay01Z, and all three must be presented in the order shown.

The first is the VitalMemoryInitPathDelay. It is used to initialize the output delay data structure. It is called exactly once per output port. Output ports may be vectored. Outputs may also be internal signals rather than ports. The possible arguments to this procedure are shown in Table 14.1.

The second is the VitalMemoryAddPathDelay. It is used to add a delay path from an input to an output. There is one call to this procedure for each input to output path. Use of this procedure is analogous to the Paths parameter of the VitalPathDelay procedure. It is used for selecting candidate paths based on the PathCondition parameter. The procedure then updates the ScheduleDataArray structure.

Table 14.2 shows the possible arguments to the VitalMemoryAddPathDelay procedure. The VitalMemoryAddPathDelay procedure is overloaded 24 ways. However, there are 64 possible parameter combinations, of which 40 will result in compiler errors that may or may not be informative. Therefore, it is recommended that you print out and read a copy of memory_p_2000.vhd if you intend to use this method. It will help in debugging your models.

The third procedure is VitalMemorySchedulePathDelay. It is used to schedule the functional output value on the output signal using the selected

Table 14.1   Arguments for VitalMemoryInitPathDelay

Name	Type	Description
		For Scalar Ports
ScheduleData	VitalMemorySchedule-DataType	Scalar form of the data structure used by VITAL_Memory path delay procedures to store persistent information for the path delay scheduling.
OutputData	STD_ULOGIC	Scalar form of the functional output value to be scheduled.
		For Vectored Ports
ScheduleDataArray	VitalMemoryScheduleData-VectorType	Vector form of the data structure used by VITAL_Memory path delay procedures to store persistent information for the path delay scheduling.
NumBitsPerSubWord	POSITIVE	Number of bits per memory subword. Optional.
OutputDataArray	STD_LOGIC_VECTOR	Vector form of the functional output value to be scheduled.

propagation delay. This procedure is overloaded for scalar and vector outputs. It can also be used to perform result mapping of the output value using the Output-Map parameter. The possible arguments to the VitalMemorySchedulePathDelay are given in Table 14.3.

One of the most desirable features of the VITAL_Memory modeling path delay procedures is their support of output-retain behavior. Many memory components exhibit an output hold time that is greater than zero but less than the new output delay. This is shown in data sheets as a waveform similar to that in Figure 14.2. This behavior can be modeled using the behavioral style, but it is a nuisance to do so. The good news is that the VITAL_Memory path delay procedures can be used in a behavioral model to take advantage of this capability without requiring the use of other features of the package. They will work in a VITAL level 0 architecture.

14.4   VITAL_Memory Timing Constraints

The VITAL_Memory package has its own versions of two timing constraint checkers: VitalMemorySetupHoldCheck and VitalMemoryPeriodPulseCheck. The VitalMemorySetupHoldCheck procedure performs the same function as the VitalSetupHoldCheck procedure, with the following enhancements:

Table 14.2 Arguments for `VitalMemoryAddPathDelay`

Name	Type	Description
ScheduleData	VitalMemorySchedule-DataType	Scalar form of the data structure used by VITAL_Memory path delay procedures to store persistent information for the path delay scheduling.
ScheduleDataArray	VitalMemoryScheduleData VectorType	Vector form of the data structure used by VITAL_Memory path delay procedures to store persistent information for the path delay scheduling.
InputSignal	STD_ULOGIC or STD_LOGIC_VECTOR	Scalar or vector input.
OutputSignalName	STRING	Name of output signal for use in messages.
InputChangeTime	TIME	Time since the last input change occurred.
PathDelay	VitalDelayType(01Z)	Path delay values used to delay the output values for scalar outputs.
PathDelayArray	VitalDelayArray-Type(01ZX)	Array of path delay values used to delay the output values for vector outputs.
ArcType	VitalMemoryArcType	Delay arc type between input and output.
PathCondition	BOOLEAN	Condition under which the delay path is considered to be one of the candidate paths for propagation delay selection.
PathConditionArray	VitalBoolArrayT	Array of conditions under which the delay path is considered to be one of the candidate paths for propagation delay selection.
OutputRetainFlag	BOOLEAN	If TRUE, output retain (hold) behavior is enabled.
OutputRetain-Behavior	OutputRetainBehavior-Type	If value is BitCorrupt, output will be set to 'X' on a bit-by-bit basis. If WordCorrupt, entire word will be 'X'.

Table 14.3 Arguments for `VitalMemorySchedulePathDelay`

Name	Type	Description
ScheduleData	VitalMemoryScheduleDataType	Scalar form of the data structure used by VITAL_Memory path delay procedures to store persistent information for the path delay scheduling.
ScheduleDataArray	VitalMemoryScheduleData-TypeVector	Vector form of the data structure used by VITAL_ Memory path delay procedures to store persistent information for the path delay scheduling.
OutputSignal	STD_ULOGIC or STD_LOGIC_VECTOR	Scalar or vector output.
OutputSignalName	STRING	Name of output signal for use in messages.
OutputMap	VitalOutputMapType	Strength mapping of output values.

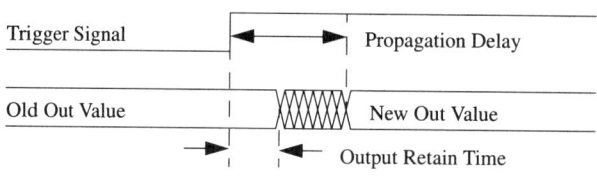

Figure 14.2 Output-retain waveform

Finer control over condition checking.

- Support for `CrossArc`, `ParellelArc`, and `Subword` timing relationships between the test signal and reference signal.

- Support for vector violation flags.

- Support for scalar and vector forms of condition in timing checks using `CheckEnabled`.

- Support of `MsgFormat` parameter to control the format of test/reference signals in the message.

- The `VitalMemoryPeriodPulseCheck` procedure is also similar to the `VitalPeriodPulseCheck` procedure, with the following differences:

- `TestSignal` is a vector rather than a scalar.

- The violation flag may be either a scalar or a vector.

- The `MsgFormat` parameter may be used to control the format of messages.

Although both of these procedures are required for accurate modeling of memories in an ASIC library environment, they are less useful for modeling off-the-shelf memory components. The `VitalMemoryPeriodPulseCheck` could be valuable should there be specification for minimum pulse width on an address bus. You will probably not find a need for the `VitalMemorySetupHoldCheck` procedure when writing a component model.

14.5   Preloading Memories

During system verification it is often desirable to be able to load the contents of a memory from an external file without going through the normal memory write process. Verilog has a system task, `$readmemh`, that can load a memory from a specified file at any time. It can be executed from the testbench if desired.

VHDL does not have an equivalent capability. However, that does not mean a memory cannot be preloaded in VHDL; it just takes a little more code. How it is done depends on the memory modeling style employed.

14.5.1   Behavioral Memory Preload

There must be a file to read. In a simple example, it may have the following format:

```
//format  : @address
//           data -> address
//           data -> address+1
@1
1234
1235
@A
55AA
```

Lines beginning with / are comments and are ignored. A line beginning with @ indicates a new address. The following lines will contain the data starting at that address and incrementing the address with each new line. Address and data do not appear on the same line. For simple cases like this one, the format is compatible with that used by the Verilog `$readmemh` task.

Upon a triggering event, usually time zero, and assuming the feature is enabled, a file like that in the example is read and its contents loaded into the memory array as specified in the file. In a model, it all starts with the declaration of a generic:

```
                                   -- memory file to be loaded
   mem_file_name       : STRING       := "km416s4030.mem";
```

The value of the generic could be passed in from the schematic or the testbench. It will specify the name of the file to read for that particular instance of the model.

Further down in the model the line

```
   FILE mem_file       : text IS mem_file_name;
```

is required to declare that mem_file is an object of type FILE. Then somewhere in the same process that defines the memory array, the preload code is placed. The following is an example of some simple memory preload code:

```
   --------------------------------------------------------------------------------
   -- File Read Section
   --------------------------------------------------------------------------------
   IF PoweredUp'EVENT and PoweredUp and (mem_file_name /= "none") THEN
      ind := 0;
      WHILE (not ENDFILE (mem_file)) LOOP
         READLINE (mem_file, buf);
         IF buf(1) = '/' THEN
            NEXT;
         ELSIF buf(1) = '@' THEN
            ind := h(buf(2 to 5));
         ELSE
            MemData(ind) := h(buf(1 to 4));
            ind := ind + 1;
         END IF;
      END LOOP;
   END IF;
```

This code waits for the triggering action, in this case an event on the signal PoweredUp. If the name of the memory load file is set to anything other than none, the memory preload code executes. It begins by initializing the index variable ind. Then it goes into a loop that will run until it reaches the end of the input file.

In this loop a line is read. If the line begins with the comment character (/) it is discarded and the next line is read. If the line begins with @ the following four characters, a hexadecimal number, are converted to a natural and assigned to ind, the array index. Otherwise, the first four characters on the line are converted to a natural and read into the memory array at the location indicated by ind, and then ind is incremented and the next line is read.

Memories with multiple words or multiple banks may be modeled. These may require slightly more complex memory preload files and file read sections. The model of a component that has four banks and is four words wide (32 bits) might have the following file read section shown in Figure 14.3.

```
--------------------------------------------------------------------
-- File Read Section
--------------------------------------------------------------------
IF PoweredUp'EVENT and PoweredUp and (mem_file_name /= "none") THEN
    ind := 0;
    WHILE (not ENDFILE (mem_file)) LOOP
        READLINE (mem_file, buf);
        IF buf(1) = '/' THEN
            NEXT;
        ELSIF buf(1) = '@' THEN
            file_bank := h(buf(2 to 2));
            ind := h(buf(4 to 8));
        ELSE
            MemData3(file_bank)(ind) := h(buf(1 to 2));
            MemData2(file_bank)(ind) := h(buf(3 to 4));
            MemData1(file_bank)(ind) := h(buf(5 to 6));
            MemData0(file_bank)(ind) := h(buf(7 to 8));
            ind := ind + 1;
        END IF;
    END LOOP;
END IF;
```

Figure 14.3   Memory preload for 4 bank memory

The corresponding preload file would have the following format:

```
// lines beginning with / are comments
// lines beginning with @ set bank (0 to 3) and starting address
// other lines contain hex data values a 32-bit value
AAAAAAAA
55555555
00030003
@1 40000
FFFF2001
FFFE2001
FFFD2003
FFFC2003
@2 00000
2000FFFF
20012001
20022001
20032001
```

This format is not compatible with that of the Verilog `$readmemh` task because it includes a method for selecting memory banks. Verilog would require a separate file and a separate call to the system task for each bank.

14.5.2   VITAL_Memory Preload

The `VitalDeclareMemory` function, if given a file name for the `MemoryLoadFile` parameter, will cause the declared memory array to be initialized during elaboration.

It requires no additional code in the model. It is the only way to preload a VITAL_LEVEL1_MEMORY model. The preload file has the following format:

```
@8  aa
a5
bf
@a  00
01
02
```

The VITAL preload format does not allow comments. Address and data may be on the same line. A file may contain data in either binary or hexadecimal but not both. The address must always be in hex.

Although it is easy to implement, it has two drawbacks. Memory can be initialized only during elaboration. In some behavioral models, memory has been initialized by a reset signal improving verification efficiency.

If multiple memory arrays are declared, each must have its own preload file. In models of memories with multiple banks, it may be necessary to manage multiple preload files.

14.6   Modeling Other Memory Types

So far our discussion of memory models has centered around SRAMs. This is because SRAMs are the simplest type of memories to model. Now we will look at some more complex memory types. These models tend to be rather long, 2,000 to 4,000 lines each, so instead of presenting the entire models, only their defining features will be discussed. The complete models can be found on the Free Model Foundry Web site.

14.6.1   Synchronous Static RAM

The first model we will examine is for a pipelined zero bus turnaround SSRAM. Its IDT part number is IDT71V65803. Compatible parts are made by Micron and Cypress. This memory type is distinguished by its fast read-to-write turnaround time. This component has two 9-bit-wide bidirectional data buses with separate write enables and a common asynchronous output enable. Memory is modeled as two arrays, each holding 9 bits of data. They are 512K words deep.

The model includes three processes. The first is used for setup and runs only once, at time zero. The second describes the functionality of the component. The third contains a generate statement that generates the required number of `Vital-PathDelay` calls to drive the output ports.

The first distinguishing feature we find in this model is a generic:

```
SeverityMode        : SEVERITY_LEVEL := WARNING;
```

It is used to control the severity of some assertion statements.

This model contains a state machine with five states. They are chip deselect (`desel`), begin read (`begin_rd`), begin write (`begin_wr`), burst read (`burst_rd`), and burst write (`burst_wr`):

```
-- Type definition for state machine
TYPE mem_state IS (desel,
                   begin_rd,
                   begin_wr,
                   burst_rd,
                   burst_wr
                   );

SIGNAL state      : mem_state;
```

In burst mode, reads and writes may be either sequential or interleaved. The interleaved order is defined by a table:

```
TYPE sequence IS ARRAY (0 to 3) OF INTEGER RANGE -3 to 3;
TYPE seqtab IS ARRAY (0 to 3) OF sequence;

CONSTANT i10 : sequence := (0, 1, 2, 3);
CONSTANT i11 : sequence := (0, -1, 2, -1);
CONSTANT i12 : sequence := (0, 1, -2, -1);
CONSTANT i13 : sequence := (0, -1, -2, -3);
CONSTANT i1  : seqtab := (i10, i11, i12, i13);

CONSTANT ln0 : sequence := (0, 1, 2, 3);
CONSTANT ln1 : sequence := (0, 1, 2, -1);
CONSTANT ln2 : sequence := (0, 1, -2, -1);
CONSTANT ln3 : sequence := (0, -3, -2, -1);
CONSTANT ln  : seqtab := (ln0, ln1, ln2, ln3);

SIGNAL Burst_Seq : seqtab;
```

The i1 constants are for the interleaved burst sequences. The ln constants are for the sequential (linear) bursts.

The burst mode for this component can be set only at power-up time. A special process is used to initialized the burst sequence:

```
Burst_Setup : PROCESS

BEGIN

   IF (LBONegIn = '1') THEN
      Burst_Seq <= i1;
   ELSE
      Burst_Seq <= ln;
   END IF;
   WAIT; -- Mode can be set only during power up

END PROCESS Burst_Setup;
```

It is run once at time zero.

This component defines four commands. They are declared in the main behavior process:

```
-- Type definition for commands
TYPE command_type is (ds,
                          burst,
                          read,
                          write
                      );
```

On the rising edge of the clock, when the component is active

```
IF (rising_edge(CLKIn) AND CKENIn = '0' AND ZZIn = '0') THEN
```

each control input is checked for a valid value:

```
ASSERT (not(Is_X(RIn)))
    REPORT InstancePath & partID & ": Unusable value for R"
    SEVERITY SeverityMode;
```

If an invalid value is found the user is notified. Depending on how the user has modified the value of SeverityMode, the assertion might stop simulation. Assuming all the control inputs are valid, the command is decoded with an IF–ELSIF statement:

```
-- Command Decode
IF ((ADVIn = '0') AND (CE1NegIn = '1' OR CE2NegIn = '1' OR
     CE2In = '0')) THEN
    command := ds;
ELSIF (CE1NegIn = '0' AND CE2NegIn = '0' AND CE2In = '1' AND
     ADVIn = '0') THEN
    IF (RIn = '1') THEN
        command := read;
    ELSE
        command := write;
    END IF;
ELSIF (ADVIn = '1') AND (CE1NegIn = '0' AND CE2NegIn = '0' AND
        CE2In = '1') THEN
    command := burst;
ELSE
    ASSERT false
        REPORT InstancePath & partID & ": Could not decode "
            & "command."
    SEVERITY SeverityMode;
END IF;
```

Model behavior is controlled by the state machine. The state machine is built using a two-deep nesting of CASE statements:

```
                    -- The State Machine
                    CASE state IS
                      WHEN desel =>
                        CASE command IS
                          WHEN ds =>
                            OBuf1 := (others => 'Z');
                          WHEN read =>
      ...
                          WHEN write =>
      ...
                          WHEN burst =>
      ...
                        END CASE;

                      WHEN begin_rd =>
                        Burst_Cnt := 0;
                        CASE command IS
                          WHEN ds =>
      ...
                          WHEN read =>
      ...
```

The outer CASE statement uses the current state value. The inner CASE statement
uses the current decoded command. The appropriate action is then taken.

Pipelining is affected using chained registers. For example, on the first clock of a read,

```
        OBuf1(8 downto 0) := to_slv(MemDataA(MemAddr),9);
```

will execute. On the second clock,

```
        OBuf2 := OBuf1;
```

and finally,

```
        IF (OENegIn = '0') THEN
          D_zd <= (others => 'Z'), OBuf2 AFTER 1 ns;
        END IF;
```

puts the output value on the zero delay output bus, ready to be used by the Vital-
PathDelay01Z procedure.

14.6.2   DRAM

DRAMs can be built with higher density and lower cost than SRAMs. That has made
them very popular for use in computers and other memory-intensive devices. Their
primary drawback is their inability to store data for more than a few tens of mil-
liseconds without being refreshed. The refresh requirement makes them more
complex than SRAM, both in their construction and their use.

The DRAM model we examine here is for Micron part number MT4LC4M16R6. It is also sourced by OKI and Samsung. This component is 64Mb memory organized as 4M words with a 16-bit-wide data bus. However, the data are also accessible as 8-bit bytes for both reads and writes. The memory is modeled as two arrays of 8-bit words.

The model includes three processes. The first describes the functionality of the component. The other two contain generate statements for generating the required VitalPathDelay calls.

The distinguishing features of this model begin with its generics. The component has a constraint for maximum time between refresh cycles. Because this time is not related to any ports, a tdevice generic is employed:

```
-- time between refresh
tdevice_REF             : VitalDelayType    := 15_625 ns;
```

It is important to initialize the generic to a reasonable default value so the model can be used without backannotation.

The next generic is for power-up initialization time. The component may not be written to until a period of time has passed after power is applied.

```
-- tpowerup: Power up initialization time. Data sheets say 100-200 us.
-- May be shortened during simulation debug.
tpowerup            : TIME        := 100 us;
```

By using a generic for this parameter, we allow the user to shorten the time during early phases of design debug.

Because a tdevice generic is used, there must be a VITAL_Primitive associated with it:

```
-- Artificial VITAL primitives to incorporate internal delays
REF : VitalBuf (refreshed_out, refreshed_in, (UnitDelay, tdevice_REF));
```

Even though this primitive is not actually used in the model, it must be present to satisfy the VITAL_Level1 requirements. In this model, tdevice_REF is used directly to time the rate of refresh cycles:

```
IF (NOW > Next_Ref AND PoweredUp = true AND Ref_Cnt > 0) THEN
    Ref_Cnt := Ref_Cnt - 1;
    Next_Ref := NOW + tdevice_REF;
END IF;
```

The component requires there be 4,096 refresh cycles in any 64 ms period. Every 15,625 nanosecond, the code decrements the value of the variable Ref_Cnt. Each time a refresh cycle occurs, the value of Ref_Cnt is incremented:

```
IF (falling_edge(RASNegIn)) THEN
    IF (CASHIn = '0' AND CASLIn = '0') THEN
        IF (WENegIn = '1') THEN
            CBR := TRUE;
            Ref_Cnt := Ref_Cnt + 1;
```

Should the value of `Ref_Cnt` ever reach zero,

```
-- Check Refresh Status
   IF (written = true) THEN
      ASSERT Ref_Cnt > 0
         REPORT InstancePath & partID &
            ": memory not refreshed (by ref_cnt)"
         SEVERITY SeverityMode;
      IF (Ref_Cnt < 1) THEN
         ready := FALSE;
      END IF;
   END IF;
```

a message is sent to the user and a flag, `ready`, is set to false. This flag will continue to warn the user if further attempts are made to store data in the component:

```
-- Early Write Cycle
ELSE
   ASSERT ready
      REPORT InstancePath & partID & ": memory is not ready for"
         & " use - must be powered up and refreshed"
      SEVERITY SeverityMode;
```

Another interesting aspect of DRAMs is that they usually have multiplexed address buses. For this component there is an 12-bit column address and an 10-bit row address, but only a single 12-bit external address bus. Our memory is modeled as two linear arrays of naturals,

```
-- Memory array declaration
TYPE MemStore IS ARRAY (0 to 4194303) OF NATURAL
                     RANGE 0 TO 255;

VARIABLE MemH        : MemStore;
VARIABLE MemL        : MemStore;
```

one for the high byte and one for the low byte.

The internal address bus is viewed as being 22 bits wide to accommodate both the row and column addresses. However, the index into the memory arrays is defined as a natural:

```
VARIABLE MemAddr     : std_logic_vector(21 DOWNTO 0)
                        := (OTHERS => 'X');
VARIABLE Location    : NATURAL RANGE 0 TO 4194303 := 0;
```

The row and column addresses must be read separately:

```
         MemAddr(21 downto 10) := AddressIn;
...
      MemAddr(9 downto 0) := AddressIn(9 downto 0);
```

then converted to a natural for use as an array index:

```
Location   :=          to_nat(MemAddr);
```

14.6.3   SDRAM

SDRAMs are DRAMs with a synchronous interface. They often include pipelining as a means of improving their bandwidth. Refresh requirements are the same as for ordinary DRAMs.

The model we examine here covers the KM432S2030 SDRAM from Samsung. The equivalent Micron part is numbered MT48LC2M32B2. This component features all synchronous inputs, programmable burst lengths, and selectable CAS latencies. It is organized as 512K words, 32 bits wide. Internal memory is divided into four addressable banks. The 32-bit output is divided into four 8-bit words that can be individually masked during read and write cycles.

To accommodate the byte masking, each memory bank is modeled as an array of four 8-bit-wide memories. This makes byte access considerably simpler than using a single 32-bit memory array. It also solves the problem of having to deal with 32-bit integers on a 32-bit machine.

This SDRAM model is composed of two processes. One models the component functionality, the other generates the `VitalPathDelay` calls.

Distinguishing features of this model begin with its generics. This component has selectable CAS latency. This means the clock-to-output delays will depend on an internal register value. To annotate two sets of delays, two generics are needed:

```
-- tpd delays
tpd_CLK_DQ2                 : VitalDelayType01Z := UnitDelay01Z;
tpd_CLK_DQ3                 : VitalDelayType01Z := UnitDelay01Z;
```

The two CAS latencies also affect the maximum clock speed of the component. Therefore, there must be two values annotated for period constraint checking:

```
-- CAS latency = 2
tperiod_CLK_posedge         : VitalDelayType     := UnitDelay;
-- CAS latency = 3
tperiod_CLK_negedge         : VitalDelayType     := UnitDelay;
```

In either of these cases the requirements could not have been met using conditional delay and constraint generics (discussed in Chapter 10).

The SDRAM has a somewhat more complex state machine than the DRAM. In this case there are 16 states described in the data sheet. The state names are as follows:

```
-- Type definition for state machine

TYPE mem_state IS (pwron,
                   precharge,
                   idle,
                   mode_set,
```

```
                          self_refresh,
                          self_refresh_rec,
                          auto_refresh,
                          pwrdwn,
                          bank_act,
                          bank_act_pwrdwn,
                          write,
                          write_suspend,
                          read,
                          read_suspend,
                          write_auto_pre,
                          read_auto_pre
                      );
```

On careful examination of the data sheets from all the vendors, it becomes apparent that each bank has its own state machine. To reduce the amount of code in the model, it was decided to treat them as an array:

```
    TYPE statebanktype IS array (hi_bank downto 0) of mem_state;
    SIGNAL statebank : statebanktype;
```

It was decided to make `statebank` a signal rather than a variable. The reason was to delay any change of state till the next delta cycle. Had a variable been used, all state changes would be instantaneous.

In modeling a component as complex as this one, it is helpful to look at the data sheets from all the vendors producing compatible products. In the case of this component, one of four vendors studied, NEC, included a state diagram in their data sheet [8]. It is shown in Figure 14.4. Having an accurate state diagram greatly facilitates modeling a complex state machine.

A signal is needed for tracking the CAS latency:

```
    SIGNAL CAS_Lat  : NATURAL RANGE 0 to 3 := 0;
```

Like the DRAM, the SDRAM excepts a number of commands. An enumerated type is defined for them:

```
    -- Type definition for commands
    TYPE command_type is (desl,
                          nop,
                          bst,
                          read,
                          writ,
                          act,
                          pre,
                          mrs,
                          ref
                      );
```

3. Simplified State Diagram

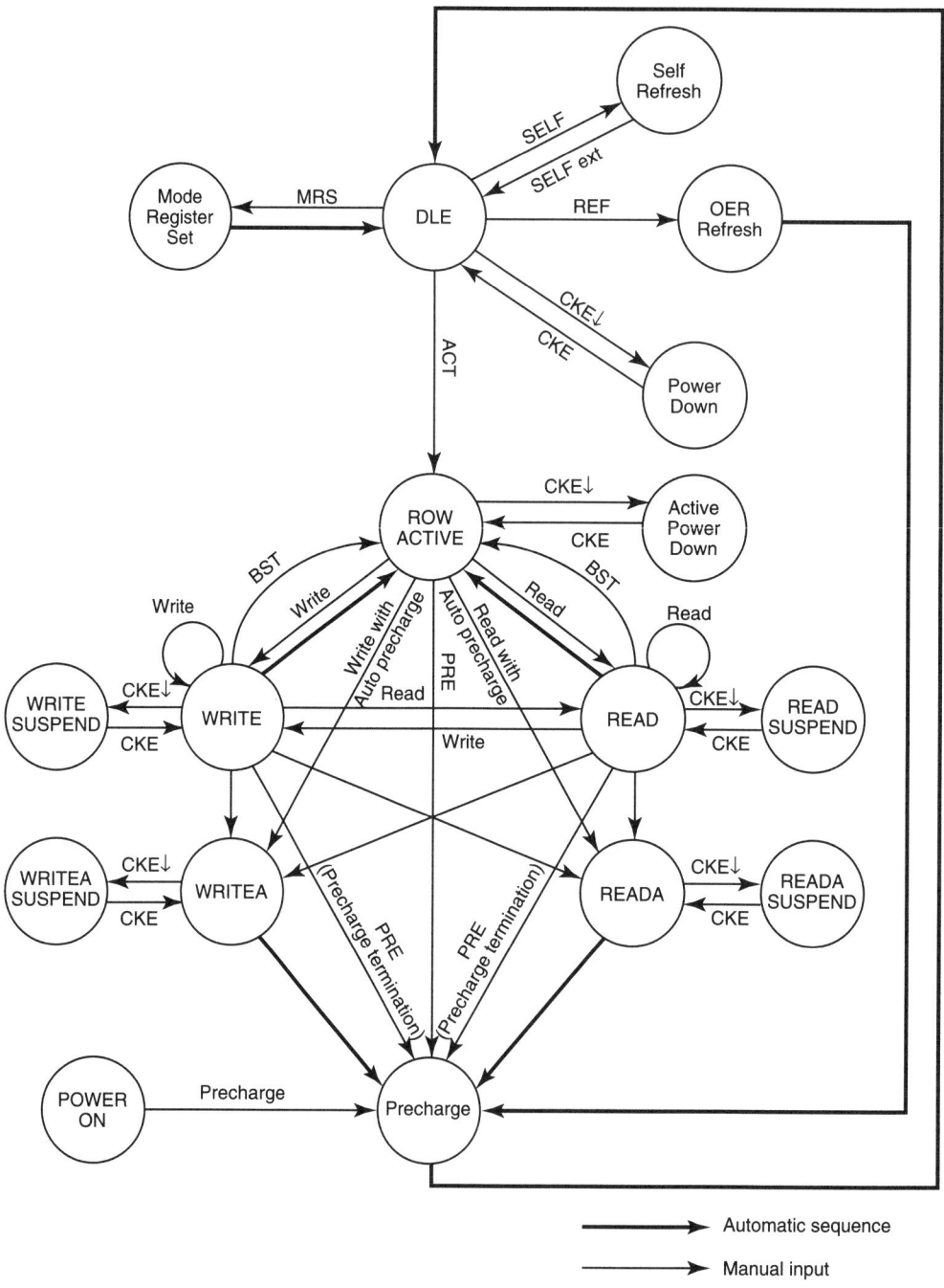

Figure 14.4 State diagram for SDRAM state machine

Memory arrays are declared in the behavior process:

```
-- Memory array declaration
TYPE MemStore IS ARRAY (0 to depth) OF INTEGER
                RANGE -2 TO 255;

TYPE MemBlock IS ARRAY (0 to 3) OF MemStore;
FILE mem_file          : text IS mem_file_name;
VARIABLE MemData0       : MemBlock;
VARIABLE MemData1       : MemBlock;
VARIABLE MemData2       : MemBlock;
VARIABLE MemData3       : MemBlock;
```

There is a separate array of four memories for each bank.

This is a programmable component. It contains a mode register that controls aspects of its behavior, such as CAS latency, burst length, and burst type:

```
VARIABLE ModeReg : std_logic_vector(10 DOWNTO 0)
                := (OTHERS => 'X');
```

As mentioned, the minimum clock period of this component varies with CAS latency. Two VitalPeriodPulseCheck calls are made. Only one is enabled at any time:

```
VitalPeriodPulseCheck (
      TestSignal       => CLKIn,
      TestSignalName   => "CLK",
      Period           => tperiod_CLK_posedge,
      PulseWidthLow    => tpw_CLK_negedge,
      PulseWidthHigh   => tpw_CLK_posedge,
      PeriodData       => PD_CLK,
      XOn              => XOn,
      MsgOn            => MsgOn,
      Violation        => Pviol_CLK,
      HeaderMsg        => InstancePath & PartID,
      CheckEnabled     => CAS_Lat = 2 );

VitalPeriodPulseCheck (
      TestSignal       => CLKIn,
      TestSignalName   => "CLK",
      Period           => tperiod_CLK_negedge,
      PulseWidthLow    => tpw_CLK_negedge,
      PulseWidthHigh   => tpw_CLK_posedge,
      PeriodData       => PD_CLK,
      XOn              => XOn,
      MsgOn            => MsgOn,
      Violation        => Pviol_CLK,
      HeaderMsg        => InstancePath & PartID,
      CheckEnabled     => CAS_Lat = 3 );
```

After the command is decoded (as in the DRAM), which bank it applies to must be determined. This is done with a CASE statement:

```
-- Bank Decode
CASE BAIn IS
   WHEN "00" => cur_bank := 0; BankString := " Bank-0 ";
   WHEN "01" => cur_bank := 1; BankString := " Bank-1 ";
   WHEN "10" => cur_bank := 2; BankString := " Bank-2 ";
   WHEN "11" => cur_bank := 3; BankString := " Bank-3 ";
   WHEN others =>
      ASSERT false
         REPORT InstancePath & partID & ": Could not decode bank"
            & " selection - results may be incorrect."
         SEVERITY SeverityMode;
END CASE;
```

Next comes the state machine itself:

```
-- The Big State Machine
IF (rising_edge(CLKIn) AND CKEreg = '1') THEN
```

It contains a FOR loop that causes four passes through the state machine code for each active clock. That is one pass for each bank:

```
banks : FOR bank IN 0 TO hi_bank LOOP
CASE statebank(bank) IS
   WHEN pwron =>
...
      IF (command = bst) THEN
         statebank(bank) <= bank_act;
         Burst_Cnt(bank) := 0;
```

The state machine code is over 800 lines. The complete model may be found and downloaded from the FMF Web site.

Finally, the output delay is determined by CAS latency:

```
VitalPathDelay01Z (
   OutSignal        => DataOut(i),
   OutSignalName    => "Data",
   OutTemp          => D_zd(i),
   Mode             => OnEvent,
   GlitchData       => D_GlitchData(i),
   Paths            => (
      1 => (InputChangeTime => CLKIn'LAST_EVENT,
            PathDelay => tpd_CLK_DQ2,
            PathCondition    => CAS_Lat = 2),
```

```
          2 => (InputChangeTime  => CLKIn'LAST_EVENT,
                PathDelay => tpd_CLK_DQ3,
                PathCondition      => CAS_Lat = 3)
      )
    );
```

14.7   Summary

The efficiency with which memory is modeled can determine whether or not a board simulation is possible within the memory resources of your workstation. The most efficient ways of modeling large memories are the Shelor method and the VITAL_memory package's `VitalDeclareMemory` function.

Functionality may be modeled using the behavioral method or the VITAL2000 method. Which you choose will depend on the complexity of the model and how comfortable you are writing tables. Some functionality may be easier to describe with one method or the other.

Path delays in memory models may be written using the path delay procedures provided by the VITAL_Timing package or the those provided by the VITAL2000 memory package. If the part being modeled has output-retain behavior, the added complexity of the memory package procedures may be worth the effort. Otherwise, you will want to use the procedures from the timing package.

The VITAL2000 memory package has its own setuphold and periodpulsewidth checks. They are of limited value for component modeling.

Adding preload capability to a memory model is not difficult and well worth the additional effort required in a behavioral model. In a VITAL2000 model the capability, though somewhat restricted, comes for free.

15 Considerations for Component Modeling

The way a model is written is influenced by the perspective of the model's author and by how he or she thinks the model will be used. Someone with a background in chip design but has never designed a board is likely to write a model that would be very good at verifying the component but unsuitable for verifying that the component is correctly designed into the system.

The key to understanding component modeling is to understand how the models will be used. The purpose of component models is not to verify components, but to verify systems.

15.1 Component Models and Netlisters

Simulation models are of little use individually. Their value comes from connecting them together in a netlist and seeing how they interact. Although one could write a netlist by hand, it would not be interesting work. It would also be prone to errors and frequent revisions.

Most vendors of schematic capture software offer tools to netlist their schematics in VHDL. These tools may be included with the schematic capture systems or may be purchased separately. In either case, the ideal is to generate a VHDL netlist from the same schematic that will be used to generate the PCB netlist. Doing this requires matching the simulation models to the requirements of the schematic capture system and its VHDL netlister.

One consideration is matching the ports in the model to the pins on the schematic symbol. It is usually permissible for the symbol to have pins that are not represented in the model. Power, ground, and voltage references are all pins that may not be needed for simulation but are required to design a printed circuit board. It is never permitted to have ports on the model that do not have corresponding pins on the schematic symbol. In other words, it is permitted to lose pins as you traverse down into the hierarchy but not as you push up through it.

Going from schematic to netlist, the facilities of the schematic capture system and netlister may be used to map unmodeled pins on the symbol to the VHDL key word OPEN. There can be no such mapping for unused ports.

A simulation model should try to be true to the logic of a component, not its physical characteristics. A 7400 NAND gate comes packaged as four gates in a 14-pin DIP (dual in-line package). A simulation model of a 7400 should contain only one NAND gate because (in most cases) a schematic symbol of a 7400 will be a single NAND gate. On the opposite end of the spectrum, a DSP (digital signal processor) model should include all the functionality of the part in a single instance because that is the way it is drawn on a schematic. However, in between there is a lot of gray area. Beginning with a 74244 tri-state buffer, decisions based on judgement and local usage are required. The 74244 can be, and usually is, drawn two different ways, as shown in Figure 15.1. One way is a single buffer with three pins. The other is a quad buffer with 9 pins. The 244 is most often placed in a schematic using the quad representation. It could be modeled that way, but if the single-buffer representation will ever be used, it must be modeled as a single buffer. Always keep in mind that it is easy to use the simple model with the more complex symbol, but it is difficult to use the complex model with the simple symbol.

Most netlisters will allow the value of a model generic to be inherited from a property or attribute of the same name attached to its schematic symbol. The inheritance is done on a per-instance basis. The generic of most importance for FMF-style models is TimingModel. It gets a value of type STRING that corresponds to a part number in the model's timing file. This part number may be the manufacturer's part number, a corporate part number, or anything else, as long as it is listed in the timing file. The TimingModel value is passed into the instance of the component in the netlist. The mk_sdf script, described in Chapter 12 and later in this chapter, reads the value and matches it to a part number in the timing file. It uses that part number to select the desired timing from the timing file and put it in the SDF file for the netlist. The TimingModel generic may also be used to control certain types of behavior in the model. An example of using TimingModel to change the functionality of a flash memory is given in Chapter 16.

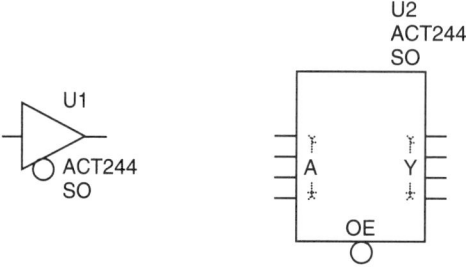

Figure 15.1   Two symbols for a 74244

15.2 File Contents

Each component in a VHDL netlist must be declared and instantiated. To do this, the netlister needs to read the entity of each model while generating a VHDL netlist from a schematic. The schematic capture tool's library must be organized so the netlister knows where to look for the model associated with each symbol. The details of how this is done will vary from tool to tool. An example for Cadence's ConceptHDL tool is given in Section 15.4. When the netlister finds and reads the model file, it expects that the first thing it comes to at the top of the file is the model's entity. Libraries, packages, and so on must not be placed ahead of the model entity.

Although VHDL is quite happy with multiple models being placed in a single file, schematic netlisters are usually less sympathetic. Include only one entity per file if you plan to netlist. Of course, there is no reason that file should not also contain the model's architecture(s).

15.3 Generics Passed from the Schematic

It is the ability to pass generics from the schematic to the model instance that makes technology-independent libraries and models possible. It also allows much smaller but richer schematic symbol and model libraries. There are two classes of generics that we are interested in passing from the schematic to the component models: timing generics and control generics.

15.3.1 Timing Generics

Most timing generics get their values from SDF backannotation. However, there are some that may be passed in through the netlister. One that is common is found in DRAMs and is usually named `tpowerup`. It is used to control how long the simulation must wait before beginning to issue commands to the memory. The model contains a default value that comes from the data sheet. This is usually a large value, on the order of 200 microseconds. The engineer has the option of adding a `tpowerup` attribute to the symbol in the schematic and setting its value to something smaller, such as 2 microseconds, while debugging a design.

15.3.2 Control Generics

Other generics have control functions. A popular generic for memory models is `mem_file`, mentioned in Chapter 14. It is attached to a symbol instance and passed into the netlist. It takes a string value that is the name of a file that contains data to be preloaded into a memory instance. At power up, or possibly at some other time, the model reads the external text file and loads the data into its memory array(s).

The most important control generic for technology-independent models is `TimingModel`. Its primary purpose is to control which of many possible sets of timing parameters for a model is selected and written into the design's SDF file. In some cases it can also be used to configure a model that represents a component with hardwired configurations. An example is a flash memory that may be purchased with a protected area at either the top or the bottom of its address range.

Many netlisters will find an attribute attached to some special object in the schematic, such as the schematic border, and annotate its value to every generic with the same name throughout the design. There are some control generics for which this may be a desirable feature. The following four generics may be updated either on an instance-by-instance basis or for the entire design. They may also be updated through the simulator.

- The `TimingChecksOn` generic controls whether timing constraint checks are performed. During the early phases of board-level verification, it may be desirable to disable timing checks while other issues are being worked on.

- The `MsgOn` generic controls the emission of text messages when a timing violation is detected. Large numbers of these messages can slow down a simulation and mask other messages that might be more useful.

- The `XOn` generic controls the generation of 'X' values driving the violation flags. The violation flags in turn may control some aspect of the model's behavior, such as the corruption of memory locations. `XOn` can be set to FALSE to disable such behavior.

- The `SeverityMode` generic can be used to control the severity level of messages issued by procedures using it. These messages can then be used to warn the user or even stop the simulation. The effect of messages of varying severity levels is controlled directly in the simulator.

15.4   Integrating Models into a Schematic Capture System

The payoff for writing component simulation models comes when they are integrated into a schematic capture system and used for board-level verification. What follows is an example of how component models can be integrated and used with one particular schematic capture system, Cadence's ConceptHDL. In the example it is assumed that Cadence's Allegro is used for PCB layout and Mentor's ModelSim is used for simulation. Although there are many other tools available and they have many differences, the basic principles are the same for all the tools. Only the details of implementation will vary.

15.4.1   Library Structure

The heart of every Computer Aided Engineering (CAE) system is its libraries. This is particularly true for board design and simulation. A library is, for our purposes, a collection of data and information regarding one or more electronic components. There are libraries of schematic symbols, PCB footprints, simulation models, and so on. Some libraries are collections of other libraries. Although the schematic libraries described in this chapter have been optimized for VHDL simulation, you can still use the Verilog RTL models of your FPGAs in board-level simulations, as described in Chapter 13.

15.4.2 Technology Independence

A key feature of a well-designed CAE library is the separation of functionality from timing. This allows for technology independence and significantly reduces the total number of schematic symbols (and models) in the library. The saving varies with the part family: for the 7400 series, the savings can be huge; for more specialized components, it may be minimal. In all cases it allows for timing customization without requiring changes to a proven model or the additttion of a copy of a schematic symbol.

15.4.3 Directories

The directory structure for our simulation-enabled CAE library starts out looking very similar to the standard ConceptHDL library, as shown in Figure 15.2. For each component library, there must also be a corresponding VHDL library (this is the ConceptHDL standard and may not be required for other systems). The VHDL libraries have three additional directories: src, TimingModel, and work. The src

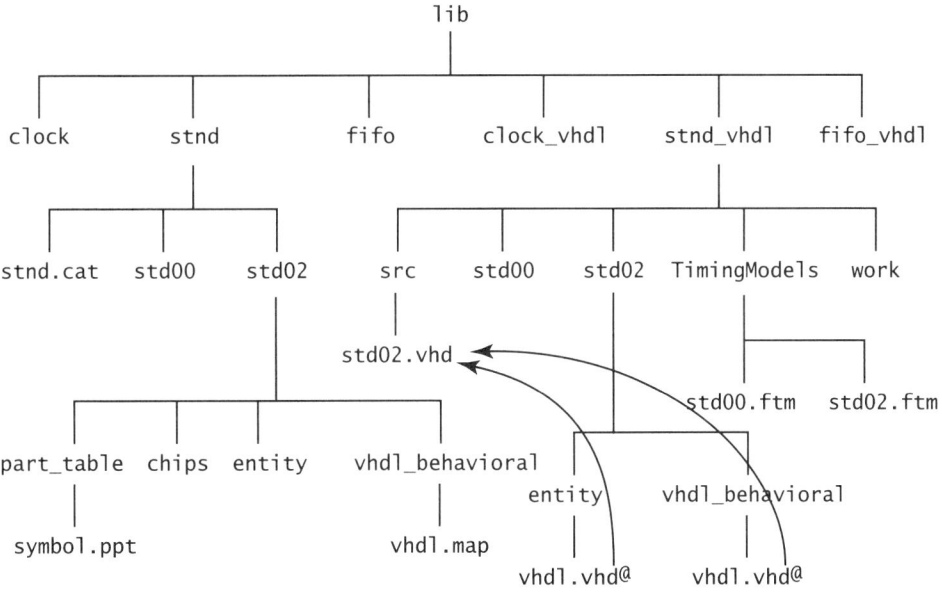

Figure 15.2 Library directory structure for ConceptHDL

```
FILE_TYPE = VHDL_MAP;
PRIMITIVE 'ACT04_DP','ACT04_SO';
  DEFAULT_MODEL = 'STD04';
  MODEL 'STD04';
    PIN_MAP
      'A'<0> = '(A)';
      '-Y'<0> = '(YNeg)';
    END_PIN;
    END_MODEL;
  END_PRIMITIVE;
  END.
```

Figure 15.3 Sample vhdl_map file

directory contains the VHDL model source code. The TimingModels directory contains the timing (.ftm) files. The work directory contains the compiled VHDL models.

It should be pointed out here that the directories found in the component library are based on the part body names but the directories found in the VHDL library are based on the model names. There is not a one-to-one mapping between the body names in the component library and the model names in the corresponding VHDL library. The mapping between the two libraries is controlled by each part's vhdl.map file.

Within the entity and vhdl_behavioral directories (in the lib_vhdl) there are links named vhdl.vhd. They point back to the model source in the src directory. These links are required for the netlister to work.

15.4.4   Map Files

Map files are required to map pin names from ConceptHDL chips.prt files to the VHDL port names listed in the entity. On a good day, they perform a simple function simply. On a bad day, it can take a lot of trial and error to figure out how they work. An example is shown in Figure 15.3.

The pin_map section of the file contains a mapping from each pin name listed in the chips.prt file to the corresponding port name in the VHDL model entity. Should there be no corresponding port in the entity, the pin is mapped to the key word OPEN:

        'VCC' = '(OPEN)';

Map files are explained in greater detail in the "Concept-HDL Digital Simulation User Guide" in cdsdoc, the Cadence online documentation library.

15.5   Using Models in the Design Process

What follows is the general design process through simulation and analysis. It discusses the various tools and the order in which they are used. This particular process is specific to the author. Your process will depend on the tools you choose but will

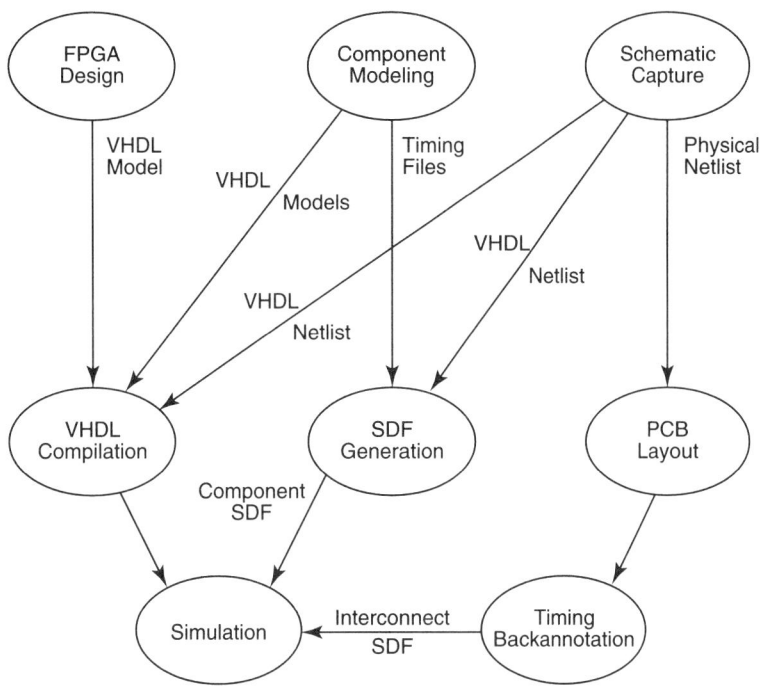

Figure 15.4   Design simulation flow

probably be similar to the one described. A diagram of this process is shown in Figure 15.4.

15.5.1   VHDL Libraries

It is assumed at this point that the VHDL libraries are in place and are precompiled for the user. This would normally be done by the CAE librarian if your company employs one; however, the user may be responsible for setting up any verilog FPGA or ASIC models in the design, as described earlier. The component libraries are technology independent and utilize the FMF-style simulation models described in this book. This reduces the effort required for library development and maintenance.

15.5.2   Schematic Entry

Schematics are drawn in ConceptHDL as they would be for any board design (with some caveats to be discussed later). The schematics may be either hierarchical or flat. Because the schematic libraries use FMF technology-independent model libraries, components must be added to the schematics using the component browser set to physical mode. Selecting parts this way causes properties to be added to the schematics that are later used to select the correct timing for each component. Other schematic capture tools have similar capabilities to the ConceptHDL component browser. Mentor's DxDesigner, for instance, has DxDataBook.

Remove Attribute

Some parts in the design are best left out of the simulation. Decoupling capacitors, for example, will add nothing to the functionality. They will just make the netlist longer. Series termination resistors also will just add clutter to your simulation, but the pull-up and pull-down resistors should stay.

ConceptHDL has an attribute that can help, called REMOVE. There are four possible values. The two most useful are LINK and EXCLUDE. If a body has REMOVE= EXCLUDE attached to it, the netlister will remove that component from the netlist all together. This is good for all those decoupling caps. If a body has REMOVE= LINK, the two nets attached (this should be used only on discrete components) will be aliased, in effect shorting them together. This is good for series terminating resistors. Be aware that aliases can have side effects in a hierarchical design if one of the nets is a port on the block.

Hierarchical Bodies

In a hierarchical design each block becomes a subdesign in the netlist. As such, it has its own entity and port list. In the ConceptHDL environment, the direction of the ports is determined by the file <block_name>/entity/vhdl.vhd. This file is created and/or modified each time the body of the block is written. If the ports don't come out with the intended direction automatically, the easiest thing to do is simply edit that file by hand. Other tools may have better ways of controlling block entities.

15.5.3   Netlisting the Design

Data are passed from schematic capture to simulation through a VHDL netlist. Depending on your tool supplier, the netlisting tool may be bundled with the schematic capture system or it may be a separate tool.

Setting Up the Netlister

Before netlisting in ConceptHDL, a number of setup steps are required:

- Ensure that for every component library in your design you also reference the corresponding VHDL library.

- Under setup –> tools –> simulation, select the simulator type that is correct for your site. If you are not using Cadence's simulator, select "Third Party VHDL." Then click on "Setup."

- Enter the VHDL package libraries you will be using. If you use FMF models you will enter "IEEE" and "FMF."

- Enter the packages you will use from the libraries. For FMF models they are IEEE.std_logic_1164.all, IEEE.VITAL_TIMING.ALL, IEEE.VITAL_PRIMITIVES.ALL, and FMF.gen_utils.all.

- If you have turned off "Create Netlist" in ConceptHDL, turn it on and check VHDL. Write all pages of your design.

Running the Netlister

Using the graphical user interface is recommended:

- From the tools menu in either ConceptHDL or the Project Manager, select "simulate."

- From the popup window that appears, enter the path to where you want the VHDL netlist and log files to go.

- Click on "Run."

Eventually, a popup will appear to tell you that netlisting was successful, or not. If it was not successful, a markers file may have been created that you can use in ConceptHDL to find where in your schematic the error occurred. However, it may be easier to read the netassembler.log file that was placed in the same directory you specified for the netlist. The most likely errors to be reported are port mismatches.

15.5.4   VHDL Compilation

Having produced a VHDL netlist, the next step is to compile it. For ModelSim, the compiler is called vcom and the command is vcom <filename>. Before the netlist can be compiled the first time, a work directory must be established to receive the compilation results. This is done with the command vlib work. If there are compiler errors, they will be written to stdout.

15.5.5   SDF Generation

Each FMF-style model has an associated timing file that describes the internal delays of a component with any required timing constraints. Some of the benefits of external timing files are reduced number of models to write and maintain and flexibility in exploring timing differences for a number of component suppliers.

SDF generation produces a file in SDF that may be used by the simulator to provide accurate timing for the simulation. Initial simulation runs may not require timing and can skip this step. Without SDF annotation, FMF-style models default to unit delays. However, some models, such as certain memories, may give misleading results when run with unit delays.

```
SET sdffile_suffix .sdf
SET use_global_timing_dir false
SET timingfile_dir TimingModels
SET timingfile_suffix .ftm
SET time_scale 1ns
SET local_path .
SET diagnostics off
SET vhdl_file vhdllink.vhd
SET lwb off
```

Figure 15.5   Sample mk_sdf.cmd file

Although SDF files may be created by hand, such work is tedious. Therefore, a tool has been created to automate the task. The SDF tool is called mk_sdf and may be obtained as a perl script from the Free Model Foundry (at no cost). It uses a command file named mk_sdf.cmd, which should reside in the working directory (a sample mk_sdf.cmd file is given in Figure 15.5). This tool reads the design's VHDL netlist or testbench (note that it does not work with uninstantiated models). It uses VHDL configuration statements, such as

```
for all : CDC339 use entity CLOCK.cdc339(VHDL_BEHAVIORAL);
```

to determine which library to search in for the timing file for each model. It also reads the TimingModel generic for each instantiation,

```
TIMINGMODEL => "CDC339DB"
```

and uses the value it finds to search the timing file for the correct section. Here CDC339DB is the real part number and is listed in the timing file cdc339.ftm. The timing file may contain timings for many parts. The timing you want to use is specified by the TimingModel generic.

The instance names in the netlist,

```
I3P_S4 : CDC339
```

are extracted for use in the SDF file.

The configuration file, called mk_sdf.cmd, provides the tool with information about its environment. The mk_sdf script is invoked with the command

```
mk_sdf [netlist_name] [sdf_file_name]
```

The mk_sdf.cmd file contains a number of **set** directives. Here are the commands available in version 2.0, the first perl version. Directives may be in any order. Except for one, diagnostics, they have no default values:

SET sdffile_suffix <.suffix> The SDF file produced will have the same name as the VHDL file read, except the suffix will be changed to whatever you specify in this command. Alternatively, the name of the SDF file may be entered as the second argument on the command line.

SET `use_global_timing_dir` <true | false> Timing files may be distributed among the CAE symbol libraries or they may all be kept in a single directory. If they are distributed it is assumed they are in a directory structure parallel to the compiled models. This directive may be set to true or false.

SET `timingfile_dir` <directory_name> The timing files will be kept in one or more directories. This directive gives the name of the directories.

SET `vendor` <modeltech | cadence> If `use_global_timing_dir` is false, `mk_sdf` will try to find the paths of the libraries. If vendor is set to `modeltech`, it will try to read the global and local `modelsim.ini` files. If vendor is set to `cadence`, it will try to read the global and local `cds.lib` files.

SET `vhdl_file` <file_name> The `vhdl_file` is the name of the netlist. It may also be entered as the first argument on the command line.

SET `diagnostics` <on | off> Diagnostics may be set **on** or **off**. The default is **off**. If set **on**, the program will write voluminous messages informing you in detail of its progress. This should be useful in debugging problems in the environment or with unexpected formatting in netlists. If even more detail is required in diagnosing problems, use the perl –d (for debug) option on the first line on the program.

Any line in the command file may be overridden from the command line. If the command file is fully and correctly set, execution may be accomplished by simply executing `mk_sdf` without any arguments.

15.5.6   Simulation

The ModelSim simulator is invoked from the command line with arguments for SDF backannotation, the name of the SDF file, and the design name: `vsim –sdfmax mydesign.sdf mydesign`. If timing is not required, the –sdfmax `mydesign.sdf` argument may be omitted for unit delay simulation. Alternatively, multiple SDF files may be read to include interconnect delays from the PCB layout tool, timing files for ASICs and FPGAs, and so on. Based on the results of the simulation, there may be a loop back to schematic capture from this point to refine the design.

15.5.7   Layout

After satisfactory simulation results are obtained, the design goes to PCB layout. If timing margins were determined to be tight, anticipated interconnect delays based on manhattan distances between pins may be computed and backannotated through SDF to check if timing constraints are likely to be met by the proposed layout.

15.5.8   Signal Analysis

Subsequent to layout and routing of the PCB, signal integrity analysis may be performed. A well-designed library will provide the names of the signal integrity models to be used in an SI tool, such as SpecctraQuest or HyperLynx. These tools are capable of computing values for various physical effects, such as crosstalk and noise margins. They can also compute accurate interconnect delays based on the characteristics of a driver's output buffer, receiver thresholds, propagation delays calculated for the board stackup, and transmission line analysis of the traces.

15.5.9   Timing Backannotation

Accurate interconnect delays, including transmission line effects, can be extracted from Allegro, the Cadence PCB layout tool, with the `writesdf` utility. Instructions for running `writesdf` appear on the popup informing you of successful netlisting.

15.5.10   Timing Analysis

Once the interconnect SDF file is generated, a final full timing simulation can be run. It would be more efficient at this point to run a static timing analyzer such as Blast or TAU from Mentor. All the data needed appear to be in the SDF files, but the few static timing analyzers available read only proprietary model formats. If your organization can support one of these tools, by all means use it. Otherwise, dynamic simulation with full timing backannotation is your best bet for timing verification.

15.6   Special Considerations

While the design-to-simulation process described here should be relatively easy and straightforward, it is possible to make it difficult. What follows are some techniques and caveats to keep in mind to avoid unnecessary complications.

15.6.1   Schematic Considerations

Schematics are the primary means of capturing board-design intent. Minor details in schematics can have substantial downstream effects.

Net Names

Try to use legal VHDL names for all nets. Names that are not legal VHDL will be changed by the netlister into legal but less readable names.

Unused Bits

Some people like to run busses into hierarchical blocks but not connect all bits of the bus inside the block. The netlister may have difficulties with this. Either connect to the block only the number of bits actually used or, for ConceptHDL users, attach a port body to each unconnected net inside the block. Other options may be available to users of other schematic tools.

Mixed Busses

Sometimes an engineer will create a bus for which some bits are inputs to a block and other bits are outputs from the block. Doing this will successfully prevent netlisting. All bits on a bus must go in the same direction. They can be inputs to the block, outputs, or bidirectional, but they must all be the same. The author does not know of any way to work around this restriction.

Port Bodies

The ConceptHDL netlister occasionally has difficulty determining the correct mode of a port on a hierarchical body. Usually the port mode can be forced to the correct value by attaching the appropriate port body to the net inside the block and/or manually editing the `entity/vhdl.vhd` file for the block. There are three usable port bodies: `INPORT`, `OUTPORT`, and `IOPORT`. They create ports of types `IN`, `OUT`, and `INOUT`, respectively.

15.6.2   Model Considerations

As noted earlier, small components that come as multiple units per package should be modeled at the lowest level they appear as schematic symbols. For example, in Figure 15.6 we have a schematic. The only components in this schematic are 74LVC2244s. There are three instances of this component. Two instances are represented as single gates. The third instance is of four gates and their common enable. This component is modeled as a single gate. When the netlister encounters the single-gate representation in the schematic, it of course adds a single instance to the netlist. When the four-gate symbol is encountered, it adds four instances to the netlist. The following shows a portion of the netlist from the schematic in Figure 15.6.

```
BEGIN
---
--- Component instances
---

   page1_I2_S1:   std244
     GENERIC MAP (
        tipd_Y => VitalZeroDelay01,
        tipd_A => VitalZeroDelay01,
        tipd_OENeg => VitalZeroDelay01,
        tpd_A_Y => UnitDelay01,
```

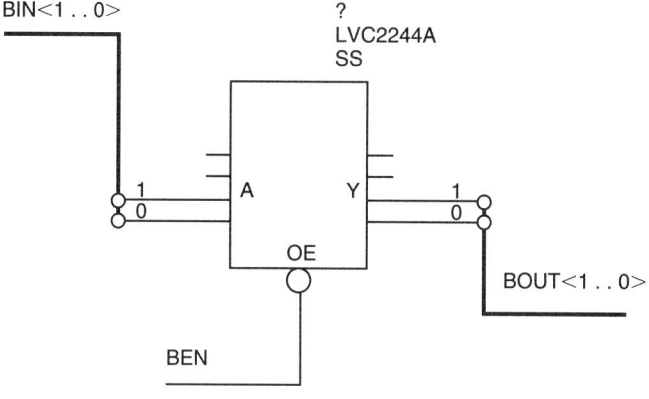

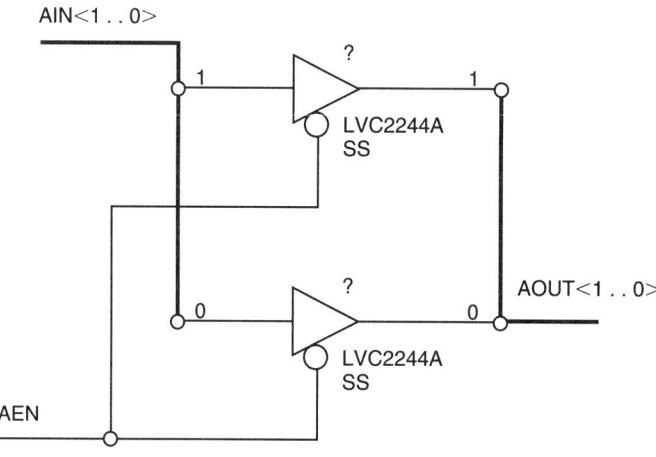

Figure 15.6   Different representations of a 74244

```
        tpd_OENeg_Y => UnitDelay01Z,
        InstancePath => "DefaultInstancePath",
        TimingModel => "LVC2244A"
    )
    PORT MAP (
       A  => open ,
       OENeg => BEN ,
       Y  => open
    );

page1_I2_S2: std244
   GENERIC MAP (
```

```
            tipd_Y => VitalZeroDelay01,
            tipd_A => VitalZeroDelay01,
            tipd_OENeg => VitalZeroDelay01,
            tpd_A_Y => UnitDelay01,
            tpd_OENeg_Y => UnitDelay01Z,
            InstancePath => "DefaultInstancePath",
            TimingModel => "LVC2244A"
        )
        PORT MAP (
          A  => open ,
          OENeg => BEN ,
          Y => open
        );

    page1_I2_S3: std244
        GENERIC MAP (
            tipd_Y => VitalZeroDelay01,
            tipd_A => VitalZeroDelay01,
            tipd_OENeg => VitalZeroDelay01,
            tpd_A_Y => UnitDelay01,
            tpd_OENeg_Y => UnitDelay01Z,
            InstancePath => "DefaultInstancePath",
            TimingModel => "LVC2244A"
        )
        PORT MAP (
          A => BIN (1),
          OENeg => BEN ,
          Y => BOUT (1)
        );

    page1_I2_S4: std244
        GENERIC MAP (
            tipd_Y => VitalZeroDelay01,
            tipd_A => VitalZeroDelay01,
            tipd_OENeg => VitalZeroDelay01,
            tpd_A_Y => UnitDelay01,
            tpd_OENeg_Y => UnitDelay01Z,
            InstancePath => "DefaultInstancePath",
            TimingModel => "LVC2244A"
        )
        PORT MAP (
          A => BIN (0),
          OENeg => BEN ,
          Y => BOUT (0)
        );
```

```
page1_I3: std244
   GENERIC MAP (
      tipd_Y => VitalZeroDelay01,
      tipd_A => VitalZeroDelay01,
      tipd_OENeg => VitalZeroDelay01,
      tpd_A_Y => UnitDelay01,
      tpd_OENeg_Y => UnitDelay01Z,
      InstancePath => "DefaultInstancePath",
      TimingModel => "LVC2244A"
   )
   PORT MAP (
      A  => AIN (0),
      OENeg => AEN ,
      Y => AOUT (0)
   );

page1_I4: std244
   GENERIC MAP (
      tipd_Y => VitalZeroDelay01,
      tipd_A => VitalZeroDelay01,
      tipd_OENeg => VitalZeroDelay01,
      tpd_A_Y => UnitDelay01,
      tpd_OENeg_Y => UnitDelay01Z,
      InstancePath => "DefaultInstancePath",
      TimingModel => "LVC2244A"
   )
   PORT MAP (
      A => AIN (1),
      OENeg => AEN ,
      Y => AOUT (1)
   );
```

As you can see, there are six instances of the std244 model in the netlist. Two of those instances have ports mapped to OPEN. The netlister was able to read a single schematic symbol and translate it to four component instantiations in the netlist.

15.7   Summary

Boards are usually designed using schematics. Board-level verification begins with the schematic-derived VHDL netlist. A component model should be tailored to work with the component's schematic representation. Every port on the model must be shown on the schematic symbol.

Model generics may be modified by attaching attributes to the symbols in the schematic. For models of the timing-independent style described in this book, the most important attribute is `TimingModel`. Other useful attributes vary by model type. To be passed from schematic to netlist the attribute name must match the generic name.

The details of how to integrate component models into a schematic capture system vary from system to system. They usually involve a mechanism to map schematic pin names to model port names and a way for the netlister to find the model entities.

After netlisting, an SDF file can be generated by running the `mk_sdf` script. The `mk_sdf.cmd` file can customize the behavior of `mk_sdf` to different design environments.

16 Modeling Component-Centric Features

There is always a new odd part to model. The semiconductor industry has produced an astounding variety of electronic devices. Just when you think you know how to model any part built, another one comes along that provides new challenges. This chapter presents just a few of the challenges you are likely to run into, and their solutions.

16.1 Differential Inputs

An increasing number of new components fall into the category of high-speed devices. One thing that characterizes many high-speed devices is differential inputs. There are several strategies for modeling differential inputs, depending on how they are used and what types of errors you would like to detect during simulation.

The simplest and least satisfying way of dealing with a differential input is to just throw away one phase and treat the other as a single-ended input. This method will fail to catch any errors related to the input's differential nature and, for that reason, is discouraged.

The next step up is to use both phases to generate a single internal signal through the use of an AND function such as

```
CLKcomb <= CLKIn AND not(CLKNegIn);
```

This requires that both phases be connected and toggled, but it can cause timing distortions under some circumstances.

A skew check, such as the one in Figure 16.1, can be used to check that the two phases of a clock or other differential signal are properly aligned:

With this, the user can be warned if the timing distortion exceeds a certain value. It would be nice to have a skew check for every differential input. Unfortunately, most data sheets do not define maximum allowable skew, so any value used in the skew check would be a guess.

Components such as differential receivers require more care in the preservation of timing in order to accurately simulate downstream circuitry. The Free Model Foundry IF490 model uses a state table for converting differential input to a single-ended signal:

```
--------------------------------------------------------------------------
-- Functionality Section
--------------------------------------------------------------------------
VitalStateTable (
    StateTable      => diff_rec_tab,
    DataIn          => (RIN_ipd, RINNeg_ipd),
    Result          => R_zd,
    PreviousDataIn  => PrevData
);
```

The state table itself is defined in the FMF `gen_utils` library and is shown in Figure 16.2.

First the inputs are checked for unknowns. An unknown on an input always results in an unknown output. Next, to recover from a previous known output, the A input is checked for a '1' or a '0'. If either value is found, it is applied to the output. If neither of the previous conditions exist, the inputs are checked for an edge on one input and an opposite value on the other. Should such values be found, a '1' or a '0' is output as appropriate. If none of the aforementioned situations apply, the output is left unchanged.

This state table should be a safe bet for converting a differential input to single ended in most non-ECL models. It works by watching both inputs for transitions. Only when the two inputs are of opposite polarity and have changed from their previous state does the table output a new value.

ECL models provide some of the most interesting examples of differential inputs. One feature that makes them interesting is they often can be made to behave as single-ended inputs by connecting one pin to a special output called vbb. vbb is a constant intermediate voltage. When tied to one pin of a differential pair, the other pin can be driven as a single-ended input. To model a component with this feature correctly requires the detection of the special voltage level. VHDL, being a digital

```
VitalOutPhaseSkewCheck (
    Signal1           => CLKIn,
    Signal1Name       =>  "CLK",
    Signal2           => CLKNegIn,
    Signal2Name       =>  "CLKNeg",
    SkewS1S2RiseFall  => tskew_CLK_CLKNeg,
    SkewS2S1RiseFall  => tskew_CLK_CLKNeg,
    SkewS1S2FallRise  => tskew_CLK_CLKNeg,
    SkewS2S1FallRise  => tskew_CLK_CLKNeg,
    CheckEnabled      => TRUE,
    HeaderMsg         => InstancePath & PartID,
    SkewData          => SD_CLK_CLKNeg,
    Trigger           => CKSKWtrg,
    XOn               => XOn,
    MsgOn             => MsgOn,
    Violation         => Sviol_CLK_CLKNeg );
```

Figure 16.1 Skew check for differential clock

```
-------------------------------------------------------------------------
-- Table for computing a single signal from a differential receiver input
-- pair.
-------------------------------------------------------------------------
CONSTANT diff_rec_tab : VitalStateTableType  := (

------INPUTS--|-PREV-|-OUTPUT----
--   A   ANeg | Aint |  Aint'  --
-------------|------|-----------
  ( 'X', '-',   '-',   'X'), -- A unknown
  ( '-', 'X',   '-',   'X'), -- A unknown
  ( '1', '-',   'X',   '1'), -- Recover from 'X'
  ( '0', '-',   'X',   '0'), -- Recover from 'X'
  ( '/', '0',   '0',   '1'), -- valid diff. rising edge
  ( '1', '\',   '0',   '1'), -- valid diff. rising edge
  ( '\', '1',   '1',   '0'), -- valid diff. falling edge
  ( '0', '/',   '1',   '0'), -- valid diff. falling edge
  ( '-', '-',   '-',   'S')  -- default
); -- end of VitalStateTableType definition
```

Figure 16.2   Differential receiver state table

hardware description language, is usually ill suited to such a task. However, this time we are lucky.

The `std_logic_1164` package defines a signal strength 'W' that is rarely used in board-level simulation. Component models can use the 'W' value for the ECL vbb output. Because there will be nothing else that ever drives a 'W', its presence on an input will be a sure indication that the pin is tied to vbb. The FMF `ecl_utils` package contains all the required declarations for modeling ECL components.

A slightly abbreviated ECL model that illustrates this is shown in Figure 16.3, a model of a D flip-flop with differential D and CLK inputs.

The first section of interest in Figure 16.3 is the port list.

```
      -- 0 denotes internal pull-down resistor, 1 pull-up
      -- (actually clamp circuit)
    PORT (
    CLK                 : IN  std_ulogic := '0';
    CLKNeg              : IN  std_ulogic := '1';
    R                   : IN  std_ulogic := '0';
    S                   : IN  std_ulogic := '0';
    D                   : IN  std_ulogic := '0';
    DNeg                : IN  std_ulogic := '1';
    Q                   : OUT std_ulogic := 'U';
    QNeg                : OUT std_ulogic := 'U' ;
    VBB                 : OUT std_ulogic := ECLVbbValue
    );
```

The component has internal pull-up and pull-down resistors. They are reflected in the model by initializing certain ports to '1' or '0'. If those ports are left

```
--------------------------------------------------------------------------------
-- File name : eclpsl29.vhd
--------------------------------------------------------------------------------
-- Copyright (C) 1999, 2002 Free Model Foundry; http://eda.org/fmf/
--
-- This program is free software; you can redistribute it and/or modify
-- it under the terms of the GNU General Public License version 2 as
-- published by the Free Software Foundation.
--
-- MODIFICATION HISTORY :
--
-- version: |  author:  | mod date: | changes made:
--   V1.0    M. Li       99 DEC 02   initial release
--   V1.1    R. Munden   02 APR 24   Fixed Dummy VPDs
--------------------------------------------------------------------------------
-- PART DESCRIPTION :
--
-- Library:       ECLPS
-- Technology:    ECL
-- Part:          ECLPSL29
--
-- Description:   Differential data and clock D Flip-Flop with Set and Reset
--
--------------------------------------------------------------------------------
LIBRARY IEEE;   USE IEEE.std_logic_1164.ALL;
                USE IEEE.VITAL_primitives.ALL;
                USE IEEE.VITAL_timing.ALL;
LIBRARY FMF;    USE FMF.ecl_utils.ALL;
                USE FMF.ff_package.ALL;


--------------------------------------------------------------------------------
--  ENTITY DECLARATION
--------------------------------------------------------------------------------
ENTITY eclpsl29 IS
    GENERIC (
        -- tipd delays: interconnect path delays
        tipd_R              : VitalDelayType01 := VitalZeroDelay01;
        tipd_S              : VitalDelayType01 := VitalZeroDelay01;
        tipd_CLK            : VitalDelayType01 := VitalZeroDelay01;
        tipd_CLKNeg         : VitalDelayType01 := VitalZeroDelay01;
        tipd_D              : VitalDelayType01 := VitalZeroDelay01;
        tipd_DNeg           : VitalDelayType01 := VitalZeroDelay01;
        -- tpd delays: propagation delays
        tpd_CLK_Q           : VitalDelayType01 := ECLUnitDelay01;
        tpd_R_Q             : VitalDelayType01 := ECLUnitDelay01;
        tpd_S_Q             : VitalDelayType01 := ECLUnitDelay01;
        -- tsetup values: setup times
        tsetup_D_CLK : VitalDelayType := ECLUnitDelay;
        -- thold values: hold times
        thold_D_CLK  : VitalDelayType := ECLUnitDelay;
        -- trecovery values: release times
        trecovery_R_CLK : VitalDelayType := ECLUnitDelay;
        trecovery_S_CLK : VitalDelayType := ECLUnitDelay;
        -- tpw values: pulse widths
        tpw_R_posedge       : VitalDelayType := ECLUnitDelay;
        tpw_S_posedge       : VitalDelayType := ECLUnitDelay;
        tpw_CLK_posedge     : VitalDelayType := ECLUnitDelay;
        tpw_CLK_negedge     : VitalDelayType := ECLUnitDelay;
        -- generic control parameters
        InstancePath        : STRING  := DefaultECLInstancePath;
        TimingChecksOn      : BOOLEAN := DefaultECLTimingChecks;
        MsgOn               : BOOLEAN := DefaultECLMsgOn;
        XOn                 : BOOLEAN := DefaultECLXOn;
        -- For FMF SDF technology file usage
        TimingModel         : STRING  := DefaultECLTimingModel
    );
```

Figure 16.3 ECL model with differential inputs

```
            -- 0 denotes internal pull-down resistor, 1 pull-up
            -- (actually clamp circuit)
        PORT (
            CLK                 : IN  std_ulogic := '0';
            CLKNeg              : IN  std_ulogic := '1';
            R                   : IN  std_ulogic := '0';
            S                   : IN  std_ulogic := '0';
            D                   : IN  std_ulogic := '0';
            DNeg                : IN  std_ulogic := '1';
            Q                   : OUT std_ulogic := 'U';
            QNeg                : OUT std_ulogic := 'U' ;
            VBB                 : OUT std_ulogic := ECLVbbValue
        );
        ATTRIBUTE VITAL_level0 of eclpsl29 : ENTITY IS TRUE;
END eclpsl29;


--------------------------------------------------------------------------------
--  ARCHITECTURE DECLARATION
--------------------------------------------------------------------------------
ARCHITECTURE vhdl_behavioral OF eclpsl29 IS
    ATTRIBUTE VITAL_level1 OF vhdl_behavioral : ARCHITECTURE IS TRUE;

    SIGNAL CLK_ipd          : std_ulogic := 'X';
    SIGNAL CLKNeg_ipd        : std_ulogic := 'X';
    SIGNAL R_ipd            : std_ulogic := 'X';
    SIGNAL S_ipd            : std_ulogic := 'X';
    SIGNAL D_ipd            : std_ulogic := 'X';
    SIGNAL DNeg_ipd          : std_ulogic := 'X';
    SIGNAL CLKint           : std_ulogic := 'X';
    SIGNAL Dint             : std_ulogic := 'X';
    SIGNAL Qint             : std_ulogic := 'X';

BEGIN

    --------------------------------------------------------------------------
    -- Wire Delays
    --------------------------------------------------------------------------
    -- wire delay block ommitted
    --------------------------------------------------------------------------
    -- Concurrent Procedures
    --------------------------------------------------------------------------
    a_1: VitalBUF (q => Q,    a => Qint, ResultMap => ECL_wired_or_rmap);
    a_2: VitalINV (q => QNeg, a => Qint, ResultMap => ECL_wired_or_rmap);


    --------------------------------------------------------------------------
    -- D inputs Process
    --------------------------------------------------------------------------
    Dinputs : PROCESS (D_ipd, DNeg_ipd)

        -- Functionality Results Variables
        VARIABLE Dint_zd        : X01;

        -- Output Glitch Detection Variables
        VARIABLE D_GlitchData   : VitalGlitchDataType;

    BEGIN

        --------------------------------------------------------------------
        -- Functionality Section
        --------------------------------------------------------------------
        Dint_zd := ECL_s_or_d_inputs_tab (D_ipd, DNeg_ipd);
```

Figure 16.3 ECL model with differential inputs *(continued)*

```
-------------------------------------------------------------------------
-- (Dummy) Path Delay Section
-------------------------------------------------------------------------
VitalPathDelay (
    OutSignal        => Dint,
    OutSignalName    => "Dint",
    OutTemp          => Dint_zd,
    GlitchData       => D_GlitchData,
    XOn              => XOn,
    MsgOn            => MsgOn,
    Paths            => (
        0 => (InputChangeTime  => D_ipd'LAST_EVENT,
              PathDelay        => VitalZeroDelay,
              PathCondition    => FALSE))

);

END PROCESS;


-------------------------------------------------------------------------
-- ECL Clock Process
-------------------------------------------------------------------------
ECLClock : PROCESS (CLK_ipd, CLKNeg_ipd)

    -- Functionality Results Variables
    VARIABLE Mode          : X01;
    VARIABLE CLKint_zd      : std_ulogic;
    VARIABLE PrevData       : std_logic_vector(0 to 2);

    -- Output Glitch Detection Variables
    VARIABLE CLK_GlitchData : VitalGlitchDataType;

BEGIN

    -------------------------------------------------------------------------
    -- Functionality Section
    -------------------------------------------------------------------------
    Mode := ECL_diff_mode_tab(CLK_ipd, CLKNeg_ipd);

    VitalStateTable (
        StateTable       => ECL_clk_tab,
        DataIn           => (CLK_ipd, CLKNeg_ipd, Mode),
        Result           => CLKint_zd,
        PreviousDataIn   => PrevData
    );


    -------------------------------------------------------------------------
    -- (Dummy) Path Delay Section
    -------------------------------------------------------------------------
    VitalPathDelay (
        OutSignal        => CLKint,
        OutSignalName    => "CLKint",
        OutTemp          => CLKint_zd,
        GlitchData       => CLK_GlitchData,
        XOn              => XOn,
        MsgOn            => MsgOn,
        Paths            => (
            0 => (InputChangeTime  => CLK_ipd'LAST_EVENT,
                  PathDelay        => VitalZeroDelay,
                  PathCondition    => FALSE))
    );
```

Figure 16.3　ECL model with differential inputs *(continued)*

```
        END PROCESS;
        ------------------------------------------------------------------
        -- Main Behavior Process
        ------------------------------------------------------------------
        VitalBehavior : PROCESS (CLKint, Dint, S_ipd, R_ipd)

             -- Timing Check Variables

-- timingcheck variables ommitted

             VARIABLE Violation     : X01 := '0';

             -- Functionality Results Variables
             VARIABLE Q_zd               : std_ulogic;
             VARIABLE PrevData           : std_logic_vector(0 to 4);

             -- Output Glitch Detection Variables
             VARIABLE Q_GlitchData   : VitalGlitchDataType;

        BEGIN

             ------------------------------------------------------------------
             -- Timing Check Section
             ------------------------------------------------------------------
             IF (TimingChecksOn) THEN

-- timingchecks ommitted

             END IF;

             ------------------------------------------------------------------
             -- Functionality Section
             ------------------------------------------------------------------
             Violation := Tviol_D_CLKint OR Pviol_CLKint OR
                          Rviol_R_CLKint OR Pviol_R OR
                          Sviol_S_CLKint OR Pviol_S;

             VitalStateTable (
                 StateTable        => DFFSR_tab,
                 DataIn            => (Violation, CLKint, Dint, S_ipd, R_ipd),
                 Result            => Q_zd,
                 PreviousDataIn    => PrevData
             );

             ------------------------------------------------------------------
             -- Path Delay Section
             ------------------------------------------------------------------
             VitalPathDelay01 (
                 OutSignal         => Qint,
                 OutSignalName     => "Qint",
                 OutTemp           => Q_zd,
                 GlitchData        => Q_GlitchData,
                 XOn               => XOn,
                 MsgOn             => MsgOn,
                 Paths             => (
                     0 => (InputChangeTime   => CLKint'LAST_EVENT,
                           PathDelay         => tpd_CLK_Q,
                           PathCondition     => TRUE),
                     1 => (InputChangeTime   => R_ipd'LAST_EVENT,
                           PathDelay         => tpd_R_Q,
                           PathCondition     => TRUE),
                     2 => (InputChangeTime   => S_ipd'LAST_EVENT,
                           PathDelay         => tpd_S_Q,
                           PathCondition     => TRUE)
                     )
             );

        END PROCESS;
END vhdl_behavioral;
```

Figure 16.3 ECL model with differential inputs *(continued)*

unconnected in the netlist, they will retain their initial values. The internal pull-up and pull-down resistors in the component are designed to work only in the case of open inputs. Likewise, initializing the ports to '1' and '0' in the model works only for open inputs.

The next section of interest is the concurrent procedures:

```
-----------------------------------------------------------------------------
-- Concurrent Procedures
-----------------------------------------------------------------------------
a_1: VitalBUF (q => Q, a => Qint, ResultMap => ECL_wired_or_rmap);
a_2: VitalINV (q => QNeg, a => Qint, ResultMap => ECL_wired_or_rmap);
```

These serve two functions. They take a single internal result, Qint, and use it to drive the differential outputs Q and QNeg. They also map the internal result to the ECL logic levels 'Z' and '1' by calling the ECL_wired_or_rmap result map in the FMF ecl_utils package. ECL outputs are open emitter, so they cannot drive a '0' output.

Each of the differential pair inputs are read in their own process. The first one is for the D inputs, while includes a function call to a table called ECL_s_or_d_inputs_tab that is defined in the ecl_utils package and is listed in Figure 16.4.

The table is called and the results assigned to Dint_zd.

```
Dint_zd := ECL_s_or_d_inputs_tab (D_ipd, DNeg_ipd);
```

The result, Dint_zd, is assigned to an internal signal, Dint, through a zero delay VitalPathDelay procedure. The VPD is required for the model to qualify as a VITAL_Level1 model.

```
CONSTANT ECL_s_or_d_inputs_tab : eclstdlogic_table := (
--
-- For the case, ECLVbbValue = 'W', table looks like this:
------------------------------------------------------------------
----|  U    X    0    1    Z    W    L    H    -        |  |
------------------------------------------------------------------
--  ( 'X',  'X', 'X', 'X', 'X', 'X', 'X', 'X', 'X' ), -- | U |
--  ( 'X',  'X', 'X', 'X', 'X', 'X', 'X', 'X', 'X' ), -- | X |
--  ( 'X',  'X', 'X', '0', 'X', 'X', '0', 'X', 'X' ), -- | 0 |
--  ( 'X',  'X', '1', 'X', 'X', '1', '1', 'X', 'X' ), -- | 1 |
--  ( 'X',  'X', 'X', 'X', 'X', 'X', 'X', 'X', 'X' ), -- | Z |
--  ( 'X',  'X', '1', '0', 'X', 'X', '1', '0', 'X' ), -- | W |
--  ( 'X',  'X', 'X', '0', 'X', 'X', '0', 'X', 'X' ), -- | L |
--  ( 'X',  'X', '1', 'X', 'X', '1', '1', 'X', 'X' ), -- | H |
--  ( 'X',  'X', 'X', 'X', 'X', 'X', 'X', 'X', 'X' )  -- | - |
--);
    '0'    => ('1' => '0', 'H' => '0', ECLVbbValue => '0', OTHERS => 'X'),
    'L'    => ('1' => '0', 'H' => '0', ECLVbbValue => '0', OTHERS => 'X'),
    '1'    => ('0' => '1', 'L' => '1', ECLVbbValue => '1', OTHERS => 'X'),
    'H'    => ('0' => '1', 'L' => '1', ECLVbbValue => '1', OTHERS => 'X'),
    ECLVbbValue => ('0' => '1', 'L' => '1', '1' => '0', 'H' => '0',
                    OTHERS => 'X'),
    OTHERS => (OTHERS => 'X')
);
```

Figure 16.4 ECL_s_or_d_inputs_tab

```
-- Table for determining whether input pair is differential or
-- single-ended. There are 3 values:
-- input, input_bar neither or both Vbb,    mode => 'X'
-- input_bar =  Vbb,                         mode => '0'
-- input   = Vbb,                  mode => '1'
-- Used as input to ECL_clk_tab: return value name convention is 'Mode'
-- Type of 'Mode' is X01
-------------------------------------------------------------------
CONSTANT ECL_diff_mode_tab : eclstdlogic_table := (
--
-- For the case, ECLVbbValue = 'W', table looks like this:
-------------------------------------------------------------------
----|  U    X    0    1    Z    W    L    H    -        |  |
-------------------------------------------------------------------
--  ( 'X', 'X', 'X', 'X', 'X', '0', 'X', 'X', 'X' ), --  | U |
--  ( 'X', 'X', 'X', 'X', 'X', '0', 'X', 'X', 'X' ), --  | X |
--  ( 'X', 'X', 'X', 'X', 'X', '0', 'X', 'X', 'X' ), --  | 0 |
--  ( 'X', 'X', 'X', 'X', 'X', '0', 'X', 'X', 'X' ), --  | 1 |
--  ( 'X', 'X', 'X', 'X', 'X', '0', 'X', 'X', 'X' ), --  | Z |
--  ( '1', '1', '1', '1', '1', 'X', '1', '1', '1' ), --  | W |
--  ( 'X', 'X', 'X', 'X', 'X', '0', 'X', 'X', 'X' ), --  | L |
--  ( 'X', 'X', 'X', 'X', 'X', '0', 'X', 'X', 'X' ), --  | H |
--  ( 'X', 'X', 'X', 'X', 'X', '0', 'X', 'X', 'X' ) --   | - |
--);

    ECLVbbValue => (ECLVbbValue => 'X', OTHERS => '1'),
    OTHERS      => (ECLVbbValue => '0', OTHERS => 'X')
```

Figure 16.5 ECL_diff_mode_tab

The next process reads the input clock pair. First the input mode must be determined. Depending on if or how **vbb** is connected, either input could be a single-ended input or they could be a differential pair. Actual use is determined by a call to **ECL_diff_mode_tab** which is listed in Figure 16.5.

The table is called and the results assigned to **Mode**.

```
Mode := ECL_diff_mode_tab(CLK_ipd, CLKNeg_ipd);
```

The result, **Mode**, can be '0', '1', or 'X'. Values '0' or '1' indicate which input is active as a single-ended input. If the value is 'X', the inputs are being driven as a differential pair. This is an input to the **ECL_clk_tab** VITAL state table (listed in Figure 16.6) that reads the inputs and outputs an internal clock variable.

```
VitalStateTable (
   StateTable       => ECL_clk_tab,
   DataIn           => (CLK_ipd, CLKNeg_ipd, Mode),
   Result           => CLKint_zd,
   PreviousDataIn   => PrevData
);
```

Once again, a **VitalPathDelay** procedure is used to assign the result to an internal signal.

The component functionality is modeled with a call to another VITAL state table that resides in the FMF **ff_package** and is listed in Figure 16.7.

```
-----------------------------------------------------------------
-- Table for computing a single signal from a differential ECL clock
-- pair. Mode is '1' or '0' when the signal is single-ended. The rest of
-- the table is self-explanatory :)
-----------------------------------------------------------------
CONSTANT ECL_clk_tab : VitalStateTableType  := (
-----------------------------------------
------INPUTS-------|-PREV---|-OUTPUT----
-- CLK CLKNeg Mode | CLKint | CLKint' --
------------------|--------|-----------
  ( '-', 'X', '1', '-', 'X'), -- Single-ended, Vbb on CLK
  ( '-', '0', '1', '-', '1'), -- Single-ended, Vbb on CLK
  ( '-', '1', '1', '-', '0'), -- Single-ended, Vbb on CLK
  ( 'X', '-', '0', '-', 'X'), -- Single-ended, Vbb on CLK_N
  ( '0', '-', '0', '-', '0'), -- Single-ended, Vbb on CLK_N
  ( '1', '-', '0', '-', '1'), -- Single-ended, Vbb on CLK_N
  -- Below are differential input possibilities only
  ( 'X', '-', 'X', '-', 'X'), -- CLK unknown
  ( '-', 'X', 'X', '-', 'X'), -- CLK unknown
  ( '1', '-', 'X', 'X', '1'), -- Recover from 'X'
  ( '0', '-', 'X', 'X', '0'), -- Recover from 'X'
  ( '/', '0', 'X', '0', '1'), -- valid ECL rising edge
  ( '1', '\', 'X', '0', '1'), -- valid ECL rising edge
  ( '\', '1', 'X', '1', '0'), -- valid ECL falling edge
  ( '0', '/', 'X', '1', '0'), -- valid ECL falling edge
  ( '-', '-', '-', '-', 'S')  -- default
); -- end of VitalStateTableType definition
```

Figure 16.6 ECL_clk_tab

```
-----------------------------------------------------------------
-- D-flip/flop with Set and Reset both active high
-----------------------------------------------------------------
CONSTANT DFFSR_tab : VitalStateTableType := (

    -- -------INPUTS-----------|PREV-|-OUTPUT--
    -- Viol CLK   D    S    R   | QI  | Q'    --
    --------------------------|-----|---------
  (  'X', '-', '-', '-', '-', '-', 'X'), -- timing violation
  (  '-', 'B', '-', 'X', '0', '1', '1'), -- set unknown
  (  '-', '/', '1', 'X', '0', '1', '1'), -- set unknown
  (  '-', '-', '-', 'X', '-', '-', 'X'), -- set unknown
  (  '-', 'B', '-', '0', 'X', '0', '0'), -- reset unknown
  (  '-', '/', '0', '0', 'X', '0', '0'), -- reset unknown
  (  '-', '-', '-', '-', 'X', '-', 'X'), -- reset unknown
  (  '-', '-', '-', '1', '1', '-', 'X'), -- both asserted
  (  '-', '-', '-', '1', '0', '-', '1'), -- set asserted
  (  '-', '-', '-', '0', '1', '-', '0'), -- reset asserted
  (  '-', 'X', '0', '0', '0', '0', '0'), -- clk unknown
  (  '-', 'X', '1', '0', '0', '1', '1'), -- clk unknown
  (  '-', 'X', '-', '0', '0', '-', 'X'), -- clk unknown
  (  '-', '/', '0', '0', '0', '-', '0'), -- active clock edge
  (  '-', '/', '1', '0', '0', '-', '1'), -- active clock edge
  (  '-', '/', '-', '0', '0', '-', 'X'), -- active clock edge
  (  '-', '-', '-', '-', '-', '-', 'S')  -- default

  ); -- end of VitalStateTableType definition
```

Figure 16.7 DFFSR_tab

```
VitalStateTable (
   StateTable        => DFFSR_tab,
   DataIn            => (Violation, CLKint, Dint, S_ipd, R_ipd),
   Result            => Q_zd,
   PreviousDataIn    => PrevData
);
```

The result, Q_zd, drives the concurrent procedure calls already examined.

16.2   Bus Hold

Bus hold is a component feature that allows an input to retain its previous value when its driver switches to high impedance. Bus hold devices incorporate a weak positive feedback buffer to maintain an input state. This buffer is connected directly to the input pin and can, in theory, hold the state of the bus for other devices connected to it. A schematic of the circuit used is shown on the left in Figure 16.8.

Modeling a bus hold circuit as shown on the left would have some disadvantages. It would require the input port to be modeled as an INOUT port. That would complicate netlisting because the schematic capture system would show the port as an input. Also, although in theory the bus hold circuit in one device might hold an entire bus, component vendors do not specify how much drive is available from one of these circuits. Therefore, it cannot be determined (from the data sheets), what will happen if one device on a multidrop bus has bus hold and the others do not.

The conservative approach (and the easy way out), is to model the bus hold circuit as not propagating back to the bus but holding only the input of which it is a part. An equivalent circuit is shown on the right side of Figure 16.8.

A model of a simple component with bus hold, the stdh125, is shown in Figure 16.9.

All of the bus hold functionality resides in the concurrent procedure calls. The signal Aint must be of type std_logic rather than the usual std_ulogic

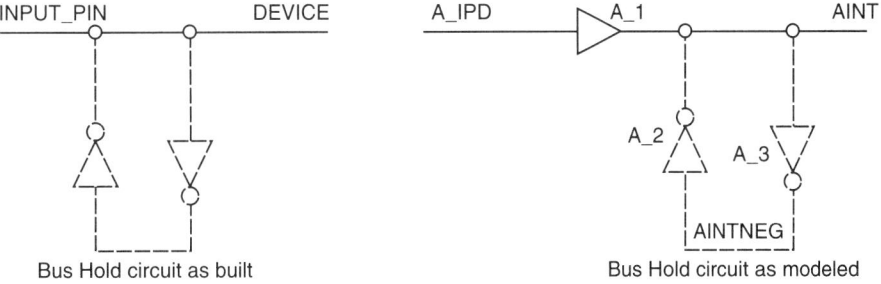

Bus Hold circuit as built   Bus Hold circuit as modeled

Figure 16.8 Bus hold circuit

```
-------------------------------------------------------------------------------
-- File Name: sdth125.vhd
-------------------------------------------------------------------------------
-- Copyright (C) 2001 Free Model Foundry; http://vhdl.org/fmf/
--
-- This program is free software; you can redistribute it and/or modify
-- it under the terms of the GNU General Public License version 2 as
-- published by the Free Software Foundation.
--
-- MODIFICATION HISTORY:
--
-- version: |  author:  | mod date: | changes made:
--   V1.0      R. Munden   01 MAR 17   initial release
--
-------------------------------------------------------------------------------
-- PART DESCRIPTION:
--
-- Library:    STNDH
-- Technology: 54/74XXXX
-- Part:       STDH125
--
-- Description: Line driver w/ 3-state output and bus hold
-------------------------------------------------------------------------------

LIBRARY IEEE;    USE IEEE.std_logic_1164.ALL;
                 USE IEEE.VITAL_timing.ALL;
                 USE IEEE.VITAL_primitives.ALL;
LIBRARY FMF;     USE FMF.gen_utils.ALL;


-------------------------------------------------------------------------------
-- ENTITY DECLARATION
-------------------------------------------------------------------------------
ENTITY sdth125 IS
    GENERIC (
        -- tipd delays: interconnect path delays
        tipd_Y             : VitalDelayType01 := VitalZeroDelay01;
        tipd_A             : VitalDelayType01 := VitalZeroDelay01;
        tipd_OENeg         : VitalDelayType01 := VitalZeroDelay01;
        -- tpd delays
        tpd_A_Y            : VitalDelayType01 := UnitDelay01;
        tpd_OENeg_Y        : VitalDelayType01Z := UnitDelay01Z;
        -- generic control parameters
        InstancePath       : STRING    := DefaultInstancePath;
        -- For FMF SDF technology file usage
        TimingModel        : STRING    := DefaultTimingModel
    );
    PORT (
        Y              : OUT   std_logic := 'U';
        A              : IN    std_logic := 'U';
        OENeg          : IN    std_logic := 'U'
    );
    ATTRIBUTE VITAL_LEVEL0 of sdth125 : ENTITY IS TRUE;
END sdth125;


-------------------------------------------------------------------------------
-- ARCHITECTURE DECLARATION
-------------------------------------------------------------------------------
ARCHITECTURE vhdl_behavioral of sdth125 IS
    ATTRIBUTE VITAL_LEVEL0 of vhdl_behavioral : ARCHITECTURE IS TRUE;

    SIGNAL A_ipd              : std_ulogic := 'U';
    SIGNAL Aint               : std_logic  := 'U';
    SIGNAL AintNeg            : std_ulogic := 'U';
    SIGNAL OENeg_ipd          : std_ulogic := 'U';
```

Figure 16.9 STDH125 driver with bus hold

```
BEGIN

    -------------------------------------------------------------------------
    -- Wire Delays
    -------------------------------------------------------------------------

    WireDelay : BLOCK
    BEGIN

        w_1 : VitalWireDelay (A_ipd, A, tipd_A);
        w_2 : VitalWireDelay (OENeg_ipd, OENeg, tipd_OENeg);

    END BLOCK;

    -------------------------------------------------------------------------
    -- Concurrent procedure calls
    -------------------------------------------------------------------------
    a_1: VitalBUF (q => Aint, a => A_ipd, ResultMap => ('U', 'Z', '0', '1'));
    a_2: VitalINV (q => AintNeg, a => Aint);
    a_3: VitalINV (q => Aint, a => AintNeg, ResultMap => ('Z', 'Z', '0', '1'));

    -------------------------------------------------------------------------
    -- Main Behavior Process
    -------------------------------------------------------------------------
    VitalBehavior : PROCESS (Aint, OENeg_ipd)

        -- Functionality Results Variables
        VARIABLE Y_zd           : std_ulogic;

        -- Output Glitch Detection Variables
        VARIABLE Y_GlitchData   : VitalGlitchDataType;

    BEGIN
        ---------------------------------------------------------------------
        -- Functionality Section
        ---------------------------------------------------------------------
        Y_zd := VitalBUFIFO (data => Aint, enable => OENeg_ipd);

        ---------------------------------------------------------------------
        -- Path Delay Section
        ---------------------------------------------------------------------
        VitalPathDelay01Z (
            OutSignal       => Y,
            OutSignalName   => "Y",
            OutTemp         => Y_zd,
            GlitchData      => Y_GlitchData,
            Paths           => (
                0 => (InputChangeTime  => Aint'LAST_EVENT,
                      PathDelay        => VitalExtendToFillDelay(tpd_A_Y),
                      PathCondition    => TRUE),
                1 => (InputChangeTime  => OENeg_ipd'LAST_EVENT,
                      PathDelay        => tpd_OENeg_Y,
                      PathCondition    => TRUE))
        );

    END PROCESS;

END vhdl_behavioral;
```

Figure 16.9 STDH125 driver with bus hold *(continued)*

because it has more than one driver and must be resolved. The combination of the result maps in a_1 and a_3 and the std_logic resolution table enable the positive feedback loop to be overridden by any new value (other than 'Z') placed on the input.

There should be some delay that requires an input pulse to exceed a certain pulse width in order to cause the bus hold to flip to a different state. However, component vendors are not supplying the value for that delay.

16.3   PLLs and DLLs

Many types of components now include phase locked loops (PLLs) or delay locked loops (DLLs) in their clock circuitry. These range from clock drivers through digital signal processors. Even memories may now include PLLs. These PLLs function as oscillators that lock to the frequency and phase of an external clock signal.

The characteristics of a PLL that need to be considered in a model are lock time, skew, and jitter. Lock time is the time required for the PLL to adjust itself to the input frequency and phase. Skew is the time or phase difference between the input and the PLL output. Jitter is the frequency stability of the PLL output. PLLs are often modeled as two processes. A typical implementation is given in Figure 16.10.

The second process here, the PLL process, behaves like an oscillator. Whenever the value of pll_out changes, the process is triggered and another change of pll_out is scheduled after a delay.

The first process here, the ADJ process, compares the period of CLK_ipd with the period of FBIN_ipd. FBIN_ipd is connected to pll_out external to these processes and may be delayed. During the time the period of FBIN_ipd is greater than that of CLK_ipd, the value of half_per is decreased. This makes pll_out oscillate faster. Should the period of FBIN_ipd become less than that of CLK_ipd, the value of half_per will be increased, making pll_out oscillate more slowly. It is important that the amount of increase be different than the amount of decrease. Otherwise, the two periods might never become equal.

Once the two periods are matched, vco_lock is set to true. The signal half_per will remain unchanged. The next step is to bring the two signals into phase alignment. This is accomplished by adding a small delay to pll_out every other cycle. The constant values chosen in the lines

```
half_per <= half_per - 50 ps;
half_per <= half_per + 60 ps;
```

determine how long the PLL model will take to match the frequency of pll_out to that of CLK_ipd. The constant value in the line

```
pll_delay <= 30 ps;
```

will determine how long it takes to achieve phase alignment. The constant value in the line

```
-------------------------------------------------------------------------
-- ADJ Process
-------------------------------------------------------------------------
ADJ : PROCESS (FBIN_ipd, CLK_ipd)

    VARIABLE fbi_period   : time := 0 ns;
    VARIABLE clk_period   : time := 0 ns;
    VARIABLE prev_clk     : time := 0 ns;
    VARIABLE prev_fbi     : time := 0 ns;
    VARIABLE toggle1      : boolean;
    VARIABLE toggle2      : boolean;

BEGIN
    ---------------------------------------------------------------------
    -- Functionality Section
    ---------------------------------------------------------------------
    IF rising_edge(CLK_ipd) THEN
        clk_period := NOW - prev_clk;
        prev_clk := NOW;
        IF FBIN_ipd = 'X' THEN
            rst_int <= '1', '0' AFTER 5 ns;
        END IF;
    END IF;

    IF (FBIN_ipd'event AND FBIN_ipd = '0') THEN
        rst_int <= '0';
        fbi_period := NOW - prev_fbi;
        prev_fbi := NOW;

        IF toggle1 AND toggle2 THEN
            IF fbi_period > clk_period THEN
                half_per <= half_per - 50 ps;
                vco_lock <= false;
            ELSIF fbi_period < clk_period THEN
                half_per <= half_per + 60 ps;
                vco_lock <= false;
            ELSE
                vco_lock <= true;
            END IF;
        END IF;
        toggle1 := not toggle1;
        IF toggle1 THEN
            toggle2 := not toggle2;
        ELSE
            pll_delay <= 0 ps;
        END IF;
    END IF;

    IF rising_edge(FBIN_ipd) AND vco_lock AND toggle1 AND toggle2 THEN
        IF (prev_clk + 150 ps) < NOW THEN
            IF pll_delay < clk_period THEN
                pll_delay <= 30 ps;
            END IF;
        END IF;
    END IF;

END PROCESS ADJ;

-------------------------------------------------------------------------
-- PLL Process
-------------------------------------------------------------------------
PLL : PROCESS (pll_out)

BEGIN

    pll_out <= TRANSPORT not pll_out AFTER pll_delay + half_per;

END PROCESS PLL;
```

Figure 16.10 PLL model

```
IF (prev_clk + 150 ps) < NOW THEN
```

will determine the accuracy of the phase alignment, also known as the skew.

This example does not attempt to introduce jitter. It approximates the behavior of a real PLL in that it takes some time to match the input clock frequency and phase. There are other PLLs in other components with different significant characteristics. The code here provides a reasonable starting point for writing PLLs.

16.4   Assertions

Assertion statements are a useful method for communicating the significant events occurring within a model to the user. They are not closely tied to modeling any particular type of component, but may be used in a variety of situations. The syntax for assertion statements is as follows:

```
[label : ] assert boolean_expression
[report expression] [severity expression];
```

If the boolean expression evaluates to false, the report expression is printed and given a severity level determined by the severity expression.

The two primary reasons for using assertion statements are to communicate error messages and progress events. An example of an error is this:

```
ASSERT (not(Is_X(AddressIn)))
    REPORT InstancePath & partID & ": Unusable value for address"
    SEVERITY SeverityMode;
```

In this code, the address bus of a memory is tested for unknown values. If an unknown value is found, a message is reported to the simulator and is assigned the severity level that corresponds to the variable SeverityMode. The severity level may be used to pause or abort the simulation.

Assertion statements may also be used to inform the user of events occurring within the model that are not errors. For example, the DLL within a QDR SRAM, once locked, will continue running at the same frequency until the input clock period slows to 30 nanoseconds:

```
IF CIn_period > 30 ns THEN
    dll_lock := false;
    ASSERT false
        REPORT "C mode DLL reseting"
        SEVERITY note;
END IF;
```

This code informs the user if that happens. The severity level is set to note. This allows the user to mask the message if he or she prefers not to see it.

Here are two more examples from a DSP model:

```
                  IF xsum = configbuf(13) THEN -- if xsum is OK initialize
...
             ASSERT false
                REPORT "PCI initialized from eeprom"
                SEVERITY note;
          ELSE -- xsum is not OK
...
             ASSERT false
                REPORT "wrong checksum - PCI not initialized from eeprom"
                SEVERITY warning;
```

These examples illustrate setting the severity level to different values for different types of messages.

Assertion statements are of particular value in testbenches, as shown in Chapter 17.

16.5 Modifying Behavior with the TimingModel Generic

The TimingModel generic is used in FMF-style models as a means of specifying to an external program a particular set of timing values to be annotated to a model instance. However, in a few cases it has also been used to specify a variant of component behavior. For example, a flash memory may be produced with a protected area at the top of its address range. Another otherwise identical component is sold with a protected area at the bottom of its address range. You could write, test, and maintain two models, one for each part, or you could write one model that could be used to simulate either part.

In the case of the flash memory, the eleventh character of the part number indicates whether the part has a boot sector at the top (AM29LV160DT-70EC) or bottom (AM29LV160DB-70EC) of its address space. Knowing that, it is easy to write a single model that reads its TimingModel generic and behaves accordingly:

```
-------------------------------------------------------------------------------
-- VarSect
-------------------------------------------------------------------------------
VarSect <= SecNum WHEN TimingModel(1 to 11) ="am29lv160mt" ELSE
   0;--WHEN TimingModel = "AM29LV160DB"

vs <= 1 WHEN TimingModel(1 to 11)="am29lv160mt" ELSE
   0;
```

16.6 State Machines

Many types of components incorporate state machines. They are found in DRAMs, FIFOs, flash memories, and processors. Writing state machines in VHDL is no problem.

State machines are usually written using CASE statements and enumerated types. For example, a state machine for a JTAG controller would begin with the following type and variable declarations:

```
TYPE tap_state_type IS (Test_Logic_Reset,
                        Run_Test_Idle,
                        Select_DR_Scan,
                        Capture_DR,
                        Shift_DR,
                        Exit1_DR,
                        Pause_DR,
                        Exit2_DR,
                        Update_DR,
                        Select_IR_Scan,
                        Capture_IR,
                        Shift_IR,
                        Exit1_IR,
                        Pause_IR,
                        Exit2_IR,
                        Update_IR
                       );
VARIABLE TAP_state  : tap_state_type;
VARIABLE prev_state : tap_state_type;
```

This is a behavioral model, so there is no need to think about state encoding. State names are all that are required. Using descriptive state names makes debugging and using the model much more pleasant. The state variables could be signals if they need to be used in multiple processes.

There must be a clock or some other signal to activate the state machine. It may just be in the sensitivity list or it may be explicit as shown later. Then there is the CASE statement itself:

```
ELSIF rising_edge(TCKIn) THEN
  CASE TAP_state IS
    WHEN Test_Logic_Reset =>
      IF TMS_nwv = '0' THEN
        TAP_state := Run_Test_Idle;
      END IF;
    WHEN Run_Test_Idle =>
      IF TMS_nwv = '1' THEN
        TAP_state := Select_DR_Scan;
      END IF;
    WHEN Select_DR_Scan =>
      IF TMS_nwv = '0' THEN
        TAP_state := Capture_DR;
      ELSIF TMS_nwv = '1' THEN
        TAP_state := Select_IR_Scan;
      END IF;
      prev_state := Select_DR_Scan;
...
```

Each **WHEN** acts to select the current state. Frequently, a condition within the **WHEN** clause will assign the next state. Complex activity can occur within a **WHEN** clause:

```
WHEN Shift_DR =>
   IF instruction = bypass OR instruction = highz THEN
      bpr := TDI_nwv;
   ELSIF instruction = idcode THEN
      FOR i IN 1 TO 31 LOOP
         IDreg(i - 1) := IDreg(i);
      END LOOP;
      IDreg(31) := TDI_nwv;
   ELSIF instruction = sample_preload OR
            instruction = extest THEN
      FOR i IN 1 TO bsr_size - 1 LOOP
         bsr(i - 1) <= bsr(i);
      END LOOP;
      bsr(bsr_size - 1) <= TDI_nwv;
   END IF;
   IF TMS_nwv = '1' THEN
      TAP_state := Exit1_DR;
   END IF;
```

Components that have state machines often have more than one. Sometimes each state machine is unique within the component, but sometimes they are duplicates. DRAMs are examples of components with multiple (usually two or four) copies of a state machine. To reduce the code size and guarantee they are identical, these state machines can be modeled in arrays.

Here are the state machine declarations from an SDRAM model:

```
-- Type definition for state machine
TYPE mem_state IS (pwron,
                   precharge,
                   idle,
                   mode_set,
                   self_refresh,
                   self_refresh_rec,
                   auto_refresh,
                   pwrdwn,
                   bank_act,
                   bank_act_pwrdwn,
                   write,
                   write_suspend,
                   read,
                   read_suspend,
                   write_auto_pre,
```

```
                         read_auto_pre
                       );

     TYPE statebanktype IS array (hi_bank downto 0) of mem_state;

     SIGNAL statebank : statebanktype;
```

In this model, the signal statebank holds the state of four state machines. On each triggering event, all state machines are examined and updated in a FOR LOOP, as shown in the following code snippets:

```
     -- The Big State Machine
     IF (rising_edge(CLKIn) AND CKEreg = '1') THEN
...
        banks : FOR bank IN 0 TO hi_bank LOOP
        CASE statebank(bank) IS
          WHEN pwron =>
...
          WHEN write =>
            IF (command = bst) THEN
               statebank(bank) <= bank_act;
...
```

Although the four state machines are each updated at each clock rising edge, they need not be in the same state.

16.7   Mixed Signal Devices

Many designs operate in both the digital and analog domains. Eventually it may be practical to simulate those designs in VHDL-AMS (analog mixed signal). Until then, it is possible to push out simulation to include many of the mixed signal components on the dividing lines between the analog and digital areas of design. These component are primarily digital-to-analog and analog-to-digital converters, but may include other mixed signal components as well.

The strategy for modeling mixed signal components is to represent analog ports as type real. Although this does not allow complete modeling of the analog characteristics of these ports, such as impedance and leakage, it is sufficient to provide a first-order approximation of the behavior of analog portions of the mixed signal components. This method does provide a good means of verifying the digital portion of the design.

An example of an analog-to-digital converter is given in Figure 16.11 with commentary to follow.

For more detail, let us begin with a comment line in the model header:

```
     -- Must be compiled with VITAL compliance checking off
```

Although this model uses the VITAL packages for timing and SDF backannotation, it is not VITAL compliant. VITAL compliance checking must be turned off or the model will not compile.

```
-------------------------------------------------------------------------------
-- File Name: ads1286.vhd
-------------------------------------------------------------------------------
-- Copyright (C) 2003 Free Model Foundry; http://eda.org/fmf/
--
-- This program is free software; you can redistribute it and/or modify
-- it under the terms of the GNU General Public License version 2 as
-- published by the Free Software Foundation.
--
-- MODIFICATION HISTORY:
--
-- version: |  author:  | mod date: | changes made:
--   V1.0    R. Munden   03 Jan 10   Initial release
--
-- Must be compiled with VITAL compliance checking off
-------------------------------------------------------------------------------
-- PART DESCRIPTION:
--
-- Library:    CONVERTERS_VHDL
-- Technology: MIXED
-- Part:       ADS1286
--
-- Description: Sampling 12-Bit A/D Converter
-------------------------------------------------------------------------------

LIBRARY IEEE;    USE IEEE.std_logic_1164.ALL;
                 USE IEEE.VITAL_timing.ALL;
                 USE IEEE.VITAL_primitives.ALL;
LIBRARY FMF;     USE FMF.gen_utils.ALL;


-------------------------------------------------------------------------------
-- ENTITY DECLARATION
-------------------------------------------------------------------------------
ENTITY ads1286 IS
    GENERIC (
        -- tipd delays: interconnect path delays
        tipd_CLK                : VitalDelayType01 := VitalZeroDelay01;
        tipd_CSNeg              : VitalDelayType01 := VitalZeroDelay01;
        -- tpd delays
        tpd_CLK_DOUT            : VitalDelayType01Z := UnitDelay01Z;
        tpd_CSNeg_DOUT          : VitalDelayType01Z := UnitDelay01Z;
        -- tsetup values: setup times
        tsetup_CSNeg_CLK        : VitalDelayType := UnitDelay;
        -- tpw values: pulse widths
        tpw_CLK_posedge         : VitalDelayType := UnitDelay;
        tpw_CLK_negedge         : VitalDelayType := UnitDelay;
        tpw_CSNeg_posedge       : VitalDelayType := UnitDelay;
        tpw_CSNeg_negedge       : VitalDelayType := UnitDelay;
        -- tperiod_min: minimum clock period = 1/max freq
        tperiod_CLK_posedge : VitalDelayType := UnitDelay;
        -- analog generics
        -- value of Vref input In Volts
        Vref              : real    := 5.00;
        -- generic control parameters
        InstancePath      : STRING    := DefaultInstancePath;
        TimingChecksOn    : BOOLEAN   := DefaultTimingChecks;
        MsgOn             : BOOLEAN   := DefaultMsgOn;
        XOn               : BOOLEAN   := DefaultXon;
        -- For FMF SDF technology file usage
        TimingModel       : STRING    := DefaultTimingModel
    );
```

Figure 16.11 Serial sampling 12-bit A/D converter

```
    PORT (
        CLK              : IN    std_ulogic := 'U';
        DOUT             : OUT   std_ulogic := 'U';
        INP              : IN    real := 0.0;
        INN              : IN    real := 0.0;
        CSNeg            : IN    std_ulogic := 'U'
    );
    ATTRIBUTE VITAL_LEVEL0 of ads1286 : ENTITY IS TRUE;
END ads1286;


-------------------------------------------------------------------------------
-- ARCHITECTURE DECLARATION
-------------------------------------------------------------------------------
ARCHITECTURE vhdl_behavioral of ads1286 IS
    ATTRIBUTE VITAL_LEVEL0 of vhdl_behavioral : ARCHITECTURE IS TRUE;

    CONSTANT partID          : STRING := "ads1286";

    SIGNAL CLK_ipd           : std_ulogic := 'U';
    SIGNAL CSNeg_ipd         : std_ulogic := 'U';

BEGIN

    -------------------------------------------------------------------------
    -- Wire Delays
    -------------------------------------------------------------------------
    WireDelay : BLOCK
    BEGIN

        w_1 : VitalWireDelay (CLK_ipd, CLK, tipd_CLK);
        w_2 : VitalWireDelay (CSNeg_ipd, CSNeg, tipd_CSNeg);

    END BLOCK;

    -------------------------------------------------------------------------
    -- Behavior Process
    -------------------------------------------------------------------------
    convert : PROCESS (CSNeg_ipd, CLK_ipd)

        -- Timing Check Variables
        VARIABLE Tviol_CSNeg_CLK : X01 := '0';
        VARIABLE TD_CSNeg_CLK    : VitalTimingDataType;

        VARIABLE PD_CLK          : VitalPeriodDataType := VitalPeriodDataInit;
        VARIABLE Pviol_CLK       : X01 := '0';

        VARIABLE PD_CSNeg        : VitalPeriodDataType := VitalPeriodDataInit;
        VARIABLE Pviol_CSNeg     : X01 := '0';
        VARIABLE Violation       : X01 := '0';

        TYPE cntdir_type IS (up, down, stop);

        CONSTANT res     : natural := 12; -- resolution in bits
        VARIABLE bitcnt  : natural := 13; -- data bit cntr
        VARIABLE reg     : std_logic_vector(res + 1 DOWNTO 0) := (others => '0');
        VARIABLE D_zd    : std_ulogic := 'Z';
        VARIABLE sample  : real;
        VARIABLE tmpref  : real;
        VARIABLE pwrdn   : boolean := true;
        VARIABLE reset   : boolean := false;
        VARIABLE convert : boolean := false;
        VARIABLE cntdir  : cntdir_type;

        -- Output Glitch Detection Variables
        VARIABLE D_GlitchData  : VitalGlitchDataType;
```

Figure 16.11 Serial sampling 12-bit A/D converter *(continued)*

```
BEGIN
    ------------------------------------------------------------------------
    -- Timing Check Section
    ------------------------------------------------------------------------
    IF (TimingChecksOn) THEN
        VitalSetupHoldCheck (
            TestSignal        => CSNeg_ipd,
            TestSignalName    => "CSNeg",
            RefSignal         => CLK_ipd,
            RefSignalName     => "CLK",
            SetupLow          => tsetup_CSNeg_CLK,
            CheckEnabled      => TRUE,
            RefTransition     => '/',
            HeaderMsg         => InstancePath & partID,
            TimingData        => TD_CSNeg_CLK,
            XOn               => XOn,
            MsgOn             => MsgOn,
            Violation         => Tviol_CSNeg_CLK
        );

        VitalPeriodPulseCheck (
            TestSignal        => CLK_ipd,
            TestSignalName    => "CLK",
            Period            => tperiod_CLK_posedge,
            PulseWidthHigh    => tpw_CLK_posedge,
            PulseWidthLow     => tpw_CLK_negedge,
            HeaderMsg         => InstancePath & partID,
            CheckEnabled      => TRUE,
            PeriodData        => PD_CLK,
            XOn               => XOn,
            MsgOn             => MsgOn,
            Violation         => Pviol_CLK
        );

        VitalPeriodPulseCheck (
            TestSignal        => CSNeg_ipd,
            TestSignalName    => "CSNeg",
            PulseWidthHigh    => tpw_CSNeg_posedge,
            PulseWidthLow     => tpw_CSNeg_negedge,
            HeaderMsg         => InstancePath & partID,
            CheckEnabled      => TRUE,
            PeriodData        => PD_CSNeg,
            XOn               => XOn,
            MsgOn             => MsgOn,
            Violation         => Pviol_CSNeg
        );

        Violation := Tviol_CSNeg_CLK OR Pviol_CLK OR Pviol_CSNeg;
    END IF;

    ------------------------------------------------------------------------
    -- Functionality Section
    ------------------------------------------------------------------------
    IF falling_edge(CSNeg_ipd) THEN    -- sample input
        reset := true;
        sample := INP - INN;
    ELSIF rising_edge(CSNeg_ipd) THEN    -- power down
        D_zd := 'Z';
        pwrdn := true;
        bitcnt :=  res + 1;
    END IF;
```

Figure 16.11 Serial sampling 12-bit A/D converter *(continued)*

```
                IF to_UX01(CSNeg_ipd) = '0' THEN
                    IF rising_edge(CLK_ipd) THEN
                        IF pwrdn AND bitcnt = res + 1 THEN      -- begin
                            convert := true;
                            reset := false;
                            pwrdn := false;
                            cntdir := down;
                        END IF;
                    ELSIF falling_edge(CLK_ipd) THEN
                        IF not pwrdn THEN
                            IF bitcnt > 12 THEN
                                D_zd := 'Z';
                            ELSE
                                D_zd := reg(bitcnt);    -- output data
                            END IF;
                            IF cntdir = down AND not convert THEN
                                bitcnt := bitcnt - 1;
                                IF bitcnt = 0 THEN
                                    cntdir := up;
                                END IF;
                            ELSIF cntdir = up THEN
                                bitcnt := bitcnt + 1;
                                IF bitcnt = res THEN
                                    cntdir := stop;
                                END IF;
                            END IF;
                        END IF;
                    END IF;
                END IF;

                IF convert THEN
                    tmpref := Vref/2.0;
                    FOR b IN (res - 1) DOWNTO 0 LOOP
                        IF sample >= tmpref THEN
                            reg(b) := '1';
                            tmpref := tmpref + tmpref/2.0;
                        ELSE
                            reg(b) := '0';
                            tmpref := tmpref/2.0;
                        END IF;
                    END LOOP;
                    convert := false;
                END IF;


                ------------------------------------------------------------------------
                -- Path Delay Section
                ------------------------------------------------------------------------
                VitalPathDelay01Z (
                    OutSignal        => DOUT,
                    OutSignalName    => "DOUT",
                    OutTemp          => D_zd,
                    GlitchData       => D_GlitchData,
                    XOn              => XOn,
                    MsgOn            => MsgOn,
                    Paths            => (
                        0 => (InputChangeTime    => CLK_ipd'LAST_EVENT,
                              PathDelay          => tpd_CLK_DOUT,
                              PathCondition      => TRUE),
                        1 => (InputChangeTime    => CSNeg_ipd'LAST_EVENT,
                              PathDelay          => tpd_CSNeg_DOUT,
                              PathCondition      => TRUE)
                    )
                );

            END PROCESS convert;
        END vhdl_behavioral;
```

Figure 16.11   Serial sampling 12-bit A/D converter *(continued)*

In the generics section, there is a value set for the reference voltage:

```
-- analog generics
-- value of Vref input In Volts
Vref              : real    := 5.00;
```

In a true analog model, this value would be measured on a pin. In a digital model, it is far easier to set it with a generic, as it should not change during the course of the simulation.

The port list shows the two analog ports as mode IN and type real:

```
PORT (
    CLK               : IN  std_ulogic := 'U';
    DOUT              : OUT std_ulogic := 'U';
    INP               : IN  real := 0.0;
    INN               : IN  real := 0.0;
    CSNeg             : IN  std_ulogic := 'U'
);
ATTRIBUTE VITAL_LEVEL0 of ads1286 : ENTITY IS TRUE;
END ads1286;
```

Also note that the model takes advantage of the VITAL_Level0 capabilities of the simulator. These capabilities include SDF backannotation. The model also has a full set of timing constraint checks.

The digital part of the functionality section is straightforward. The analog section is an area of interest:

```
IF convert THEN
    tmpref := Vref/2.0;
    FOR b IN (res - 1) DOWNTO 0 LOOP
        IF sample >= tmpref THEN
            reg(b) := '1';
            tmpref := tmpref + tmpref/2.0;
        ELSE
            reg(b) := '0';
            tmpref := tmpref/2.0;
        END IF;
    END LOOP;
    convert := false;
END IF;
```

Conversions are performed by successive approximation. Each pass through the loop determines the value of one bit, beginning with the most significant bit. It starts by setting tmpref to one-half the full range voltage. If the sample voltage exceeds tmpref, the first bit (msb) is set to '1', otherwise it is set to '0'. The value of tmpref is then halved and the value of the second bit is determined. The process continues until all 12 bits have been set. In a model such as this, the analog (real) inputs would probably be supplied directly from the testbench.

16.8   Summary

There always seems to be another odd feature to learn how to model. Most of them seem to be cases of an analog circuit that found its way into a digital component. With a little imagination, methods can be devised for modeling them.

Differential inputs can be modeled using tables to detect changes on each input and determine the correct output. Bus hold inputs can be modeled by incorporating weak positive feedback loops, although each model will hold only its own inputs. PLLs and DLLs are both modeled as synchronizing oscillators. Mixed signal devices are modeled using ports of type real. Should VHDL-AMS become more popular, more complete modeling of the analog portions of these devices will become practical.

It is sometimes practical to use the `TimingModel` generic to configure a model to simulate specific members of a component family.

New modeling challenges are certain to present themselves to the intrepid modeler.

17 Testbenches for Component Models

At least half the work of creating a component model is in the verification. The purpose of component models is the verification of board- or system-level designs, but the component models must first be verified themselves. This is done using a testbench. Writing the testbench is often as much work as writing the model, sometimes even more. The quality of the testbench may determine the quality of the model.

The topic of testbenches is a large one. Books have been devoted to it. This chapter will discuss only a few aspects of particular interest to those modeling components.

17.1 About Testbenches

The purpose of a testbench is to exercise a model and test it for correct operation. It exercises the model by applying stimuli to its inputs and tests it by reading its outputs and comparing them to the known correct responses.

17.1.1 Tools

A testbench contains a significant amount of boilerplate code. This is code that is predetermined based on the component entity and requires no thought or creativity to write. The testbench for a 3-port gate model has about 80 lines of such code. For a 54-port device, the line count exceeds 400 just to declare and instantiate the device under test. Rather than bear the tedium of writing such boilerplate, use a perl script to do it for you. You can write your own or download one called `mktb` from the Free Model Foundry Web site.

A testbench generator should read the model and write a testbench that includes an empty entity, a component declaration, a configuration statement, a signal declaration for every port in the model, and a component instantiation. Once the boilerplate has been generated, you can concentrate on the creative side of testbench writing.

17.2   Testbench Styles

Testbenches serve not only as a means to verify models, but also as an aid to their development. The range of complexity of testbenches parallels the range of complexity of the models they test. Even the most basic model should have a testbench. In the sections that follow we examine the various testbench styles and the different types of component models they are best suited for.

17.2.1   The Empty Testbench

Even the simplest model should have a testbench. An empty testbench merely instantiates the device under test. It is useful for testing backannotation and verifying that the timing file matches the component entity. At a minimum you can drive the model's input(s) using the force commands in the simulator, and manually inspect the resulting waveform for correct operation.

To use such a testbench, all that is required is to automatically generate the testbench, set the `TimingModel` generic in the component instantiation, run `mk_sdf` (the SDF generator discussed in Chapter 12) against the testbench, and start the simulator and backannotate the generated SDF file. If backannotation is successful, everything in the timing file matches the model.

17.2.2   The Linear Testbench

The linear testbench is characterized by its uninterrupted start-to-finish flow. It assigns values to interface signals, waits, and then checks for correct results. An example testbench process for an open collector nand gate is given in Figure 17.1.

There are two key items to note about this code: The time delays should be long enough to work with the slowest timing that will be simulated (or use a generic instead of a hard-coded value), and testbenches should always be self-checking. Self-checking is implemented by using assertion statements, covered in detail later in this chapter.

The self-checking feature may seem unimportant in a short testbench like that shown here, but as models get more complex and their testbenches get longer, self-checking becomes indispensable. Without it, a change made to a model to correct an error in one test might create an error in a previous test that would go unnoticed. It is not practical to verify a complex model merely by examining the simulator waveform output.

17.2.3   The Transactor Testbench

Most components of interest have buses and operate on bus cycles. Bus cycles often require controlling several signals in a specific sequence. Rather than the linear coding style, procedure calls are used. Each procedure call results in a transaction on the bus. The main stimulus process becomes much shorter and more readable:

```
Stim : PROCESS
BEGIN
    T_A <= '0';
    T_B <= '0';
    WAIT for 50 ns;
    ASSERT (T_YNeg = 'Z')
        REPORT "YNeg is " & to_bin_str(T_YNeg) & "should be Z"
        SEVERITY ERROR;
    T_A <= '1';
    T_B <= '0';
    WAIT for 50 ns;
    ASSERT T_YNeg = 'Z'
        REPORT "YNeg is " & to_bin_str(T_YNeg) & "should be Z"
        SEVERITY ERROR;
    T_A <= '1';
    T_B <= '1';
    WAIT for 50 ns;
    ASSERT T_YNeg = '0'
        REPORT "YNeg is " & to_bin_str(T_YNeg) & "should be 0"
        SEVERITY ERROR;
    T_A <= '0';
    T_B <= '1';
    WAIT for 50 ns;
    ASSERT T_YNeg = 'Z'
        REPORT "YNeg is " & to_bin_str(T_YNeg) & "should be Z"
        SEVERITY ERROR;
    WAIT;
END PROCESS Stim;
```

Figure 17.1   Linear-style testbench code

```
test <= write;
Command(Instruct => write, length => 4);
Command(Instruct => write, length => 4, bsel0 => false,
    wr_addr_strt => 4, data3_strt => 13);
test <= cmode_rd;
Command(Instruct => rd, length => 4);
```

In this code there is a signal, named **test**. Test is of an enumerated type and the value it is assigned is the name of the test that is being started. When viewing a waveform, a glance at the current value of test tells you where you are in the simulation.

The procedure **Command** is defined in the testbench. More on writing transactor-based testbenches is provided in Section 17.4.

17.3   Using Assertions

One of the most effective means of comparing an actual output to an expected output is the assertion statement. The assertion statement allows a testbench to be self-checking, which is essential. The formal syntax for an assertion statement is as follows:

```
[label : ] assert boolean_expression
[ report expression ]
[ severity expression ] ;
```

Assertion statements may be placed in either concurrent or sequential code. When encountered, the boolean expression is evaluated. If it evaluates to TRUE, the simulator moves to the next statement. If it evaluates to FALSE, the simulator reports the fact. If there is a report clause, it will be included in the report. If there is a severity clause, it will set the severity of the assertion. Without a severity clause the severity level will be error. Although both the report and severity clauses are optional, report should always be used and severity, for reasons of clarity, should be explicitly given.

Unlike the rest of VHDL, the report clause in an assertion statement is sensitive to carriage returns. The lines in multiline reports must be linked with &s. Otherwise, a newline character indicates the end of the report clause.

An example assertion is as follows:

```
ASSERT (T_B = '0')
    REPORT "B is " & to_bin_str(T_B) & "should be 0"
    SEVERITY ERROR;
```

Here we are checking that T_B is '0'. If it is not, we report what its value is and what it should be and say it is an error.

Most simulators will allow suppressing assertions below a specified severity. For example, the assertion

```
ASSERT to_nat(DQ(7 downto 0)) /= D_lo
    REPORT "READ: OK - "&
    to_int_str(DQ(7 downto 0))
    SEVERITY note;
```

will emit an assert message every time it encounters a correct result. This could be useful at some point in the design process but annoying at others. By setting the severity to note, the message can be masked by the simulator while more severe assertions are still reported.

Assertion statements can use boolean expressions of arbitrary complexity. It is sometimes useful to use a generic to set the severity level.

17.4   Using Transactors

Transactor-style testbenches are used with any component that requires several signals to work together in a coordinated fashion. SDRAMs, processors, and flash memories are examples of such components.

Although they may vary greatly from one testbench to another, transactors are built as procedures. These procedures are usually in the testbench file but may be placed in a separate package so they can be used with multiple testbenches.

```
Stim: PROCESS

    VARIABLE datain : NATURAL;
    VARIABLE addrin : NATURAL;
    VARIABLE outreg : std_logic_vector(11 DOWNTO 0);

    PROCEDURE s1
        (datain : IN NATURAL := 0;
         addrin : IN NATURAL := 0)
    IS
    BEGIN
        outreg(11 DOWNTO 8) := to_slv(addrin, 4);
        outreg(7 DOWNTO 0) := to_slv(datain, 8);
        FOR i IN 11 DOWNTO 0 LOOP
            T_SDI <= outreg(i);
            WAIT FOR 50 ns;
            T_CLK <= 'H';
            WAIT FOR 50 ns;
            T_CLK <= 'L';
        END LOOP;
        WAIT FOR 10 ns;
        T_LD <= 'H';
        WAIT FOR 90 ns;
        T_LD <= 'L';

    END s1;
```

Figure 17.2   Simple transactor process

However, if a procedure is placed in a separate package, it will not be able to directly assign any signals.

The procedures can span the full range of complexity. For starters, let us begin with a simple one shown in Figure 17.2.

This procedure is from the testbench of a component with a serial interface. Each transaction requires clocking in 4 bits of address followed by 8 bits of data followed by a load strobe. The procedure is called from an otherwise linear flow testbench:

```
s1(addrin => 1, datain => 128);
WAIT FOR 10 ns;
s1(addrin => 2, datain => 0);
```

Each time s1 is called it executes the equivalent of 64 statements of linear concurrent code.

A more common scenario involves testing a model of a component that accepts specific instructions on its input bus. In this case it is convenient to be able to issue instructions by name, so a procedure is developed to enable that.

The code in Figure 17.3 is from the testbench for a JPEG CODEC.

In this code relatively simple bus cycles are generated by calling the procedure Host with the name of an instruction. Data required for the instruction are passed implicitly by setting variables prior to the procedure call. In the code in Figure 17.3, the lines

```
Stim : PROCESS

-- Type Definitions
TYPE Instruction IS (START,
                     RESET,
                     Idle,
                     ISR,
                     READ,
                     LOAD,
                     SLEEP
                     );

    VARIABLE Instruct          : Instruction;
    VARIABLE ADDRint           : std_logic_vector(1 downto 0);
    VARIABLE DATAint           : std_logic_vector(7 downto 0);

PROCEDURE Host
    (Instruct  : IN Instruction)
IS
BEGIN
    CASE Instruct IS
        WHEN ISR =>
            T_CSNeg <= '0', '1' AFTER 360 ns;
            T_WRNeg <= '0' AFTER 6 ns, '1' AFTER 130 ns;
            T_ADDR <= "10", "11" AFTER 130 ns, "ZZ" AFTER 350 ns;
            T_Data <= "00001000", "ZZZZZZZZ" AFTER 130 ns;
            T_RDNeg <= '0' AFTER 170 ns, '1' AFTER 350 ns;
        WHEN Read =>
            T_CSNeg <= '0', '1' AFTER 190 ns;
            T_RDNeg <= '0' AFTER 6 ns, '1' AFTER 180 ns;
            T_ADDR <= "11", "ZZ" AFTER 180 ns;
        WHEN Load =>
            T_CSNeg <= '0', '1' AFTER 140 ns;
            T_WRNeg <= '0' AFTER 6 ns, '1' AFTER 120 ns;
            T_ADDR <= ADDRint, "ZZ" AFTER 130 ns;
            T_Data <= Dataint, "ZZZZZZZZ" AFTER 130 ns;
        WHEN Reset =>
            T_RESETNeg <= '0', '1' AFTER 600 ns;
        WHEN Sleep =>
            T_SLEEPNeg <= '0', '1' AFTER 600 ns;
        WHEN Others => NULL;
    END CASE;

END Host;

BEGIN
    WAIT FOR 200 ns;
    Host(RESET);
    WAIT UNTIL T_RESETNeg = '1';
    WAIT FOR 400 ns;
    Host(Sleep);
    WAIT UNTIL T_SLEEPNeg = '1';
    WAIT FOR 3700 ns;
    -- Load Register 2 Code Interface
    DATAint := "00000000"; ADDRint := "01";
    Host(Load);
    WAIT FOR 150 ns;
    DATAint := "00000010"; ADDRint := "10";
    Host(Load);
    WAIT FOR 150 ns;
    DATAint := "00000001"; ADDRint := "11";
    Host(Load);
    WAIT FOR 150 ns;
...
```

Figure 17.3 Instruction-based transactor

```
                           -- Read Register 1
                           DATAint := "00000001"; ADDRint := "10";
                           Host(Load);
                           WAIT FOR 150 ns;
                           T_CBUSYNeg <= '1';
                           BusyLoop : WHILE T_DATA(7) /= '0' LOOP
                               Host(Read);
                               WAIT UNTIL T_ACKNeg = '0';
                               IF T_DATA(7) = '0' THEN
                                   LoadComplete <= '1';
                                   EXIT BusyLoop;
                               END IF;
                               WAIT FOR 500 ns;
                           END LOOP BusyLoop;
                           T_PVALID <= '1';
                           WAIT FOR 800 ns;
          ...
```

Figure 17.3   Instruction-based transactor *(continued)*

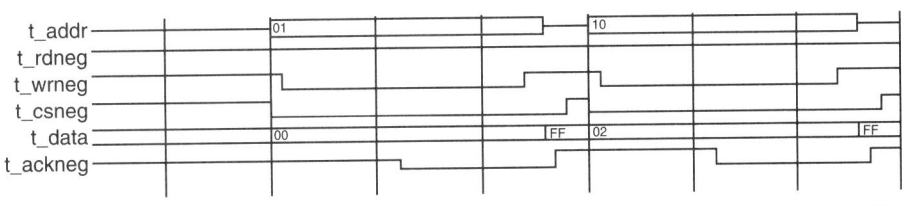

Figure 17.4   Transactor-produced stimulus

```
          DATAint := "00000000"; ADDRint := "01";

          Host(Load);

          WAIT FOR 150 ns;

          DATAint := "00000010"; ADDRint := "10";

          Host(Load);
```

produce the stimulus shown in Figure 17.4.

17.5   Testing Memory Models

The two things that set memory models and their testbenches apart from most
other models is their complexity and their need for (sometimes large amounts of)
data storage. The following code is from the testbench for a dual-port synchronous
SRAM. The memory-specific code begins with the process declarations:

```
          Stim: PROCESS

              -- Memory array declaration
              TYPE MemStore IS ARRAY (0 to 32767) OF INTEGER
                            RANGE -2 TO 255;
```

```
TYPE Instruction IS (rdLa,     -- read left port using address bus
                     wrtLa,    -- write left port using address bus
                     wrtLc,    -- write left port using counter
                     rdLc,     -- read left port using counter
                     rdRpa,    -- read right port pipeline address
                     rdRpc,    -- read right port pipeline counter
                     rdRfa,    -- read right port FT address
                     rdRfc,    -- read right port FT counter
                     wrtRa,    -- write right port using address bus
                     wrtRpa,   -- wrt rt port pipeline using address bus
                     matchwr,  -- wrt lt port rd rt port FT using address bus
                     desel
                    );
VARIABLE data    : NATURAL;
VARIABLE datain  : NATURAL;
VARIABLE MemData    : MemStore;
```

The first type declaration is for a testbench storage array that will mirror the
memory storage array in the model. That is followed by another type declaration
that enumerates the instructions that will be issued to the model.

Next is the definition of the procedure **Command**. It begins with a declaration of
its interface:

```
PROCEDURE Command
   (Instruct    : IN Instruction;
      length    : IN NATURAL := 0;
      addr_strt : IN NATURAL := 0;
      data_strt : IN NATURAL := 0)
   IS
```

It can accept up to four arguments. Instruct is the name of the instruction to
execute, length is the number of times to loop within the instruction, `addr_strt`
is a starting address, and `data_strt` is a starting data value.

The first instruction defined is an address-controlled write to the left side of the
dual-port memory:

```
BEGIN
   CASE Instruct IS
      WHEN wrtLa =>
         data := data_strt;
         T_RWL  <= '0';
         WAIT FOR 30 ns;
         T_ADSLNeg <= '0';
         FOR i IN addr_strt TO addr_strt + length LOOP
            T_AL  <= To_slv(i,15);
            T_IOL <= To_slv(data,8);
```

```
                    MemData(i) := data;
                    data := data + 1;
                    WAIT FOR 30 ns;
                END LOOP;
                T_IOL <= (others => 'Z');
```

During this instruction, every time data are written to the memory model a copy is also written to the same location in the testbench memory. The two memory arrays should always have the same contents.

The next instruction is an address-controlled read from the left side of the memory:

```
        WHEN rdLa =>
            T_RWL <= '1';
            T_ADSLNeg <= '0';
            FOR i IN addr_strt TO addr_strt + length LOOP
                T_AL <= To_slv(i,15);
                WAIT FOR 30 ns;
                IF (i > addr_strt) THEN
                    datain := To_Nat(T_IOL);
                    ASSERT MemData(i - 1) = datain
                        REPORT "expected IOL = " & to_hex_str(MemData(i - 1))
                            & " got " & to_hex_str(datain)
                        SEVERITY error;
                END IF;
            END LOOP;
```

Here, every time data are read from a location in the model memory, the data are compared to the same location in the testbench memory. The comparison is done using an assertion statement. If the two do not match, a message is emitted giving the expected and received values.

The **Command** procedure is called from the main body of the testbench with arguments:

```
        Test <= WrLa;
        Command(Instruct => wrtLa, addr_strt => 0, data_strt => 8,
                length => 16);
        Test <= RdLPa;
        Command(Instruct => rdLa, addr_strt => 0, length => 16);
```

In this code fragment a command is issued to write to the left side of the memory using address control. The start address is 0, the starting data value is 8, and the number of consecutive locations to be written is 16. The following command is a read back from the same 16 locations.

A less complex but more tedious method simplifies the **Command** procedure at the expense of the main body. The following is from a testbench for an SDRAM model:

```vhdl
Stim      :          PROCESS

TYPE Instruction IS (nop,
                     act,
                     read,
                     writ,
                     bst,
                     pre,
                     mrs,
                     ref
                    );

PROCEDURE Command
     (Instruct : IN Instruction)
IS
BEGIN
     CASE Instruct IS
        WHEN nop =>
           T_CSNeg <= '1';
        WHEN act =>
           T_CSNeg <= '0', '1' AFTER 30 ns;
           T_RASNeg <= '0', '1' AFTER 30 ns;
           T_CASNeg <= '1', '1' AFTER 30 ns;
           T_WENeg <= '1', '1' AFTER 30 ns;
        WHEN read =>
           T_CSNeg <= '0', '1' AFTER 30 ns;
           T_RASNeg <= '1', '1' AFTER 30 ns;
           T_CASNeg <= '0', '1' AFTER 30 ns;
           T_WENeg <= '1', '1' AFTER 30 ns;
           WHEN writ =>
           T_CSNeg <= '0', '1' AFTER 30 ns;
           T_RASNeg <= '1', '1' AFTER 30 ns;
           T_CASNeg <= '0', '1' AFTER 30 ns;
           T_WENeg <= '0', '1' AFTER 30 ns;
        WHEN bst =>
           T_CSNeg <= '0', '1' AFTER 30 ns;
           T_RASNeg <= '1', '1' AFTER 30 ns;
           T_CASNeg <= '1', '1' AFTER 30 ns;
           T_WENeg <= '0', '1' AFTER 30 ns;
        WHEN pre =>
           T_CSNeg <= '0', '1' AFTER 30 ns;
           T_RASNeg <= '0', '1' AFTER 30 ns;
           T_CASNeg <= '1', '1' AFTER 30 ns;
           T_WENeg <= '0', '1' AFTER 30 ns;
        WHEN ref =>
```

```
                        T_CSNeg <= '0', '1' AFTER 30 ns;
                        T_RASNeg <= '0', '1' AFTER 30 ns;
                        T_CASNeg <= '0', '1' AFTER 30 ns;
                        T_WENeg <= '1', '1' AFTER 30 ns;
                    WHEN mrs =>
                        T_CSNeg <= '0', '1' AFTER 30 ns;
                        T_RASNeg <= '0', '1' AFTER 30 ns;
                        T_CASNeg <= '0', '1' AFTER 30 ns;
                        T_WENeg <= '0', '1' AFTER 30 ns;
                END CASE;

        END Command;
```

Here the Command procedure is very clean and understandable. However, the main body of the testbench has become much less readable:

```
Test <= reg_set; -- burst = 4; sequential; CAS Lat = 2;
T_A0 <= 'L';
T_A1 <= 'H';
T_A2 <= 'L';
T_A3 <= 'L';
T_A4 <= 'L';
T_A5 <= 'H';
T_A6 <= 'L';
T_A7 <= 'L';
T_A8 <= 'L';
T_A9 <= 'L';
T_A10 <= 'L';
T_A11 <= 'L';
T_A12 <= 'L';
Command(mrs); -- 110400
WAIT FOR 40 ns;
Command(nop);
WAIT FOR 80 ns;

Test <= bank_sel; -- bank 0
T_A0 <= 'L';
T_A1 <= 'L';
T_A2 <= 'L';
T_A3 <= 'L';
T_A4 <= 'L';
T_A5 <= 'L';
T_A6 <= 'L';
T_A7 <= 'L';
T_A8 <= 'L';
T_A9 <= 'L';
```

```
T_A10  <=  'L';
T_A11  <=  'L';
T_A12  <=  'L';
T_DM  <=  'L';
Command(act); --
WAIT FOR 65 ns;

Test  <=  write; -- burst = 4
T_A0  <=  'L';
T_A1  <=  'L';
T_A2  <=  'L';
T_A3  <=  'H';
T_A4  <=  'L';
T_A5  <=  'L';
T_A6  <=  'L';
T_A7  <=  'L';
T_A8  <=  'L';
T_A9  <=  'L';
T_A10  <=  'L';
T_A11  <=  'L';
T_A12  <=  'L';
WAIT FOR 10 ns;
Command(writ); --
T_DQS  <=  '0';
WAIT FOR 52 ns;
T_DQ  <=  "01010101"; -- 85
WAIT FOR 2 ns;
T_DQS  <=  '1';
WAIT FOR 20 ns;
T_DQ  <=  "01010000"; --
--WAIT FOR 2 ns;
T_DQS  <=  '0';
WAIT FOR 15 ns;
T_DQ  <=  "00010101"; --
WAIT FOR 5 ns;
T_DQS  <=  '1';
WAIT FOR 15 ns;
T_DQ  <=  "01011000"; --
WAIT FOR 5 ns;
T_DQS  <=  '0';
WAIT FOR 20 ns;
T_DQ  <=  "ZZZZZZZZ";
T_DQS  <=  'Z';
Command(nop);
```

```
WAIT FOR 70 ns;
Command(nop);

Test <= read; -- burst = 4
T_A0 <= 'L';
T_A1 <= 'L';
T_A2 <= 'L';
T_A3 <= 'H';
T_A4 <= 'L';
T_A5 <= 'L';
T_A6 <= 'L';
T_A7 <= 'L';
T_A8 <= 'L';
T_A9 <= 'L';
T_A10 <= 'L';
T_A11 <= 'L';
Command(read); --
WAIT FOR 40 ns;
Command(nop);
WAIT UNTIL rising_edge(T_DQS);
WAIT FOR 1 ns;
ASSERT (T_DQ = "01010101")
   REPORT "expected data = " & to_int_str("01010101") &
     " got " & to_int_str(T_DQ)
   SEVERITY error;
WAIT FOR 20 ns;
ASSERT (T_DQ = "01010000")
   REPORT "expected data = " & to_int_str("01010000") &
     " got " & to_int_str(T_DQ)
   SEVERITY error;
WAIT FOR 20 ns;
ASSERT  (T_DQ = "00010101")
   REPORT "expected data = " & to_int_str("00010101") &
     " got " & to_int_str(T_DQ)
   SEVERITY error;
WAIT FOR 20 ns;
ASSERT (T_DQ = "01011000")
   REPORT "expected data = " & to_int_str("01011000") &
     " got " & to_int_str(T_DQ)
   SEVERITY error;
WAIT FOR 75 ns;
```

Although the Command procedure is driving the control pins to send instructions
to the model under test, all the addresses and data are being set in the main body.
Also, all the results checking is done in the main body. The advantage of this

method is that it is faster and less work to write enough code to begin testing the model. However, that must be balanced against the increased difficulty of debugging and maintaining the testbench over time.

17.6   Summary

Testbenches are an integral part of model development and a significant portion of the total effort. Certain parts of the testbench are boilerplate and should be generated programmatically rather than manually.

There are different styles of testbenches that lend themselves to different type of component models. All testbenches should be self-checking. As you write your testbench, keep in mind that it will not be done as quickly as you estimated and you are likely to want to reuse it in the future. A little extra effort in the beginning will yield large dividends later on.

Books have been published on the subject of testbenches. Although they are usually oriented to RTL design, they are worth reading even if you are only writing component models.

INDEX

2-input nand gate, synthesizable, 16
12-bit A/D converter, serial sampling, 289–292

A/D converter, serial sampling 12-bit, 289–292
abstraction, levels of, 6–7
adding timing to RTL code, 191–208
additions, VITAL, 19–25
algorithms, state table, 99
alternatives, conditional delay, 150–152
analyses
    signal, 262
    timing, 262
annotation. *See* Backannotation
arguments for `VitalMemoryTable`, 105
arrays, memory, 209–211
ASIC design flow. *See* FPGA/ASIC design flow
assertions, 284–285, 297–298
attributes, VITAL, 20–21

backannotating path delays, 88–89
backannotation, 60
    and hierarchy, 185–187
backannotation, timing, 262
backannotation, timing files and, 179–187
    anatomy of timing files, 179–182
    backannotation and hierarchy, 185–187
    custom timing sections, 183–184
    generating SDF files, 184–185
    generating timing files, 184
    importing timing values, 183
behavior, modifying with `TimingModel` generic, 285
behavioral memory preload, 235–237
behavioral (Shelor) method, 211–223
BFMs (bus functional models), 7
blocks
    negative constraint, 65
    signal delay, 66
    wire delay, 63–65
board-level verification, 3–14
    definition of model, 5–10
    design methods and models, 10
    getting models, 13–14

models fit in FPGA/ASIC design flow, 10–13
    need for models, 3–5
bus functional models (BFMs), 7
bus hold, 279–282

CAE (Computer Aided Engineering), 254–255
calls
    procedure, 70
    VITAL primitive, 21–22
    `VitalOutPhaseSkewCheck`, 118
    `VitalOutPhaseSkewCheck` procedure, 118
    `VitalPeriodPulseCheck`, 113
    `VitalPeriodPulseCheck` procedure, 112
    `VitalRecoveryRemovalCheck`, 115
    `VitalRecoveryRemovalCheck` procedure, 115
    `VitalSetupHoldCheck`, 109
    `VitalSetupHoldCheck` procedure, 109
calls, `VitalPeriodPulseCheck` procedure
    parameters of Mode IN, 113–114
    parameters of Mode INOUT, 114
    parameters of Mode OUT, 114
cells, 50
    RTL produces, 8
characteristics, negative setup timing, 159
checks
    conditional timing, 153–156
    in-phase skew, 117
    memory timing, 42
    out-of-phase skew, 118
    purpose of timing constraint, 107
    recovery/removal, 114–117
    `RecoveryRemoval`, 161
    setup/hold, 108–112
    `SetupHold`, 161
    skew, 117–121
    timing, 55–57
circuit delays, 52–55
clocks, modeling delays in designs with internal, 206–207
CMOS (Complimentary Metal Oxide on Silicon), 74
code, adding timing to RTL, 191–208
compilation, VHDL, 259

component-centric features, modeling, 269–294
  assertions, 284–285
  bus hold, 279–282
  differential inputs, 269–279
  mixed signal devices, 288–294
  modifying behavior with `TimingModel` generic,
    285
  PLLs and DLLs, 282–284
  state machines, 285–288
component modeling, considerations for,
    251–267
  component models and netlisters, 251–252
  file contents, 253
  generics passed from schematic, 253–254
  integrating models into schematic capture
    system, 254–256
  model considerations, 263–266
  schematic considerations, 262–263
  special considerations, 262–266
  using models in design process, 256–262
component models
  and netlisters, 251–252
  VHDL packages for, 35–45
component models, testbenches for, 295–308
  about testbenches, 295
  assertions, 297–298
  testbench styles, 296–297
  testing memory models, 301–308
  transactors, 298–301
components
  modeling, 125–146
  VITAL model of nand gate, 28–29
Computer Aided Engineering (CAE), 254–255
concurrent procedure section, 70
conditional delays
  alternatives, 150–152
  in SDF, 149–150
  timing table for part with, 148
conditional delays and timing constraints,
    147–156
  conditional delay alternatives, 150–152
  conditional delays in SDF, 149–150
  conditional delays in VITAL, 147–149
  conditional timing checks in VITAL, 153–156
  mapping SDF to VITAL, 152–153
conditional timing checks in VITAL, 153–156
constraint checks, purpose of timing, 107
constraints
  conditional delays and timing, 147–156
  how simulators handle negative, 176–177
  modeling negative, 158–176
  models of component negative, 162–174
  negative setup/hold, 158
  negative timing, 157–178
  `RecoveryRemoval` checks with negative, 161
  `SetupHold` checks with negative, 161
  timing, 107–122
  VITAL_Memory timing, 232–235
  workings of negative, 157–158
construction
  memory table, 102–103
  state table, 97–98
  truth table, 92
constructs, example usage of, 57
control generics, 253–254

conversions, FMF, 45
converter, serial sampling 12-bit A/D, 289–292

D flip-flops, 98, 100, 101
D register, `StateTable` for, 41
data flow, simulation, 13
DDR (Double Data Rate) DRAMs, 209
declarations, 36–37, 37–38, 40
declarative section, 66–67
delay alternatives, conditional, 150–152
delay blocks
  signal, 66
  wire, 63–65
delay generic, nand gate with, 18
delay locked loops (DLLs), 282
delay procedures, path, 76–82
delay sections, path, 69
delay types
  and glitches, 73–74
  VITAL, 19–20
delays
  backannotating path, 88–89
  circuit, 52–55
  conditional, 147–156, 149–150
  device, 83–88
  distributed, 75
  interconnect, 25–27, 89–90
  model, 18–19
  pin-to-pin, 75–76
  timing table for part with conditional, 148
  transport and inertial, 73–74
  VITAL nand gate model with interconnect, 26
  VITAL_Memory path, 231–232
delays, modeling, 73–90, 206–207
  backannotating path delays, 88–89
  delay types and glitches, 73–74
  device delays, 83–88
  distributed delays, 75
  generates and VPDs, 83
  interconnect delays, 89–90
  path delay procedures, 76–82
  pin-to-pin delays, 75–76
  VPDs, 82–83
design flow, models fit in FPGA/ASIC, 10–13
design methods and models, 10
design, netlisting, 258–259
design processes, using models in, 256–262
  layout, 261
  netlisting design, 258–259
  schematic entry, 257–258
  SDF generation, 259–261
  signal analysis, 262
  simulation, 261
  timing analysis, 262
  timing backannotation, 262
  VHDL compilation, 259
  VHDL libraries, 257
design/verification flow, 11–13
designs with internal clocks, modeling delays in,
    206–207
device delays, 83–88
device under test (DUT), 7
devices, mixed signal, 288–294
DFFCEN state table, 134
differential inputs, 269–279

ECL model with, 272–275
directories, 255–256
distributed delays, 75
DLLs (delay locked loops), 282
DLLs, PLLs and, 282–284
DRAMs, 241–244
DRAMs, DDR (Double Data Rate), 209
duplicated outputs, models with, 84–85
DUT (device under test), 7

ECL model with differential inputs, 272–275
empty testbench, 296
entry, schematic, 257–258
expressions in SDF, operators for, 151
eXtensible Markup Language (XML), 47

features, modeling component-centric, 269–294
files
    generating SDF, 184–185
    generating timing, 184
    map, 256
    sample SDF, 48
    sample vhdl_map, 256
    timing, 179–187
files, anatomy of timing, 179–182
    body, 181
    FMFTIME, 181–182
    header, 179–181
    separate timing specifications, 182
files, SDF, 47–52
    cells, 50
    headers, 48–50
    timing specifications, 50–52
flip-flops, anatomy of, 125–137
    architecture, 129–131
    B side, 135–137
    entity, 125–129
    functionality section, 133–134
    path delay, 134–135
    VITAL process, 131–133
flip-flops, oversimplified D, 98
flow
    design/verification, 11–13
    simulation data, 13
FMF (Free Model Foundry), 42
    conversions, 45
    ff_package, 44
    gen_utils and ecl_utils, 43
    packages, 42–45
FMFTIME, 181–182
formatting, nand model with improved, 17
FPGA/ASIC design flow, models fit in, 10–13
FPGA/ASIC design flow, models fitting in,
    design/verification flow, 11–13
Free Model Foundry (FMF), 42
functionality
    memory, 41
    modeling memory, 211–231
    section, 68–69
functions, 37
    and procedures, 40–41

gate component, VITAL model of nand, 28–29
gate model, VITAL nand, 26
gates, 39

basic VITAL nand, 20
nand, 18
synthesizable 2-input nand, 16
VITAL nand, 23
generate statement, VPD inside, 86
generating
    SDF files, 184–185
    timing files, 184
generation, SDF, 259–261
generics
    control, 253–254
    modifying behavior with TimingModel, 285
    nand gates with delay, 18
    timing, 253
generics passed from schematic, 253–254
    control generics, 253–254
    map files, 256
    timing generics, 253
glitches, delay types and, 73–74

headers, 48–50
hierarchy, backannotation and, 185–187
HL (high to low transitions), 20
hold, bus, 279–282
hold checks. See Setup/hold checks
hold constraints. See Negative setup/hold
    constraints

improved formatting, nand model with, 17
in-phase skew check, 117
independence, technology, 255
inertial delays, transport and, 73–74
inputs
    differential, 269–279
    ECL model with differential, 272–275
instruction based transactor, 300–301
integrating models into schematic capture system,
    254–256
    directories, 255–256
    library structure, 254–255
    technology independence, 255
interconnect delays, 25–27, 89–90
interfaces, standard, 17–18
internal clocks, modeling delays in designs with,
    206–207
iteration times, relative, 5

keywords, SDF, 62

latches, anatomy of, 137–146
    architecture, 140–146
    entity, 138–140
LATNDFF state table, 145
level 0 guidelines, 59–62
levels, comparison of VITAL, 21
LH (low to high transitions), 20
libraries, VHDL, 257
library structure, 254–255
linear testbench, 296

machines, state, 285–288
map files, 256
mapping
    SDF to netlist, 61
    SDF to VITAL, 152–153

mapping *(continued)*
  SDF to VITAL symbol, 153
memories, modeling, 209–249
  memory arrays, 209–211
  modeling memory functionality, 211–231
  modeling miscellaneous memory types, 238–249
  preloading memories, 235–238
  VITAL_Memory path delays, 231–232
  VITAL_Memory timing constraints, 232–235
memories, preloading, 235–238
  behavioral memory preload, 235–237
  VITAL_Memory preload, 237–238
memory arrays, 209–211
  Shelor Method, 210
  VITAL_Memory package, 210–211
memory functionality, 41
memory functionality, modeling, 211–231
  behavioral (Shelor) method, 211–223
  Shelor Method, 211–223
  VITAL2000 method, 223–231
memory models, testing, 301–308
memory preload, behavioral, 235–237
memory tables, 101–105
  allowed symbols for, 103
  construction, 102–103
  memory table construction, 102–103
  memory table symbols, 101–102
  memory table usage, 103–105
  for simple SRAMs, 104
  symbols, 101–102
  usage, 103–105
memory timing
  checks, 42
  specification, 42
memory types, modeling miscellaneous, 238–249
  DRAMs, 241–244
  SDRAMs, 244–249
  synchronous static RAM, 238–241
method, behavioral (Shelor), 211–223
methods
  design, 10
  VITAL2000, 223–231
Methods, Shelor, 210, 211–223
mixed signal devices, 288–294
model delays, 18–19
modeling
  memories, 209–249
  memory functionality, 211–231
  negative constraints, 158–176
modeling components with registers, 125–146
  anatomy of flip-flop, 125–137
  anatomy of latch, 137–146
modeling, considerations for component, 251–267
  component models and netlisters, 251–252
  file contents, 253
  generics passed from schematic, 253–254
  integrating models into schematic capture
    system, 254–256
  model considerations, 263–266
  schematic considerations, 262–263
  special considerations, 262–266
  using models in design process, 256–262
modeling delays, 73–90
  backannotating path delays, 88–89
  delay types and glitches, 73–74

in designs with internal clocks, 206–207
device delays, 83–88
distributed delays, 75
generates and VPDs, 83
interconnect delays, 89–90
path delay procedures, 76–82
pin-to-pin delays, 75–76
VPDs, 82–83
models
  anatomies of VITAL, 59–71
  component, 251–252
  design methods and, 10
  with duplicated outputs, 84–85
  ECL, 272–275
  fit in FPGA/ASIC design flow, 10–13
  getting, 13–14
  integrating, 254–256
  nand, 17
  structure of VITAL, 64
  testing memory, 301–308
  VHDL packages for component, 35–45
  VITAL, 28–29
  VITAL nand gate, 26
models, definitions of, 5–10
  levels of abstraction, 6–7
  model types, 7–9
  technology-independent models, 9–10
models, need for, 3–5
  prototyping, 3
  simulation, 4–5
models, technology-independent, 9–10
models, testbenches for component, 295–308
  about testbenches, 295
  assertions, 297–298
  testbench styles, 296–297
  testing memory models, 301–308
  transactors, 298–301
models, tour of simple, 15–32
  finishing touches, 27–31
  formatting, 15–17
  interconnect delays, 25–27
  model delays, 18–19
  standard interfaces, 17–18
  VITAL additions, 19–25
models, using in design process, 256–262
  layout, 261
  netlisting design, 258–259
  schematic entry, 257–258
  SDF generation, 259–261
  signal analysis, 262
  simulation, 261
  timing analysis, 262
  timing backannotation, 262
  VHDL compilation, 259
  VHDL libraries, 257
models, using timing constraint checks in VITAL,
  108–121

nand gate component, VITAL model of, 28–29
nand gate model, VITAL, 26
nand gates
  basic VITAL, 20
  with delay generic, 18
  synthesizable 2-input, 16
  VITAL, 23

nand models with improved formatting, 17
NCC (Negative Constraint Calculation), 177
NCC, timing values before and after, 177
negative constraint block, 65
Negative Constraint Calculation (NCC), 177
negative constraints
    how simulators handle, 176–177
    modeling, 158–176
    models of component, 162–174
    `RecoveryRemoval` checks with, 161
    `SetupHold` checks with, 161
    workings of, 157–158
negative setup/hold constraints, 158
negative setup timing characteristics, 159
negative timing constraints, 157–178
    how simulators handle negative constraints,
        176–177
    modeling negative constraints, 158–176
    ramifications, 177–178
    workings of negative constraints, 157–158
netlist mapping, SDF to, 61
netlisters, component models and, 251–252
netlisting design, 258–259

off-the-shelf (OTS), 10
operators for expressions in SDF, 151
OTS (off-the-shelf), 10
out-of-phase skew check, 118
outputs, model with duplicated, 84–85

packages
    FMF, 42–45
    VHDL, 35–45
    VITAL_Memory, 210–211
parameters for `VitalInPhaseSkewCheck`, 119
    parameters of Mode IN, 119–120
    parameters of Mode INOUT, 120
    parameters of Mode OUT, 120–121
parameters to `VitalRecoveryRemovalCheck`
        procedure, 115–117
    parameters of Mode IN, 115–117
    parameters of Mode OUT, 117
parameters to `VitalSetupHoldCheck` procedure,
    110
    parameters of Mode IN, 110–111
    parameters of Mode INOUT, 112
    parameters of Mode OUT, 112
path delays
    backannotating, 88–89
    procedures, 76–82
    sections, 69
    VITAL_Memory, 231–232
PCB (printed circuit board), 12
period/pulsewidth checks, 112–114
pessimism, reducing, 100–101
phase locked loops (PLLs), 282
pin-to-pin delays, 75–76
PLLs and DLLs, 282–284
PLLs (phase locked loops), 282
prefixes, VITAL, 62
preload
    behavioral memory, 235–237
    VITAL_Memory, 237–238
preloading memories, 235–238
primitive call, VITAL, 21–22

primitives
    procedure calls to VITAL, 70
    user-defined, 39
    VITAL, 70
printed circuit board (PCB), 12
procedure calls
    to VITAL primitives, 70
    `VitalOutPhaseSkewCheck`, 118
    `VitalRecoveryRemovalCheck`, 115
    `VitalSetupHoldCheck`, 109
procedure calls, `VitalPeriodPulseCheck`, 112
    parameters of Mode IN, 113–114
    parameters of Mode INOUT, 114
    parameters of Mode OUT, 114
procedures, 38–39
    functions and, 40–41
    `VitalRecoveryRemovalCheck`, 114
    `VitalSetupHoldCheck`, 108
procedures, parameters to
        `VitalRecoveryRemovalCheck`, 115–117
    parameters of Mode IN, 115–117
    parameters of Mode OUT, 117
procedures, parameters to
        `VitalSetupHoldCheck`, 110
    parameters of Mode IN, 110–111
    parameters of Mode INOUT, 112
    parameters of Mode OUT, 112
processes, VITAL, 22–23
prototyping, 3
pulsewidth checks. *See* Period/pulsewidth
    checks

RAM, synchronous static, 238–241
RC (resistance capacitance), 74
recovery/removal checks, 114–117
`RecoveryRemoval` checks with negative
        constraints, 161
register transfer level (RTL), 6
registers, modeling components with, 125–146
    anatomy of flip-flop, 125–137
    anatomy of latch, 137–146
registers, `StateTable` for D, 41
removal checks. *See* Recovery/removal checks
RTL code, adding timing to, 191–208
    basic wrapper, 192–205
    caveats, 207–208
    modeling delays in designs with internal clocks,
        206–207
    using VITAL to simulate RTL, 191–192
    wrapper for Verilog RTL, 206
RTL (register transfer level), 6
    produces cells, 8
    using VITAL to simulate, 191–192
    wrapper for Verilog, 206

schematic capture system, integrating models
        into, 254–256
    directories, 255–256
    library structure, 254–255
    technology independence, 255
schematic entry, 257–258
schematics, generics passed from, 253–254
    control generics, 253–254
    map files, 256
    timing generics, 253

SDF
  conditional delays in, 149–150
  generation, 259–261
  keywords, 62
  mapping to VITAL, 152–153
  to netlist mapping, 61
  operators for expressions in, 151
  to VITAL symbol mapping, 153
SDF capabilities, 52–57
  circuit delays, 52–55
  timing checks, 55–57
SDF constructs, example usage of, 57
SDF files, 47–52
  cells, 50
  generating, 184–185
  headers, 48–50
  sample, 48
  timing specifications, 50–52
SDF, introduction to, 47–58
  SDF capabilities, 52–57
  SDF files, 47–52
SDRAMs (Synchronous Dynamic RAMs), 209,
    244–249
sections
  concurrent procedure, 70
  declarative, 66–67
  example of VITAL process declarative, 67
  functionality, 68–69
  path delay, 69
  timing check, 67–68
serial sampling 12-bit A/D converter, 289–292
setup/hold checks, 108–112
SetupHold checks with negative constraints, 161
SGML (Standard Generalized Markup Language),
    179
Shelor Method, 210, 211–223
signal analysis, 262
signal delay blocks, 66
signal devices, mixed, 288–294
simple model, tour of, 15–32
simple SRAMs, memory tables for, 104
simulation, 4–5
  data flow, 13
simulators, how they handle negative constraints,
    176–177
skew checks, 117–121
  in-phase, 117
  out-of-phase, 118
specifications
  memory timing, 42
  separate timing, 182
  timing, 50–52
SRAMs, memory tables for simple, 104
SSRAMs (synchronous static RAMs), 209
Standard Generalized Markup Language (SGML),
    179
standard interfaces, 17–18
state machines, 285–288
state tables, 97–99
  advantages of truth and, 91
  algorithms, 99
  construction, 97–98
  DFFCEN, 134
  LATNDFF, 145
  state table algorithm, 99

state table construction, 97–98
state table symbols, 97
state table usage, 98–99
symbols, 97
usage, 98–99
statements, VPD inside generate, 86
StateTable for D register, 41
Static RAM, synchronous, 238–241
STD_LOGIC_1164, 35–37
  functions, 37
  type declarations, 36–37
structures, library, 254–255
styles, testbench, 296–297
symbol mapping, SDF to VITAL, 153
symbols
  allowed for memory tables, 103
  memory table, 101–102
  state table, 97
  VITAL table, 92–93
  VITALMemory table, 102
Synchronous Dynamic RAMs (SDRAMs), 209,
    244–249
synchronous static RAMs (SSRAMs), 238–241
system, integrating models into schematic
    capture, 254–256

table symbols
  VITAL, 92–93
  VITALMemory, 102
tables
  advantages of truth and state, 91
  allowed symbols for memory, 103
  DFFCEN state, 134
  LATNDFF state, 145
  timing, 148
tables, memory, 101–105
  memory table construction, 102–103
  memory table symbols, 101–102
  memory table usage, 103–105
tables, state, 97–99
  state table algorithm, 99
  state table construction, 97–98
  state table symbols, 97
  state table usage, 98–99
tables, truth, 92–97
  truth table construction, 92
  truth table usage, 93–97
  VITAL table symbols, 92–93
tables, vital, 91–106
  advantages of truth and state tables, 91
  memory tables, 101–105
  reducing pessimism, 100–101
  state tables, 97–99
  truth tables, 92–97
technology independence, 255
technology-independent models, 9–10
testbench styles, 296–297
  empty testbench, 296
  linear testbench, 296
  transactor testbench, 296–297
testbenches
  empty, 296
  linear, 296
  transactor, 296–297
testbenches, about, 295

tools, 295
testbenches for component models, 295–308
    about testbenches, 295
    assertions, 297–298
    testbench styles, 296–297
    testing memory models, 301–308
    transactors, 298–301
timing
    analysis, 262
    backannotation, 262
timing, adding to RTL code, 191–208
    basic wrapper, 192–205
    caveats, 207–208
    modeling delays in designs with internal clocks,
        206–207
    using VITAL to simulate RTL, 191–192
    wrapper for Verilog RTL, 206
timing characteristics, negative setup, 159
timing check section, 67–68
timing checks, 55–57
    conditional, 153–156
    memory, 42
timing constraint checks in VITAL models, using,
        108–121
    period/pulsewidth checks, 112–114
    recovery/removal checks, 114–117
    setup/hold checks, 108–112
    skew checks, 117–121
timing constraint checks, purpose of, 107
timing constraints, 107–122
    purpose of timing constraint checks, 107
    using timing constraint checks in VITAL
        models, 108–121
    violations, 121
    VITAL_Memory, 232–235
timing constraints, conditional delays and,
        147–156
    conditional delay alternatives, 150–152
    conditional delays in SDF, 149–150
    conditional delays in VITAL, 147–149
    conditional timing checks in VITAL, 153–156
    mapping SDF to VITAL, 152–153
timing constraints, negative, 157–178
    how simulators handle negative constraints,
        176–177
    modeling negative constraints, 158–176
    ramifications, 177–178
    workings of negative constraints, 157–158
timing files, anatomy of, 179–182
    body, 181
    FMFTIME, 181–182
    header, 179–181
    separate timing specifications, 182
timing files and backannotation, 179–187
    anatomy of timing files, 179–182
    backannotation and hierarchy, 185–187
    custom timing sections, 183–184
    generating SDF files, 184–185
    generating timing files, 184
    importing timing values, 183
timing files, generating, 184
timing generics, 60–61, 253
timing sections, custom, 183–184
timing specifications, 50–52
    memory, 42

    separate, 182
timing table for part with conditional delays, 148
timing values
    before and after NCC, 177
    importing, 183
TimingModel generic, modifying behavior with,
        285
transactor testbenches, 296–297
transactors, 298–301
    instruction based, 300–301
transport and inertial delays, 73–74
truth and state tables, advantages of, 91
truth tables, 92–97
    construction, 92
    truth table construction, 92
    truth table usage, 93–97
    usage, 93–97
    VITAL table symbols, 92–93
type declarations, 36–37
types
    delay, 73–74
    VITAL delay, 19–20

UDPs (user-defined primitives), 39
Unit Under Test (UUT), 186
usage
    memory table, 103–105
    state table, 98–99
    truth table, 93–97
user-defined primitives (UDPs), 39
UUT (Unit Under Test), 186

values, importing timing, 183
verification, board-level, 3–14
verification flow. *See* Design/verification flow
Verilog RTL, wrapper for, 206
VHDL compilation, 259
VHDL libraries, 257
VHDL packages for component models, 35–45
    FMF packages, 42–45
    STD_LOGIC_1164, 35–37
    VITAL_Primitives, 39–41
    VITAL_Timing, 37–39
violations, 121
VITAL
    attributes, 20–21
    conditional delays in, 147–149
    conditional timing checks in, 153–156
    delay types, 19–20
    mapping SDF to, 152–153
VITAL additions, 19–25
    VITAL attributes, 20–21
    VITAL delay types, 19–20
    VITAL primitive call, 21–22
    VITAL processes, 22–23
    VitalPathDelays, 24–25
VITAL levels, comparison of, 21
VITAL models
    of nand gate component, 28–29
    structure of, 64
VITAL models, anatomy of, 59–71
    backannotation, 60
    concurrent procedure section, 70
    declarative section, 66–67
    functionality section, 68–69

VITAL models, anatomy of *(continued)*
  level 0 guidelines, 59–62
  level 1 guidelines, 63–70
  negative constraint block, 65
  path delay section, 69
  processes, 65–69
  timing check section, 67–68
  timing generics, 60–61
  VITAL primitives, 70
  `VitalDelayTypes`, 61–62
  wire delay block, 63–65
VITAL models, using timing constraint checks in, 108–121
  period/pulsewidth checks, 112–114
  recovery/removal checks, 114–117
  setup/hold checks, 108–112
  skew checks, 117–121
VITAL nand gate
  basic, 20
  model with interconnect delays, 26
  using `VitalPathDelay`, 23
VITAL prefixes, 62
VITAL primitives, 70
  calls, 21–22
  procedure calls to, 70
VITAL process declarative section, example of, 67
VITAL processes, 22–23
VITAL symbol mapping, SDF to, 153
VITAL table symbols, 92–93
VITAL tables, 91–106
  advantages of truth and state tables, 91
  memory tables, 101–105
  reducing pessimism, 100–101
  state tables, 97–99
  truth tables, 92–97
VITAL, using to simulate RTL, 191–192
VITAL2000 method, 223–231
`VitalDelayTypes`, 61–62
`VitalInPhaseSkewCheck`, parameters for, 119, 119–120
  parameters of Mode IN, 119–120
  parameters of Mode INOUT, 120
  parameters of Mode OUT, 120–121
VITAL_Memory, 41–42
  package, 210–211

path delays, 231–232
preload, 237–238
timing constraints, 232–235
VITALMemory table symbols, 102
`VitalMemoryTable`, arguments for, 105
`VitalOutPhaseSkewCheck`
  call, 118
  procedure call, 118
`VitalPathDelay`, VITAL nand gate using, 23
`VitalPathDelay` (VPD), 24–25, 82–83
`VitalPeriodPulseCheck`
  call, 113
  procedure call, 112
`VitalPeriodPulseCheck` procedure call
  parameters of Mode IN, 113–114
  parameters of Mode INOUT, 114
  parameters of Mode OUT, 114
VITAL_Primitives, 39–41
`VitalRecoveryRemovalCheck`
  call, 115
  procedure, 114
`VitalRecoveryRemovalCheck` procedure, parameters to, 115–117
  parameters of Mode IN, 115–117
  parameters of Mode OUT, 117
`VitalSetupHoldCheck`
  procedure, 108
  procedure call, 109
`VitalSetupHoldCheck` procedure, parameters to, 110
  parameters of Mode IN, 110–111
  parameters of Mode INOUT, 112
  parameters of Mode OUT, 112
VITALSignalDelay, component with, 66
VITAL_Timing, 37–39
VPD inside generate statement, 86
VPD (`VitalPathDelay`), 22, 24–25, 82–83

wire delay block, 63–65
wrappers
  basic, 192–205
  for Verilog RTL, 206

XML (eXtensible Markup Language), 47

ZBT (Zero Bus Turnaround), 209